Artificial Intelligence for Future Networks

IEEE Press
445 Hoes Lane
Piscataway, NJ 08854

Artificial Intelligence for Future Networks

Edited by

Mohammad A. Matin
Department of Electrical and Computer Engineering
North South University
Bangladesh

Sotirios K. Goudos
Department of Physics
Aristotle University of Thessaloniki
Greece and the Director of the ELEDIA@AUTH lab member of the ELEDIA Research Center Network

George K. Karagiannidis
Department of Electrical and Computer Engineering
Aristotle University of Thessaloniki
Greece and the Head of the Wireless Communications and Information Processing (WCIP) Group

IEEE Press
WILEY

Published by John Wiley & Sons, Inc., Hoboken, New Jersey.
Published simultaneously in Canada.

Library of Congress Cataloging-in-Publication Data Applied for:

Hardback ISBN: 9781394227921

Cover Design: Wiley
Cover Image: © zf L/Getty Images

Set in 9.5/12.5pt STIXTwoText by Straive, Chennai, India

Mohammad A. Matin dedicates the book to his wife Momtaz Begum who inspires him and their sons Zabeer Ahmed and Zawad Ahmed.

Sotirios K. Goudos dedicates the book to his wife Athena who inspires him and their daughters Mary and Mandy.

George K. Karagiannidis dedicates the book to Kostis.

Contents

About the Editors

Mohammad A. Matin (Senior Member, IEEE) is a professor in the Department of Electrical and Computer Engineering at North South University (NSU). He has published over 150 peer-reviewed journal and conference papers. He is the author/editor of 18 academic books and 22 book chapters. He has received several prizes and scholarships including the Best Student Prize (Loughborough University), Commonwealth Scholarship, and Overseas Research Scholarship (ORS) conferred by the Committee of Vice Chancellors and Principals (CVCP) in the United Kingdom. He is currently serving as a member of the editorial boards for several international publications *including IEEE Communications Magazine, IET Wireless Sensor Systems.*

Sotirios K. Goudos (Senior Member, IEEE) is a professor in the Department of Physics at the Aristotle University of Thessaloniki and the director of the ELEDIA@AUTH lab member of the ELEDIA Research Center Network. He has published and presented more than 350 technical papers in scientific journals and international conferences. He is the founding editor-in-chief of the *Telecom* open-access journal (MDPI publishing). He is currently serving as *associate editor for IEEE Transactions on Antennas and Propagation, IEEE ACCESS, and IEEE Open Journal of the Communication Society*. He was honored as an IEEE Access Outstanding Associate Editor for 2019, 2020, 2021, 2022, and 2023.

George K. Karagiannidis (Fellow, IEEE) is currently a professor in the Department of Electrical and Computer Engineering of Aristotle University of Thessaloniki, Greece, and head of the Wireless Communications & Information Processing (WCIP) Group. He has published and presented more than 700 technical papers in scientific journals and international conferences. He was the past editor of several IEEE journals, and from 2012 to 2015, he was the editor-in-chief *of IEEE Communications Letters. From September 2018 to June 2022, he served as associate editor-in-chief of the IEEE Open Journal of Communications Society*. Currently, *he is the editor-in-chief of IEEE Transactions on Communications*. He received three prestigious awards: The 2021 IEEE ComSoc RCC Technical Recognition Award, the 2018 IEEE ComSoc SPCE Technical Recognition Award, and the 2022 Humboldt Research Award from the Alexander von Humboldt Foundation. He is one of the highly cited authors across all areas of electrical engineering, recognized from Clarivate Analytics as a Highly Cited Researcher in the nine consecutive years 2015–2023.

List of Contributors

Aly S. Abdalla
Department of Electrical and Computer Engineering
Mississippi State University
Mississippi State, MS
USA

Manuel Arrebola
Department of Electrical Engineering
Group of Signal Theory and Communications
Universidad de Oviedo
Gijon, Asturias
Spain

Mehdi Bennis
Centre for Wireless Communications
University of Oulu
Oulu
Finland

Yuanzhu Chen
School of Computing
Queen's University
Kingston, Ontario
Canada

Christos Christodoulou
Department of Electrical and Computer Engineering
University of New Mexico
Albuquerque, New Mexico
USA

Merouane Debbah
KU 6G Research Center
Khalifa University of Science and Technology
Abu Dhabi
United Arab Emirates

Octavia A. Dobre
Faculty of Engineering and Applied Science
Memorial University, St. John's
Newfoundland and Labrador
Canada

Ahmet M. Elbir
Department of Electrical and Electronics Engineering
Istinye University
Istanbul
Turkey

Carlo Fischione
School of Electrical Engineering and Computer Science
KTH Royal Institute of Technology
Stockholm
Sweden

Gábor Fodor
School of Electrical Engineering and Computer Science
KTH Royal Institute of Technology
Stockholm
Sweden

and

Radio Network Algorithms
Ericsson Research, Ericsson AB
Kista
Sweden

Sotirios K. Goudos
Department of Physics
Aristotle University of Thessaloniki
Greece

and

Director of the ELEDIA@AUTH lab member of the ELEDIA Research Center Network

Arjun Gupta
Department of Electrical and Computer Engineering
University of New Mexico
Albuquerque, New Mexico
USA

Yongming Huang
National Mobile Communications Research Laboratory
Southeast University
Nanjing
China

George K. Karagiannidis
Department of Electrical and Computer Engineering
Aristotle University of Thessaloniki
Greece

and

Head of the Wireless Communications and Information Processing (WCIP) Group

Maokun Li
Department of Electronic Engineering
Beijing National Research Center for Information Science and Technology (BNRist)
Tsinghua University
Beijing
China

and

ELEDIA Research Center (ELEDIA@TSINGHUA – Tsinghua University)
Beijing
China

Jesús A. López-Fernández
Department of Electrical Engineering
Group of Signal Theory and Communications
Universidad de Oviedo
Gijon, Asturias
Spain

Vuk Marojevic
Department of Electrical and Computer Engineering
Mississippi State University
Mississippi State, MS
USA

Eduardo Martinez-de-Rioja
Department of Signal Theory and Communications and Telematic Systems and Computing
Universidad Rey Juan Carlos
Fuenlabrada, Madrid
Spain

Manel Martinez-Ramón
Department of Electrical and Computer Engineering
University of New Mexico
Albuquerque, New Mexico
USA

Christos Masouros
Department of Electronic & Electrical Engineering
University College London
London
UK

Andrea Massa
ELEDIA Research Center (ELEDIA@UniTN – University of Trento)
DICAM – Department of Civil Environmental, and Mechanical Engineering
Trento
Italy

and

ELEDIA Research Center (ELEDIA@ TSINGHUA – Tsinghua University)
Beijing
China

and

ELEDIA Research Center (ELEDIA@UESTC – UESTC)
School of Electronic Science and Engineering
Chengdu
China

and

School of Electrical Engineering
Tel Aviv University
Tel Aviv
Israel

Mohammad A. Matin
Department of Electrical and Computer Engineering
North South University
Dhaka
Bangladesh

Marco Salucci
ELEDIA Research Center
(ELEDIA@UniTN – University of Trento), DICAM – Department of Civil, Environmental, and Mechanical Engineering
Trento
Italy

Wei Shi
School of Information Technology
Carleton University
Ottawa, Ontario
Canada

Oscar Stenhammar
School of Electrical Engineering and Computer Science
KTH Royal Institute of Technology
Stockholm
Sweden

and

Radio Network Algorithms
Ericsson Research, Ericsson AB
Kista
Sweden

Bo Tang
Department of Electrical and Computer Engineering
Worcester Polytechnic Institute
Worcester, MA
USA

Álvaro F. Vaquero
Department of Mathematics
Universidad de Oviedo
Oviedo, Asturias
Spain

Jayakrishnan Vijayamohanan
Department of Electrical and Computer Engineering
University of New Mexico
Albuquerque, New Mexico
USA

Xu Wang
School of Computing
Queen's University
Kingston, Ontario
Canada

Jianjun Zhang
College of Computer Science and Technology
Nanjing University of Aeronautics and Astronautics
Nanjing
China

Qiyang Zhao
Technology Innovation Institute
Abu Dhabi
United Arab Emirates

Hang Zou
Technology Innovation Institute
Abu Dhabi
United Arab Emirates

Acknowledgments

We express our sincere thanks to the authors of different chapters, whose invaluable contributions helped us in completing this book. The editors are grateful to numerous reviewers for their helpful recommendations and comments during the review process. We are also grateful to the Wiley team, especially Sandra Grayson, Becky Cowan, Kavipriya Ramachandran, Sathishwaran Pathmanabhan, and Kavin Shanmughasundaram, for their kind assistance during this endeavor.

Best regards,

Mohammad A. Matin
North South University, Bangladesh

Sotirios K. Goudos
Aristotle University of Thessaloniki, Greece

George K. Karagiannidis
Aristotle University of Thessaloniki, Greece

1

Intelligent Beam Prediction and Tracking

Christos Masouros[1], Jianjun Zhang[2], and Yongming Huang[3]

[1] *Department of Electronic & Electrical Engineering, University College London, London, UK*
[2] *College of Computer Science and Technology, Nanjing University of Aeronautics and Astronautics, Nanjing, China*
[3] *National Mobile Communications Research Laboratory, Southeast University, Nanjing, China*

CHAPTER MENU

1.1 Introduction

Because of abundant spectrum resources at high-frequency band, that enable to achieve ultrahigh-speed data transmission (DT), high-frequency communications, e.g., millimeter wave or even Terahertz communications, have attracted extensive interest from academia, industry, and government [1]. For high-frequency communications, the transmitter and/or receiver are often equipped with large-scale antenna arrays, i.e., massive multiple-input multiple-output (MIMO), to achieve high array gains to overcome signal attenuation of high-frequency band. However, the use of pencil-like highly directional beams makes channel state information (CSI) acquisition via beam alignment (BA) very challenging. First, acquiring CSI in a mobile network is particularly challenging since the wireless channel often varies rapidly. Second, in contrast to the fully digital transceiver, where the pilot

Artificial Intelligence for Future Networks, First Edition.
Edited by Mohammad A. Matin, Sotirios K. Goudos, and George K. Karagiannidis.

transmission scheme can be utilized to acquire CSI [2, 3], channel estimation is more complicated in the hybrid antenna array architecture, which has been used widely in practice, because we cannot extract the actual received signals on all antennas simultaneously. Last but not the least, the large dimension of massive MIMO inevitably results in large and even unaffordable pilot overhead, even if the pilot transmission-based method can be used.

To tackle this challenging issue, the two-step precoding and combining based scheme is widely used in practical systems, e.g., standardized in IEEE 802.11ad/802.15.3c [4, 5]. Let **F**, **H**, and **W** represent the precoding matrix, channel matrix, and combing matrix, respectively. The precoding (combining) matrix is assumed to be decomposed as $\mathbf{F} = \mathbf{F}_a\mathbf{F}_d$ ($\mathbf{W} = \mathbf{W}_a\mathbf{W}_d$), where $\mathbf{F}_a$ and $\mathbf{F}_d$ ($\mathbf{W}_a$ and $\mathbf{W}_d$) denote the analog and digital parts, respectively. To reduce the dimension of the CSI estimate problem, beam training is first performed between the transmitter and receiver to obtain the analog precoder $\mathbf{F}_a$ and combiner $\mathbf{W}_a$. Then, the effective or equivalent channel $\mathbf{W}_a^{\mathrm{H}}\mathbf{H}\mathbf{F}_a$ can be estimated, based on which the digital parts, i.e., the precoder $\mathbf{F}_d$ and combiner $\mathbf{W}_d$, can be designed in the analog domain via a variety of methods, e.g., the heuristic methods or optimization-based algorithms [6–8]. Note that since the size of the effective channel matrix $\mathbf{W}_a^{\mathrm{H}}\mathbf{H}\mathbf{F}_a$ is much smaller than that of the original channel matrix **H**, the pilot overhead in the second step is relatively low. It is observed that the remaining difficulty lies in how to design an efficient beam training scheme to find the optimal $\mathbf{F}_a$ and $\mathbf{W}_a$.[1]

Initially, beam training (also referred to as beam sounding) is implemented via the exhaustive and hierarchical search [5, 9, 10]. Compared to the exhaustive search, whose sounding overhead is N with N denoting the size of the training codebook, the sounding overhead of the hierarchical search is $4\log_2 N$ for the typical binary tree search-based implementation, which is smaller than that of the exhaustive search scheme. For this reason, along with the advantage of easy implementation, the hierarchical search-based scheme has been adopted in several IEEE standards, such as IEEE 802.15.3c and IEEE 802.11aj. Note the performance of the hierarchical search-based algorithms heavily depends on the codebook used. In fact, besides the demand for multi resolution, namely, various widths of main lobes, other properties, such as flat main lobe and side lobe, narrow transition band, and high-power efficiency of power amplifier, are also very important and should be well addressed [9, 11]. In general, the research on hierarchical search often boils down to sounding codebook design [5, 9–14].

1 In industry, beam training or tracking involving beam measurements is also referred to as beam sounding. Since the term "training" has been used in machine learning (ML) to refer to learning, from now on, we use the term "sounding" to refer to training in communications so as to avoid confusion.

The advantage of the exhaustive or hierarchical search-based methods is that they can be applied to an arbitrary scenario because they are nonadaptive methods and thus independent of external environments. However, the beam sounding overhead is almost always very large, especially for a large-scale antenna array and/or a rapidly changing environment. In fact, on the one hand, as the scale of the antenna array increases, the beam width decreases accordingly, which thus increases the sounding overhead. On the contrary other hand, the coherence time or period becomes shorter in a rapidly fluctuating environment. Hence, much of the precious time resource is spent on beam sounding, and the proportion of time resources used for DT is very small. This phenomenon is particularly pronounced for the highly varying communication scenarios, e.g., unmanned aerial vehicle (UAV) communication.

To avoid frequent searches, beam tracking is invoked to reduce the sounding overhead. The complete process of BA operation in a relatively long time consists of two phases. First, initial BA is performed in the first stage to find the optimal beam or beam pair via the exhaustive or hierarchical search, which involves a large beam sounding overhead, as mentioned before. Then, the beam tracking technique is invoked in the second phase to enable efficient search. Compared to the initial BA, the number of beams used for tracking is relatively small, e.g., maybe only one beam is used for sounding. Note that if the tracking fails, which is inevitable, the initial BA operation is invoked again to reinitialize the beam tracking.

The key to beam tracking is beam prediction, i.e., to predict a beam subspace that contains the real beam. In practice, two types of metrics are closely related to beam prediction. The first one is the success rate and prediction efficiency, i.e., the beam subspace predicted should contain the real optimal beam, and meanwhile, the beam subspace should be as small as possible. The second one is the complexities of beam prediction, including both sample complexity and inference complexity. To balance these indicators, various methods have been proposed, the core of which is to exploit temporal and spatial correlations of wireless channels. The most important step toward beam prediction is to construct an appropriate prediction model. Overall, there are mainly two ways to construct a prediction model, i.e., the traditional manual fashion and the recent automatic fashion. The classical and representative manual method to construct a prediction model is the Kalman filtering- or Bayesian filtering-based Beam prediction and tracking (BPT) algorithms [15–22]. Machine learning (ML) methods are used to automatically construct prediction models, typically, in the data-driven manner [23–29].

The Kalman filtering and Bayesian filtering methods address the issue of prediction model construction by building a dynamical model that characterizes the underlying physical system. Specifically, two stochastic differential equations (SDEs), referred to as state-space and measurement equations in literature, are first established. As long as the two SDEs are available, the well-known Kalman filter or Bayesian filter can be invoked to perform real-time inference or prediction. For example, both the extended Kalman filter- and Bayesian filter-based beam tracking algorithms are proposed in Liu et al. [15] and Yuan et al. [18] for the dual-functional radar and communication systems. For the distributed millimeter-wave massive MIMO problem, a monopulse beam tracking method based on the unscented Kalman filter is designed in [21], which shows to achieve good robustness as well as generalization ability.

An important and appealing advantage of the Kalman filtering based methods is that they have low computational complexity. In particular, the scaling of computational complexity for the Kalman filter is linear $\mathcal{O}(L)$ (where L is the number of samples), as opposed to the cubic scaling $\mathcal{O}(L^3)$ for Gaussian process (GP) regression-based BPT algorithms. Note that this advantage is attributed to the fact that the underlying state-space system model is exploited. However, since the prediction model is obtained via manual derivation manner, it may fail in complicated scenarios or environments. To tackle this issue, recently a novel hybrid model and data-driven-based approach, referred to as hybrid data-induced Kalman filtering (HDIKF), has been proposed by Zhang et al. [30, 31].

In contrast to the Kalman filtering-based designs, ML typically addresses the issue of prediction modeling by employing the data-driven mode. In fact, it is well known that a powerful ability of ML is that it can automatically extract meaningful patterns and further derive directly an appropriate model from the observed data. According to the underlying ML theory and methods, the ML-based beam prediction methods fall into two categories, i.e., the (deep) reinforcement learning (RL)-based algorithms [26–29, 32, 33] and supervised learning (SL)-based algorithms [13, 23–25, 34–38].

The RL-based BPT solutions can further fall into two subcategories, including the state-free RL-based solutions [26–29] and RL-based solutions incorporating state design [32, 33]. The state-free algorithms, which are mainly implemented via stochastic bandits, cannot characterize and exploit complex environment information, which may fail to exploit useful spatial or other channel correlations. In contrast, the RL-based solutions that characterize the channel correlation, variation, and other information via dedicated state designs when formulating the BPT problem as an RL counterpart can achieve better performance in complicated scenarios. Compared to other ML-based methods, a salient advantage of the RL-based methods is that they can collect training samples via interacting with the environment, which facilitates an online implementation [33]. However, since the mathematical foundation of RL is the Markov decision process, they have a low convergence rate, which may fail to achieve good performance in the short term.

Another major category of ML-based beam prediction solutions is the SL-based algorithms [13, 23–25, 34–38]. The SL-based BPT solutions circumvent the difficulty of manually building models, thanks to the data-driven design paradigm. But it should be noted that one drawback of the SL-based solutions is that the number of training samples required is often very large so as to achieve satisfactory performance. Moreover, rapidly fluctuating wireless environments can also invalidate the deterministic prediction models established. Among the SL-based algorithms, GP regression-based algorithms have attracted considerable attention [35–38], thanks to the advantages of nonparametric modeling and uncertainty calibration. Unfortunately, the scaling of their computational complexity is typically $\mathcal{O}(L^3)$. Since the wireless channels vary quickly and L can be very large, the large complexity becomes an obstacle to real-time applications.

We have already briefly outlined the main BPT techniques and methods, from the conventional nonlearning solutions to the state-of-the-art learning-based methods. We have also briefly analyzed the pros and cons of different BPT methods. In the rest of this chapter, we will first highlight the difference between generic ML applications and prediction modeling in wireless communications. Then, from the perspective of the modern stochastic process (SP), we will propose a novel design methodology whose core is to identify and exploit the analytic properties (e.g., stochastic continuity and stochastic smoothness) of the sample functions of the underlying beam process. Based on the methodology, we further design several efficient BPT solutions, all of which enjoy good small sample performance, low computational complexity and robustness to environment variation. To further enhance system performance, we present an important technique closely related to BPT, namely, beam width optimization (BWO) and propose several efficient ML-based algorithms. Finally, the simulation results are provided to evaluate different BPT techniques.

1.2 Challenge of Beam Prediction Modeling in Wireless Communications

Although tremendous ML prediction models and algorithms have been developed in the past two decades in various fields, such as computer vision, natural language processing, and machine translation, this does not mean that the problem of beam prediction modeling can be easily addressed. In fact, because of the characteristics of wireless communications, the ML models and algorithms practicable are very limited, and this problem is far from being resolved. Next, we highlight and analyze the differences and challenges of beam prediction modeling in wireless communications via comparing them with the typical ML applications or tasks. As illustrated in Figure 1.1, the main challenges are abbreviated as USMC, i.e., various types of uncertainties, small-sample performance, prediction modeling, and training/inference complexities.

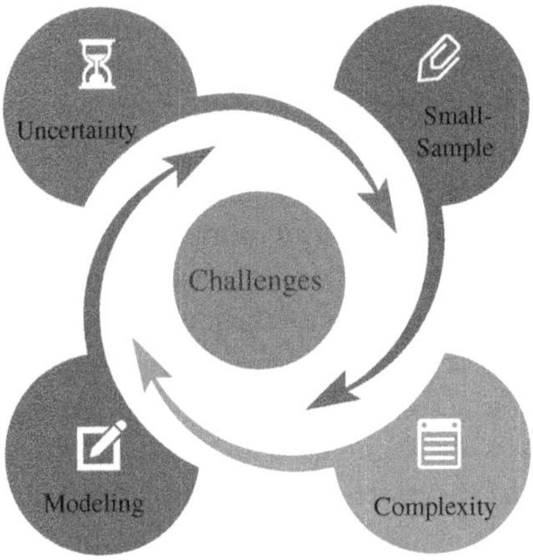

Figure 1.1 Main challenges of BPT in wireless communications.

Uncertainty (U): Roughly speaking, the uncertainties in beam prediction stem from epistemic uncertainty and aleatoric uncertainty. On the one hand, both the formation and variation mechanisms of beam direction are very complex, which results in epistemic uncertainty. On the other hand, due to the rapid fluctuation of wireless channel environments, the available training samples are very limited, which further increases the epistemic uncertainty. The aleatoric uncertainty is derived from the natural randomness inherent in a beam prediction task. In practice, the training samples inevitably contain noise, which thus requires us to build a prediction model based on the noisy samples. For example, because the resolution of an antenna array is limited, it is impossible to obtain an accurate beam direction. Similarly, when we make predictions in the phase of inference, we also need to consider the uncertainty.

Small-sample (S): In contrast to many ML applications, e.g., image recognition, where enormous training images are available with the help of Internet, it is often expensive to collect training samples for a wireless communication ML task. As a result, a few public datasets can be used by the researchers. What is harder is that the wireless communication system is time-varying, which implies that the collected training samples and the trained prediction model can be easily outdated. This is different from a typical image recognition task. Therefore, the small-sample performance is very desired and important in wireless communications, which excludes many powerful but data-hungry prediction models and algorithms.

Modeling (M): The modeling challenge is three-fold. First, as mentioned earlier, the time-varying characteristic of wireless communications makes the ML model obtained outdated easily. Second, the time-varying feature also increases the difficulty of model training. Besides the challenge of sample collection, the task of training a new model has to be completed in a very short time, which requires more computing resources. Third, modeling in wireless communications must take into account the issue of scalability. In fact, scalability is a fundamental (and also essential and important) feature of wireless communications, which, however, does not exist in most other ML applications. Specifically, due to various factors and/or demands, e.g., the design goal of maximizing system transmission rate, the need of meeting the quality of service of some users or just caused by a system scheduling algorithm, the prediction model setting, e.g., the dimension of input or output layer of a prediction model, is time-varying, which greatly increases the challenge of ML modeling.

Complexity (C): As a data-driven paradigm, building ML models often requires huge computing and storage resources, such as graphics processing unit (GPU). However, in contrast to classical ML applications, where abundant computing and storage resources are available, most mobile communication systems are often resource-constrained, caused by the power supply, device cost, etc. Moreover, because of the time-varying feature, the prediction model has to be updated at a high frequency, and thus, the real-time requirement is much higher than other ML tasks. All these factors determine that the complexities of model learning and online inference, in terms of both sample complexity and computation complexity should be sufficiently low, which may be in conflict with the deep learning model having lots of trainable parameters. When it comes to considering uncertainty, it becomes harder.

In a nutshell, due to the above reasons, it is nontrivial to design efficient prediction models and training and inference algorithms for a wireless communication system. We have to take into account at least a part of the above factors. Next, we address and alleviate the issues by identifying and exploiting an important type of nonparametric priors existing extensively in many practical communication systems.

1.3 Prior Identification – Perspective of Function Space

To address these issues, first and foremost, we shall tackle the small-sample challenge since as the number of samples decreases, the other challenges can be greatly alleviated. Note that the training samples critically depend on environments, which could be very expensive to obtain even a small amount of them. Therefore, it is of great importance to identify and exploit the priors from both the communication system and external environment. To this end, an effective and systematic approach is proposed to identify and exploit the priors. In this section, we elaborate on the theoretical part of this approach [39]. Later, specific examples are presented to show how to use this approach to design efficient algorithms.

1.3.1 Perspective of Function Space

The existing BPT solutions mainly exploit the probabilistic properties of beam directions or beam direction variations. For example, the RL-based algorithms essentially exploit the transition probability of beam directions. However, the beam directions or beam direction variations also have rich and useful analytic properties, which unfortunately have been ignored in the past BPT designs. If they can be well identified and exploited, the desired small-sample performance can be achieved.

To identify and exploit these analytic properties, we need to inspect the beam directions and beam direction variations from the perspective of function space. Specifically, we first adopt a dynamic view by considering the discrete beam directions as the sampling points of a continuous-time or -space function, which is referred to as beam direction function. Second, we analyze, characterize, and model the beam direction function. Note that the beam direction function is essentially random, which can be regarded as an outcome sampled from a set of functions, i.e., function space. For a practical scenario, due to various factors, e.g., specific deployments or surroundings of base stations or mobile users, the function space is endowed with a probability structure that characterizes the randomness.[2] Finally, we attach equal importance to both the analytic properties and probabilistic properties.

Mathematically, a function space endowed with a probability structure, in fact, constitutes an SP. In literature, an SP is described as a collection of random variables $\{X_t \mid t \in T\}$, all defined on a common probability space $\{\Omega, \mathcal{E}, \mathbb{P}\}$, where Ω is the sample space, $\mathcal{E}$ is a σ-algebra on Ω, and $\mathbb{P}$ is a probability measure on $\mathcal{E}$. T is often an interval of the real line, which enables to think of the process as evolving in time almost irresistible. The random variable X_t depends on both t and the point ω in Ω at which it is evaluated. As a function of two variables, it is written as $X(\omega, t)$. For a fixed t, $X(\cdot, t)$ is a measurable mapping from Ω into $\mathbb{R}$.

The above definition and interpretation of SP, which underlie most BPT solutions, however, fail to identify and exploit important prior information. In contrast, we emphasize particularly on the second perspective of SP, i.e., the perspective of the sample path of SP, which attracted people's attention and became popular due to Kolmogrov's work. Specifically, for a fixed ω, the function $X(\omega, \cdot)$ is called a sample path of the SP. If all the sample paths lie in some fixed collection $\mathscr{X}$, the process X can be thought of as a map from Ω into $\mathscr{X}$, i.e., a random element of $\mathscr{X}$. Based on this perspective, each sample path of X is a single point in $\mathscr{X}$. For example, if a process indexed by $[0, 1]$ has continuous sample

2 In general, we cannot explicitly describe the probability structure. Otherwise, we can obtain an analytical prediction model. However, it can be implicitly characterized and exploited via the data-driven mode.

paths, it defines a random element of the classical function space $C[0, 1]$ of all continuous and real-valued functions on $[0, 1]$.

The motivation and advantage that SP is utilized to characterize and describe beam direction functions are twofold. First, many important priors, which cannot be exploited by conventional and even learning-based methods, can be identified. Second, efficient BPT algorithms can be derived based on this perspective. Before proceeding, several typical examples, and the priors that cannot be exploited by existing methods, are provided to intuitively illustrate our perspective.

1.3.2 Useful Priors for Beam Process Modeling

In practice, thanks to specific deployments (e.g., high-speed train and motorway), specific and/or limited movement speeds (e.g., the low speed of human beings while typical speeds on motorway), surroundings of base stations and mobile users (e.g., buildings and trees), electromagnetic environments, radio spectrum (e.g., Sub-6G or Terahertz), weather (to affect high-frequency communications), etc., there exist a variety of useful priors. Moreover, for a specific scenario (Figure 1.2), some factors are dominated, while the others are marginal, which can thus further strengthen and highlight a specific prior.

1.3.2.1 High-speed Train Communication

In practice, the base stations are deployed along the rails at regular distances, and the velocities of high-speed trains are within a reasonable range [40]. As a result, the beam directions observed by multiple base stations present an inexact periodicity [39].

1.3.2.2 Indoor Environment

A typical example is a mobile user moving randomly within a room. As a consequence, it is difficult to predict the beam for the next time slot accurately because of the randomness. But fortunately, the difference in beam directions at two adjacent time slots is small, as shown in Figure 1.3a.

1.3.2.3 City Street Environment

A typical scenario is considered, where a user moves along a street and maybe changes to another one or occasionally stops for a moment. It can be observed from Figure 1.3b that the beam direction function is flat in some intervals. Because of the blockage, it may also be discontinuous.

As mentioned earlier, to identify these priors, we need to adopt the Kolmogrov's perspective of modern SP, i.e., beam directions within an interval are regarded as the samples of a random function $b(t)$ $(t \in [0, T])$, and more importantly, each single function $b(t)$ is further regarded as a random realization of an SP, which

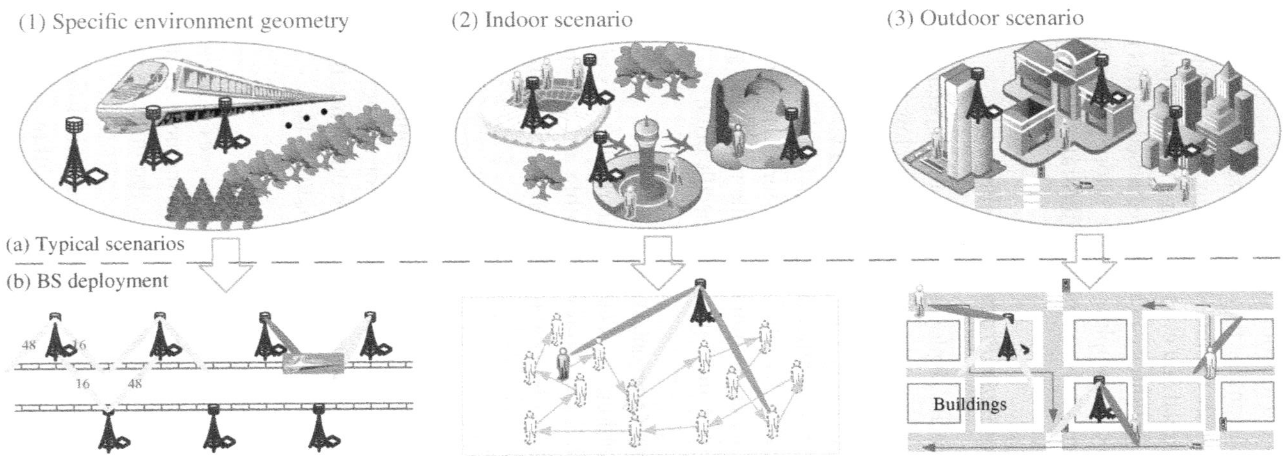

Figure 1.2 (a) Typical communication scenarios – lines with arrows in part (b) represent motion trajectories.

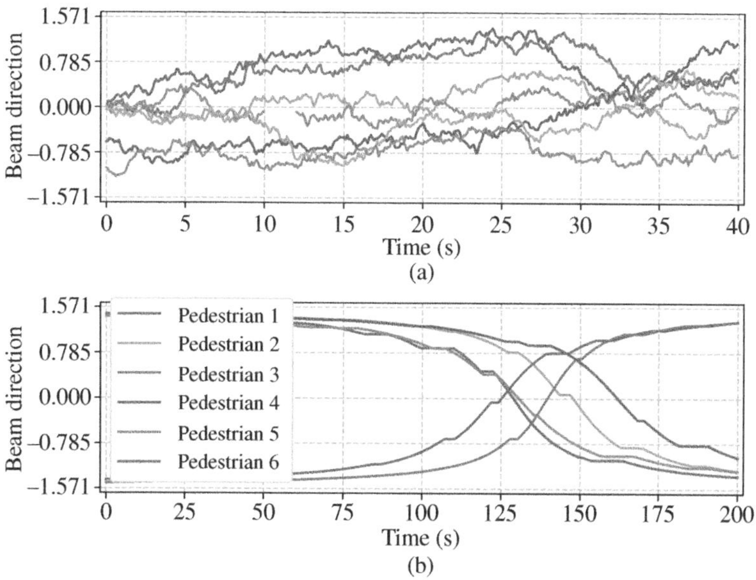

Figure 1.3 Beam direction functions of city block and indoor environments. (a) Beam directions of indoor users and (b) beam directions of pedestrians.

is referred to as beam process and denoted by B – the (function) space of sample paths.[3] We investigate and exploit two types of important properties or priors, i.e., sample path property and statistical regularity property.

- *Sample path property (or prior)*: A sample path property is referred to the analytic property of random sample path $B(\omega, \cdot)$, whose existence is as follows. For example, due to the limited movement speed as well as slow-varying environ ment, $B(\omega, \cdot)$ is a continuous function, although it may be fairly random, e.g., it behaviors like the Brownian motion. This sample-path property is referred to as beam change continuity. Another property or prior is that since the movement speed of a mobile user is limited and beam sounding is operated with an appropriate frequency, the jump or variation of $B(\omega, \cdot)$ is bounded, although $B(\omega, \cdot)$ can be very complex, e.g., $B(\omega, \cdot)$ is discontinuous due to the blockage effect. This sample-path property is referred to as the bounded jump or variation property.

3 A beam direction is a continuous quantity, while a beam index is a discrete quantity (i.e., an integer) obtained via quantization with a codebook. Here, $B(\omega, \cdot)$ represents a continuous and random quantity (i.e., beam direction function within $[0, T]$). In practice, the beam index is used instead. Note also that the first part of the statement, i.e., the discretization of a continuous function, does not belong to the Kolmogrov's perspective, which is, however, required to elaborate on the SP based modeling methodology.

- *Statistic regularity property (or prior)*: Although a single sample path, e.g., $B(\omega, \cdot)$, may be too random to predict its behavior, a large number of sample paths will present some regularity. This property is referred to as the statistic regularity (SR). In fact, for a specific scenario, i.e., the high-speed train communication, a set of sample paths $\{B(\omega_1, \cdot), \dots, B(\omega_K, \cdot)\}$ can be regarded as multiple realizations of a probability distribution, i.e., each $B(\omega_k, \cdot)$ is sampled as per a common distribution. The property related to the probability distribution is, in fact, the SR property or prior. The existence of SR is due to the fact that the surrounding environment is relatively stable.

The two types of properties are also observed from different timescales. Specifically, since the SR property can only be obtained from multiple sample paths or long-term observations, it belongs to long-term property or prior. In contrast, as the sample path property may be represented based on a single sample path or short-term observations, it therefore belongs to short-term property or prior. For the above scenarios, the beam direction functions have the aforementioned properties. For example, the beam direction function of an indoor environment has the property of beam continuity. In some cases, there is no obvious SR seen from a small number of sample paths, it presents significant SR if many sample paths are available.

It is not difficult to understand that although these priors are useful and exist extensively in practice, it is challenging to incorporate them into the existing frameworks or models. The reason for this is twofold. On the one hand, they involve considerable degrees of randomness. On the other hand, the dimension of function space underlying these priors is infinite. The two factors invalidate many deterministic and/or parametric techniques. Compared to the prior identification, we are more concerned with the use of these priors to improve efficiency, e.g., to reduce the sounding overhead.

1.4 Methodology from Stochastic Process

The perspective of function space not only provides useful insights but also offers efficient methodologies to model and characterize the beam process and predict the beam direction. As mentioned before, a mathematical object incorporating both function (or analytic) structure and probabilistic structure is the SP. Next, we further provide two methodologies to model the beam process and predict beam direction, i.e., SP theory and SDE. We first investigate the beam modeling methodology based on the SP theory, which is both natural and intuitive as per the function space view.

There are various SPs, such as the Poisson process, Markov chain, and Brownian motion. As an example, we take the GP to illustrate the methodology. In practice,

GP is also preferred by researchers and engineers. The reason for this is twofold. First, compared with many SPs, GP has explicit and analytical conditional density expression, which can avoid inference intractability. Second, GP is very flexible, which can be further extended and enhanced via kernel design and deep neural networks (DNNs). A GP describing beam directions is denoted by B. Since the mean function $m(t)$ of GP is often assumed to be zero function, a GP is characterized completely by the covariance function $k(t, t')$ (also referred to as kernel), which is defined as

$$k(t, t') = \mathbb{E}[(B(t) - m(t))(B(t') - m(t'))]. \tag{1.1}$$

Refer to Rasmussen and Williams [41] for more details of GP.

The key to GP modeling is that the nonparametric prior is encoded into the GP kernel. We still take high-speed train communication as an example to illustrate how to realize this goal [39]. In view that: (i) the beam direction of the high-speed train has the properties of inexact periodicity and continuity; and (ii) the spectral mixture kernel $k_{\text{SM}}(\cdot, \cdot)$ and squared exponential kernel $k_{\text{SE}}(\cdot, \cdot)$, respectively, encode periodicity [42] and continuity [41], the GP kernel is designed as a linear combination of the two kernels (along with a noise term) i.e.,

$$k(t, t') = c_1 k_{\text{SM}}(t, t') + c_2 k_{\text{SE}}(t, t') + \sigma_w^2 \delta_{t,t'}. \tag{1.2}$$

Since the representative ability of a simple kernel is limited and the selection or design of the kernel is heuristic, it may fail to encode complex priors into the GP kernel, which may lead to performance loss in complex scenarios. Hence, it is essential to strengthen the representative ability of the GP kernel and optimize the parameters of the kernel automatically. To this end, the (deep) neural network can be incorporated into a simple kernel so as to yield a more powerful kernel. A more flexible learning model is shown in Figure 1.4. To bring the advantages of

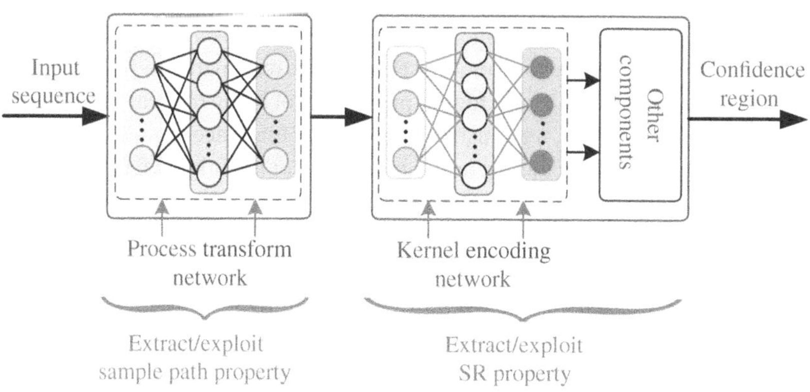

Figure 1.4 Network structure tailored for beam prediction oriented GP learning.

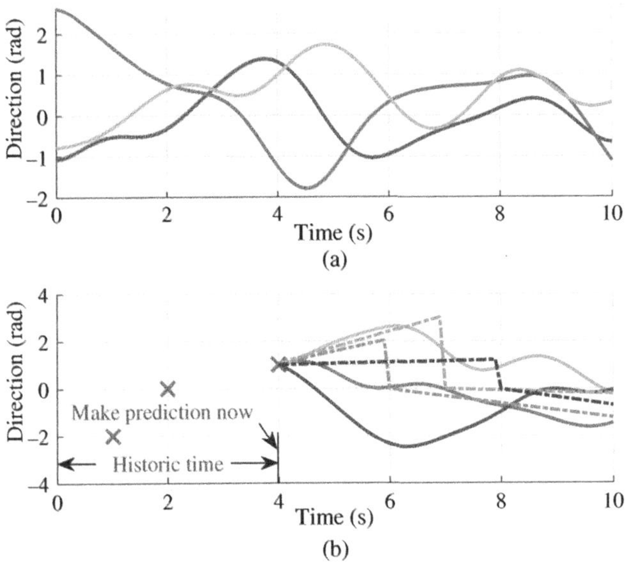

Figure 1.5 A visualization of SP inference that incorporates the nonparametric prior (e.g., smoothness). (a) Prior dataset and (b) inference instance.

the structure into full play, it is vital to tailor an efficient training method. Refer to Zhang et al. [39] for more details.

We further tackle the inference issue, which includes qualitative inference and quantitative inference. First, to obtain an intuitive understanding, we use a simple inference example to qualitatively demonstrate the essence of incorporating a nonparametric prior into SP inference. As illustrated in Figure 1.5, given a prior dataset (i.e., three sample paths) in Figure 1.5a and three data points (labeled by "+") in Figure 1.5b, the nonparametric SP inference paradigm is based on the fact that the prior sample paths are smooth and their variations are also small while the three-dotted curves are nonsmooth. Hence, it is reasonable to infer that the solid curves in Figure 1.5b have a higher probability of being the real sample path than the dotted ones. Next, we consider quantitative inference.

The basic principle of predicting or inferring $b(t_*)$ for an unknown time t_* based on a set of observations or samples $\mathcal{D} = \{(t_i, y_i) \,|\, y_i = b(t_i) + w_i, w_i \sim \mathcal{N}(0, \sigma_w^2), i = 1, \ldots, n\}$ is as follows. In contrast to other ML models or methods, GP regression is based on the Bayesian inference, which generates a probability distribution for the quantity of interest rather than a simple point estimate. Hence, it can provide a quantitative characterization of prediction uncertainty. Given the above $\mathcal{D}$, the conditional distribution of $b_* = b(t_*)$ is given by

$$p(b_*|\mathcal{D}, t_*) \sim \mathcal{N}\left(\mu(t_*), \sigma^2(t_*)\right), \tag{1.3}$$

$$\mu(t_*) = \mathbf{k}_*^{\mathrm{T}}(\mathbf{K} + \sigma_w^2\mathbf{I})^{-1}\mathbf{y}, \tag{1.4}$$

$$\sigma^2(t_*) = k_{**} - \mathbf{k}_*^{\mathrm{T}}(\mathbf{K} + \sigma_w^2\mathbf{I})^{-1}\mathbf{k}_*, \tag{1.5}$$

where the (i, j)th element of $\mathbf{K}$ and the ith element of $\mathbf{k}_*$ are calculated as $k(t_i, t_j)$ and $k(t_i, t_*)$, respectively. k_{**} is calculated as $k(t_*, t_*)$.

Next, we proceed to use SDE to model beam direction. Although the choice of SP, such as GP, to model the beam direction is very natural, it may not be intuitive to do this via SDE. In fact, the SDE theory has shown that the solution of an SDE is an SP. Moreover, SDE has a close relationship with dynamical system theory, which enriches our research means, e.g., it can better describe the variation of beam direction. The relationship between SP and SDE, as well as further extension of SDE, will be discussed in Section 1.6.

In practice, the system dynamics is almost always characterized by a system of SDEs, which can be written as

$$d\boldsymbol{x}(t) = \boldsymbol{f}(\boldsymbol{x}, t)dt + \boldsymbol{\sigma}(t)d\boldsymbol{\xi}(t), \quad \boldsymbol{x}(0) = \mathbf{x}_0, \tag{1.6}$$

where $\boldsymbol{x}(t)$ and $\boldsymbol{\xi}(t)$ represent the system state and Brownian motion, respectively. $\boldsymbol{f}(\boldsymbol{x}, t)$ and $\boldsymbol{\xi}(t)$ (modulated by $\boldsymbol{\sigma}(t)$) characterize the deterministic and stochastic parts of system evolution, respectively. From a control point of view, the drift term $\boldsymbol{f}(\boldsymbol{x}, t)$ controls the system to achieve a good predictive accuracy, while the diffusion term $\boldsymbol{\sigma}(t)$ characterizes model uncertainty in a stochastic environment [43].

A prominent advantage of the SDE-based prediction model is that it is good at quantifying uncertainty, which is often an urgent need for many real-life applications, especially BPT. As mentioned earlier, there are two types of uncertainties, i.e., aleatoric uncertainty (natural randomness inherent in a prediction task) and epistemic uncertainty (model uncertainty caused by lack of observation data). In many tasks, it is important to separate the two sources of uncertainties, while many existing methods suffer from the drawback of conflating the two types of uncertainties. Compared to existing models, the SDE-based models are able to separate the two sources of uncertainties in prediction, by explicitly modeling aleatoric uncertainty and epistemic uncertainty.

Similar to other ML models, modeling and learning are of great importance and have to be well addressed. There are mainly two methods to address these issues, including the direct method and the indirect method. The direct method models the drift term and diffusion term via parametric models directly, typically via the DNN. As for the training of an SDE-based model, it can be addressed via expectation maximization, variational Bayes, and other methods. As SDE can also be solved via various numerical or simulation methods, such as the Euler–Maruyama, Taylor series approximation, and Runge–Kutta methods, numerical- or simulation-based training and inference algorithms have been developed [43, 44].

The indirect method exploits the relationship between the stochastic dynamical model and the probabilistic ML model. Here, the probabilistic ML model is referred to the familiar ML model, such as DNN, Bayesian neural network, linear basis function model, GP and their combinations. Compared to a dynamical model, these classical ML models have been studied deeply, and both training and inference are very mature, stable, and reliable. Therefore, we can convert the stochastic dynamical model into a probabilistic one and utilize relevant methods to train the equivalent prediction model. We can also separate the training and inference so as to achieve the advantages from both fields. An example will be provided in Section 1.6 to illustrate this method.

To show the superiority and effectiveness of the SP-based methodology, we take two typical nonparametric priors, i.e., stochastic continuity and stochastic smoothness, as examples and employ the above methods to design efficient algorithms. The two priors are of infinite dimension, which thus cannot be described by classical parametric methods. We highlight that the ideas and algorithms themselves in the two examples are important as well. We first elaborate on the beam index difference method, which exploits the stochastic continuity.

1.5 Stochastic Continuity – Beam Index Difference

In this section, we identify and exploit the nonparametric prior, i.e., stochastic continuity, to design efficient BPT algorithms. The well-known Weierstrass approximation theorem asserts that any continuous function can be approximated by a polynomial function of infinite dimension, which shows that the continuity prior is also of infinite dimension. To tackle this issue, we propose the beam index difference technique.

1.5.1 Beam Index Difference Technique

Since the rate of change of beam direction heavily depends on the variation of the environment, the top priority within the beam adjustment strategy is to sense the variation of the environment. Let θ denote the angle of departure or arrival, i.e., AoD or AoA. Without loss of generality, we take AoD as an example. The rate of AoD change v_{AoD} can be defined as

$$v_{\text{AoD}} = \frac{\Delta\theta}{\Delta t} = \frac{\theta_t - \theta_{t-\Delta t}}{\Delta t}, \tag{1.7}$$

where $\Delta\theta$ denotes the difference of AoDs in two successive time slots and Δt equals to the duration of one time slot. This definition is similar to that of the velocity. Since Δt is fixed for a specific communication system, v_{AoD} reflects the rate of change of the AoD or the environment. By quantizing θ via a beam codebook, the difference of beam indexes also reflects the variation of the environment. Here is an example.

Example 1.1 ***(Beam Index Difference)***
Let $C = \{\mathbf{f}_0, \dots, \mathbf{f}_{31}\}$ be the codebook used by the base station (BS). Without loss of generality, it is assumed that $\mathbf{f}_{10}$ is used for DT in the current time slot. If the environment varies slowly, in the next time slot beam $\mathbf{f}_{11}$ may be used, i.e., the difference of the beam indices is $11 - 10 = 1$. On the other hand, if the environment varies quickly, the beam used in the next time slot may be $\mathbf{f}_7$, i.e., the difference is $7 - 10 = -3$.

Let the beam index of current time slot and the difference of beam indexes estimated be i and Δi, respectively. The beam for the next time slot is $i + \Delta i$. But Δi is a random variable, as the beam direction function is a random function. In fact, from the SP point of view, we must attach importance to the probability structure of the underlying beam process. To this end, instead of predicting a single beam, we predict a beam subspace that contains the real beam direction so as to alleviate misalignment. Then, local beam sweeping is invoked to find the beam. We also provide an example.

Example 1.2 ***(Beam Sweeping Subspace)***
It is assumed that $\mathbf{f}_{10}$ and $\mathbf{f}_{12}$ are the optimal beams in the current time slot t and the next time slot $t + 1$, respectively. Given the optimal beam $\mathbf{f}_{10}$ in the current time slot t, in the next time slot $t + 1$ the problem is to find the corresponding optimal beam $\mathbf{f}_{12}$. If only one beam is chosen to sweep the beam space, the chosen beams may be $\mathbf{f}_{11}$ or $\mathbf{f}_{13}$, which leads to misalignment. To tackle this issue, we can construct a beam subspace, e.g. $\{\mathbf{f}_{12}, \mathbf{f}_{13}\}$ or $\{\mathbf{f}_{11}, \mathbf{f}_{12}\}$, and then sweep the subspace.

Remark 1.1 Here, narrow beams are used to sweep, which is in contrast to the conventional beam search scheme, where the multiresolution beams are utilized. In general, the use of narrower beams may lead to larger beam sounding overhead. The rationale behind the design philosophy is born of the (LPBC) or stochastic continuity – in practice, the beam in the current timeslot is more likely to switch to one of its neighbor beams in the next timeslot, since AoD/AoA often changes continuously and the amplitude is limited by the maximal speed of mobile users. It is referred to Section 1.7 for more details about the LPBC. Moreover, a narrow beam often has a large array gain, which can better combat the large path loss.

1.5.2 BPT Solution via Beam Index Difference

In view that wireless communication is a process of interacting constantly with an ever-changing external environment, RL is chosen to design efficient BPT algorithms. To reveal important characteristics of beam index difference, we consider the simple stochastic bandit model [45, 46]. To employ the stochastic bandit model, we formulate the BPT problem as a stochastic bandit problem (a simplified Markov decision process problem), i.e., to define the action space and reward. In fact, the definition of action has already been discussed in detail, which, for clarity, is formally defined below.

Definition 1.1 Each action is defined by a pair of integers (u, b), where u denotes the difference or offset of the indices of the optimal beams in two adjacent time slots and b denotes the number of beams used to sweep the beam space. The action space is denoted by

$$\mathcal{B} = \{(u_1, b_1), (u_2, b_2), \dots, (u_L, b_L)\}, \tag{1.8}$$

where L denotes the size of the action space.

Without loss of generality, it is assumed that $\mathbf{f}_c$ is the optimal beam in the current time slot t and (u, b) is the action chosen for the next time slot $t + 1$, the set of beams used to sweep the beam space in the next time slot $t + 1$, denoted by $\mathcal{F}_{u,b}$, is given by

$$\mathcal{F}_{u,b} = \{\mathbf{f}_{c+u}, \mathbf{f}_{c+u+1}, \dots, \mathbf{f}_{c+u+b-1}\}. \tag{1.9}$$

Note that if the beam index i in $\mathbf{f}_i$ satisfies $i < 0$ or $i \geq M$, the beam used is $\mathbf{f}_{i \bmod M}$. In this chapter, any beam index is understood similarly if not explicitly stated. As shown in Figure 1.6, an action corresponds to a subset of codebook.

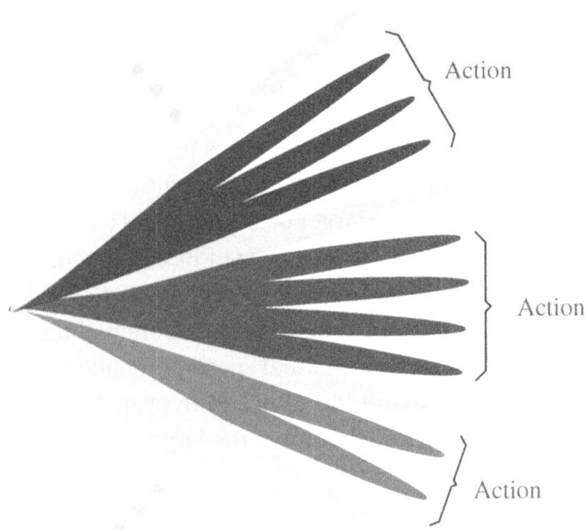

Figure 1.6 The actions are defined via beam subsets, which may overlap.

Remark 1.2 The first component u in action (u, b) is designed to sense the average rate of change of the environment. Intuitively, adjusting the beam index largely for a fast-changing environment can offset the rapid change of the environment, which helps to reduce the beam training overhead. The second component b is introduced to measure the variance of the rate of change of the environment. Intuitively, if the environment changes more irregularly, more beams (i.e., a larger b) should be swept in order to avoid misalignment. The introduction of b helps to improve the robustness of designed algorithms.

Next, we present an effective BPT algorithmic framework. Let $\mathbf{h}$ denote the channel vector between the transmitter and receiver. For an action (u, b) and beam $\mathbf{f}_i \in \mathcal{F}_{u,b}$, the received signals are given by

$$\mathbf{y}_i = \sqrt{P}\mathbf{h}^{\mathrm{H}}\mathbf{f}_i\mathbf{s} + \mathbf{n}_i,$$

where P is the transmit power, $\mathbf{s}$ is the pilot sequence, and $\mathbf{n}_i \sim \mathcal{CN}(\mathbf{0}, \mathbf{I})$ denotes the noise vector. The mean power of the received signals is calculated as $Y_i = |\mathbf{1}^{\mathrm{T}}\mathbf{y}_i|^2/L_p$, where L_p denotes the pilot length. The beam used for data transmission in the next time slot $t+1$ is $\mathbf{f}_{i^\star_{u,b}}$, with $i^\star_{u,b} = \mathrm{argmax}_i\{Y_i \mid \mathbf{f}_i \in \mathcal{F}_{u,b}\}$. Let T_0 denote the duration of transmitting one pilot signal. The reward corresponding to action (u, b), denoted by $R_{u,b}$, is defined as

$$R_{u,b} = (1 - bT_0/T_S)\log(1 + P|\mathbf{h}^{\mathrm{H}}\mathbf{f}_{i^\star_{u,b}}|^2). \tag{1.10}$$

Note that bT_0 in (1.10) characterizes the sounding overhead/time required to find the optimal beam. In practice, it is sufficient to estimate $|\mathbf{h}^{\mathrm{H}}\mathbf{f}_{i^\star_{u,b}}|$ to compute $R_{u,b}$, without estimating other system parameters, e.g., the path gain.

Let $\mathcal{F}_t$ denote the set of training beams in time slot t. The optimization goal is to maximize expected cumulative effective achievable rate (EAR) from time slot 1 to time slot n, i.e.,

$$\max_{\mathcal{F}_t} \mathbb{E}\left(\sum_{t=1}^{n} R_{\mathrm{eff},t}\right),$$

where $R_{\mathrm{eff},t}$ denotes the EAR for time slot t [46]. Based on the action and reward defined, we can present the BPT framework, as shown in Table 1.1. As an example, a specific BPT algorithm implemented based on the well-known upper confidence bound (UCB) policy is provided in Algorithm 1. The formulas used to update the bandit information are given by

$$\begin{aligned} T_{u,b} &\leftarrow T_{u,b} + 1 \\ \overline{\mu}_{u,b} &\leftarrow \overline{\mu}_{u,b} + (R_{u,b} - \overline{\mu}_{u,b})/T_{u,b}. \end{aligned} \tag{1.11}$$

It is referred to Zhang et al. [26] for more details about (1.11).

Table 1.1 BPT framework via stochastic bandit.

1 **input:** codebook C and action space $\mathcal{B} = \{(u_i, b_i)\}$
2 **find** out an initial optimal beam from C
3 **initialize** parameters and bandit information of the chosen stochastic bandit learning policy
4 **repeat** for each time slot
 (1) choose an optimal action $(u^\star, b^\star)$ and play the chosen action
 (2) choose an optimal beam $\mathbf{f}^\star$ within $\mathcal{F}_{u^\star,b^\star}$
 (3) transmit data with the optimal beam $\mathbf{f}^\star$
 (4) update bandit information of the policy

Algorithm 1 UCB-based BPT Algorithm

1: **input:** codebook C and $\mathcal{B} = \{(u_i, b_i)\}$ of size K
2: **find** an initial optimal beam from codebook C
3: **play** each action once and update bandit information according to (1.11); and let $n = K + 1$
4: **repeat** for each time slot
 (1) compute $\text{UCB}_{u_i,b_i}(n)$ for each (u_i, b_i) and let $(u^\star, b^\star) = \arg\max_{(u_i,b_i)} \text{UCB}_{u_i,b_i}(n-1)$
 (2) play the chosen action $(u^\star, b^\star)$
 (3) choose an beam in $\mathcal{F}_{u^\star,b^\star}$ to transmit data
 (4) update $T_{u^\star,b^\star}$ and $\bar{\mu}_{u^\star,b^\star}$ according to (1.11)
 (5) update counter: $n \leftarrow n + 1$

We further consider how to extend the proposed algorithm to the MIMO case. The key is still to design the action space and reward. Specifically, the reward is similarly defined as EAR. The (overall) action space $\mathcal{B}$ is defined as $\mathcal{B} = \mathcal{B}_T \times \mathcal{B}_R$, where $\mathcal{B}_T$ and $\mathcal{B}_R$ represent the action spaces for the transmitter (Tx) and receiver (Rx), respectively. With the optimal beam and action chosen by Tx (Rx) in the current time slot t available, the set of beams used for sweeping the beam space in the next time slot $t + 1$, denoted by $\mathcal{F}_{t,\text{T}}$ ($\mathcal{F}_{t,\text{R}}$), can be constructed according to (1.9). Then, the set of sounding beams for time slot $t + 1$ is $\mathcal{F}_t = \mathcal{F}_{t,\text{T}} \times \text{F}_{t,\text{R}}$.

The above method can also be extended to the multiuser case. However, it can only exploit the temporal correlation of channel variation, while ignoring the spatial correlation, which is also important in the multiuser communication scenario and should be paid attention to. To tackle this issue, the concept of beam image, as well as more efficient BPT solutions, has been proposed in Zhang et al. [32, 33]. In particular, an interactive learning design framework (ILDF) has been tailored for the BPT problem, which is demonstrated in Figure 1.7. The details are omitted here

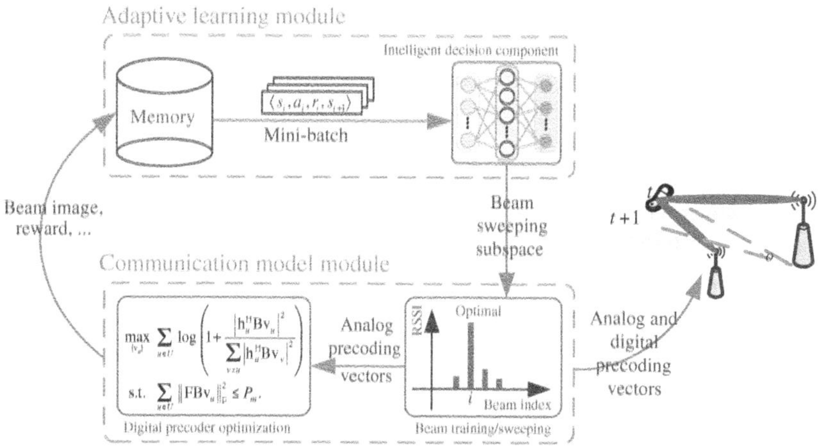

Figure 1.7 The principle of the interactive learning design framework.

due to space limitation. It is referred to Zhang et al. [32, 33] for more details. Note that till now, only the stationary environment is assumed and considered. But, we may encounter a nonstationary environment in practice. An effective method that applies to the nonstationary environment has been proposed in Zhang et al. [26].

1.5.3 Theoretical Analysis for Beam Index Difference

In this subsection, important characteristics and meaningful insights are analyzed and revealed for the beam index difference technique by deriving the lower bounds of the expected cumulative reward for the previous algorithm. All proofs are omitted to save space, which can be found in [26]. To simplify the analysis and derivation, a few assumptions are made below. However, the conclusions derived hold for general cases.

Assumption 1.1 The number of channel paths is assumed to be 1. The rationality of the assumption is that the line of sight (LOS) link is often dominated by a high-frequency channel with an LOS link [47]. The channel path gain α is further assumed to be a constant α_C so as to reveal the impact on system performance caused by the change of beam direction.

Assumption 1.2 The size of the codebook C is equal to the number of antennas, i.e., $M = N$. Similar to Alkhateeb et al. [48], the AoD is assumed to take value from a uniform grid of N points:

$$\phi \in \{-1 + 2i/M \mid i = 0, 1, \dots, M - 1\}. \tag{1.12}$$

Assumption 1.3 The reward for each action is drawn independently from a fixed distribution that depends on the action but not on the time slot. Moreover, if an optimal action, e.g. (u, b) is chosen, the probability that the optimal beam is within $\mathcal{F}_{u,b}$ is 1. Otherwise, it cannot be optimal. The reward for each action has been normalized to be in $[0, 1]$.

For a time-varying channel, it is more desired to evaluate finite time or long-term performance, i.e., expected cumulative effective achievable rate (ECEAR) $E_n = \mathbb{E}\left(\sum_{t=1}^{n} R_{\text{eff},t}\right)$. With E_n available, it is possible to characterize the asymptotic average ECEAR, i.e., $n^{-1}E_n$ or its limit. For this purpose, we require the following useful lemma [49].

Lemma 1.1 For an action (u, b), the expected EAR and the number of times that (u, b) has been chosen within n rounds are denoted by $\mu_{u,b}$ and $T_{u,b}(n)$, respectively. Then, E_n can be decomposed as

$$E_n = \sum_{(u,b)\in\mathcal{B}} \mu_{u,b}\mathbb{E}(T_{u,b}(n)). \tag{1.13}$$

According to the above lemma, we can obtain that it is sufficient to analyze $\{\mu_{u,b}\}$ and $\{\mathbb{E}(T_{u,b}(n))\}$ to characterize E_n. Let $\mathbf{f}_{\text{opt}}$ denote the optimal beam for a given time slot. Under Assumptions 1.1–1.3, $\mu_{u,b}$ can be calculated as

$$\begin{aligned}\mu_{u,b} &= \mathbb{P}(\mathbf{f}_{\text{opt}} \in \mathcal{F}_{u,b})\mathbb{E}(R_{u,b}|\mathbf{f}_{\text{opt}} \in \mathcal{F}_{u,b}) + \mathbb{P}(\mathbf{f}_{\text{opt}} \notin \mathcal{F}_{u,b})\mathbb{E}(R_{u,b}|\mathbf{f}_{\text{opt}} \notin \mathcal{F}_{u,b})\\ &\overset{(*)}{=} \mathbb{P}(\mathbf{f}_{\text{opt}} \in \mathcal{F}_{u,b})\mathbb{E}(R_{u,b}|\mathbf{f}_{\text{opt}} \in \mathcal{F}_{u,b}),\end{aligned}$$

where $(*)$ is due to $R_{u,b} = 0$ if $\mathbf{f}_{\text{opt}} \notin \mathcal{F}_{u,b}$. Due to the third assumption, $\mu_{u,b}$ can be rewritten as $\mu_{u,b} = \mathbb{E}(R_{u,b}|\mathbf{f}_{\text{opt}} \in \mathcal{F}_{u,b})$ if (u, b) is an optimal action. For convenience, let $\mu_{u,b}^{\text{suc}} = \mathbb{E}(R_{u,b}|\mathbf{f}_{\text{opt}} \in \mathcal{F}_{u,b})$, which can be bounded below.

Lemma 1.2 Under Assumptions 1.1–1.3, $\mu_{u,b}^{\text{suc}}$ can be, respectively, upper bounded and lower bounded by

$$\begin{aligned}&\mu_{u,b}^{\text{suc}}(\alpha_C, P) \leq r_{u,b}(\alpha_C, P)\\ &\mu_{u,b}^{\text{suc}}(\alpha_C, P) \geq \max\left\{0, r_{u,b}(\alpha_C, P)\left(1 - 0.5(b-1)e^{-\frac{1}{4}\lambda^\star}\right)\right\},\end{aligned} \tag{1.14}$$

where $r_{u,b}(\alpha_C, P)$ and $\lambda^\star$ are, respectively, given by

$$r_{u,b}(\alpha_C, P) = (1 - bT_0/T_S)\log(1 + PN|\alpha_C|^2\beta^{-1}) \tag{1.15}$$

$$\lambda^\star = 2L_p PN|\alpha_C|^2\beta^{-1}. \tag{1.16}$$

It can be verified that for a sufficiently large $\lambda^\star$ (e.g., when the transmit power P is large), the term $\exp(-\frac{\lambda^\star}{4})$ tends to zero. As a result, $\mu_{u,b}^{\text{suc}}$ can be safely approximated by $r_{u,b}$ in this case, i.e.,

$$\mu_{u,b}^{\text{suc}}(\alpha_C, P) \approx r_{u,b}(\alpha_C, P).$$

Hence, the upper and lower bounds nearly coincide.

Next, we analyze and bound the ECEAR performance metric, whose key idea is as follows. Let $(u^\star, b^\star)$ be an optimal action, which is further assumed to be unique. Then, set $\mathcal{B}_{\text{nop}} \triangleq \mathcal{B}\backslash\{(u^\star, b^\star)\}$ collects all nonoptimal actions. Since E_n is the sum of expected cumulative rewards of the optimal action $(u^\star, b^\star)$ and nonoptimal actions $\{(u, b) \in \mathcal{B}\backslash(u^\star, b^\star)\}$ and, more importantly, the optimal action $(u^\star, b^\star)$ dominates E_n, a lower bound can be obtained by ignoring the contributions from all nonoptimal actions, i.e.,

$$E_n = \mu_{u^\star,b^\star}\mathbb{E}[T_{u^\star,b^\star}(n)] + \sum_{(u,b)\in\mathcal{B}_{\text{nop}}} \mu_{u,b}\mathbb{E}[T_{u,b}(n)] \geq \mu_{u^\star,b^\star}\mathbb{E}[T_{u^\star,b^\star}(n)]. \tag{1.17}$$

The upper and lower bounds of $\mu_{u^\star,b^\star}$ have been characterized in Lemma 1.2. To lower bound E_n, it is sufficient to lower bound $\mathbb{E}[T_{u^\star,b^\star}(n)]$, which can be calculated as

$$\mathbb{E}[T_{u^\star,b^\star}(n)] = n - \sum_{(u,b)\in\mathcal{B}_{\text{nop}}} \mathbb{E}[T_{u,b}(n)]. \tag{1.18}$$

The following task is to upper bound $\mathbb{E}(T_{u,b}(n))$ for each nonoptimal action, which depends on a specific strategy. Two typical strategies, i.e., the ε-greedy and UCB strategies are considered here. We first consider the ε-greedy strategy under the assumption $\varepsilon_t = \varepsilon_0\,(t \geq 1)$ with $\varepsilon_0 \in (0, 1)$ a constant. As per Lemma 1.1, it is sufficient to bound $\{\mathbb{E}[T_{u,b}(n)]\}$ to characterize E_n. The following lemma provides an upper bound for a nonoptimal action (u, b).

Lemma 1.3 If $\varepsilon_t = \varepsilon_0\,(t \geq 1)$ with $\varepsilon_0 \in (0, 1)$, for any nonoptimal action (u, b), $\mathbb{E}[T_{u,b}(n)]$ is upper bounded by

$$\mathbb{E}[T_{u,b}(n)] \leq \frac{\varepsilon_0 n}{K} + 2(1-\varepsilon_0)\sum_{t=1}^{n}\left(\frac{t}{2K}\exp\left(-\frac{t}{10K}\right)\right.$$
$$\left. + \frac{2}{\Delta_{u,b}^2}\exp\left(-\frac{\lfloor t/(2K)\rfloor \Delta_{u,b}^2}{2}\right)\right), \tag{1.19}$$

where $\Delta_{u,b} = \mu_{u^\star,b^\star} - \mu_{u,b}$ characterizes the expected EAR gap between the optimal action and nonoptimal action (u, b).

Although the upper bound in (1.19) consists of two terms, i.e., the linear term (in terms of n) and exponential-sum term, the latter is asymptotically negligible.

In fact, it is independent of the number of rounds n [26]. Specifically, $\mathbb{E}[T_{u,b}(n)]$ can be upper bounded by

$$\mathbb{E}[T_{u,b}(n)] \leq \frac{\varepsilon_0 n}{K} + 2(1-\varepsilon_0)\left(10K + 1 + \frac{8K}{\Delta_{u,b}^4}\right). \tag{1.20}$$

Based on (1.20), as well as Lemmas 1–3, the lower bound of ECEAR for the ε-greedy based BPT algorithm can be obtained immediately.

Theorem 1.1 Under Assumptions 1.1–1.3 and $\varepsilon_t = \varepsilon_0\,(t \geq 1)$, a lower bound of ECEAR for the BPT algorithm in Table 1.1 with the ε-greedy strategy is given by

$$E_n \geq \mu_{u^\star,b^\star}(\alpha_C, P)\cdot\Bigg((1-\varepsilon_0+\varepsilon_0/K)n - 2(1-\varepsilon_0)\sum_{(u,b)\in\mathcal{B}_{\text{nop}}}\left(10K+1+8K\Delta_{u,b}^{-4}\right)\Bigg). \tag{1.21}$$

Moreover, the asymptotic average ECEAR satisfies

$$\lim_{n\to\infty}\frac{E_n}{n} = (1-\varepsilon_0+\varepsilon_0/K)\mu_{u^\star,b^\star}(\alpha_C, P) \tag{1.22}$$

$$\approx (1-\varepsilon_0+\varepsilon_0/K)(1-b^\star T_0/T_S)\cdot\log(1+P|\alpha_C|^2\beta^{-1}) \tag{1.23}$$

$$\geq (1-\varepsilon_0+\varepsilon_0/K)(1-b_{\max} T_0/T_S)\cdot\log(1+P|\alpha_C|^2\beta^{-1}), \tag{1.24}$$

where $b_{\max} = \max\{b\,|\,(u,b)\in\mathcal{B}\}$.

The parameter ε_0 measures the degree of exploration. A larger ε_0 implies more explorations, which, however, decreases the ECEAR performance. To better balance the exploration and exploitation, an adaptive exploration strategy, e.g., $\varepsilon_n = C/\log(n+1)$ or $\varepsilon_n = C/(n+D)$, can be used.

Remark 1.3 The values $b^\star$ and $b_{\max}$ in Theorem 1.1 depend on the specific environment, i.e., the proposed BPT algorithm is an environment-adaptive algorithm, which can achieve good performance in a variety of environments.

Remark 1.4 Theorem 1.1 shows that ECEAR E_n or asymptotic average $\lim_{n\to\infty}E_n/n$ is independent of the average rate of change of environment (i.e., the component $u^\star$), which implies that the beam index difference technique can offset the rapid change in environment. Hence, the sounding overhead can still be small even though the environment varies rapidly. Theorem 1.1 also shows that the ECEAR E_n and its asymptotic average $\lim_{n\to\infty}E_n/n$ depend on the variance of the change of the environment. If the variance of change of the environment (i.e., $b^\star$ in (1.23)) is larger, the sounding overhead is larger, e.g., the environment varies more irregularly.

Next, we consider another classical strategy, i.e., the UCB strategy. Similar to the ε-greed strategy, the first step is also to bound $\{\mathbb{E}[T_{u,b}(n)]\}$ of a nonoptimal action (u, b) for the UCB-based BPT algorithm. The following lemma provides an upper bound of $\mathbb{E}[T_{u,b}(n)]$ for a nonoptimal action (u, b).

Lemma 1.4 Under Assumptions 1.1–1.3, $\mathbb{E}[T_{u,b}(n)]$ of a nonoptimal action (u, b) for the UCB strategy is bounded by

$$\mathbb{E}[T_{u,b}(n)] \leq 1 + \frac{8c}{\Delta_{u,b}^2(c-2)} + \frac{8\left(\log(n^{c/2}) + \sqrt{\pi \log(n^{c/2})} + 1\right)}{\Delta_{u,b}^2}. \tag{1.25}$$

In contrast to the ε-greedy strategy, whose $\mathbb{E}[T_{u,b}(n)]$ for a nonoptimal action (u, b) is of order $\mathcal{O}(n)$, $\mathbb{E}[T_{u,b}(n)]$ of the UCB strategy is of order $\mathcal{O}(\log n)$. Hence, it is expected that the UCB-based BPT algorithm can achieve a better ECEAR performance, as shown in the following theorem.

Theorem 1.2 Under Assumptions 1.1–1.3, the ECEAR of BPT solution in Table 1.1 with the UCB strategy is given by

$$E_n \geq \mu_{u^\star,b^\star}(\alpha_C, P) \cdot \left(n - K + 1 - \frac{8c}{c-2} \sum_{(u,b)\in\mathcal{B}_{\text{nop}}} \Delta_{u,b}^{-2} \right.$$
$$\left. 8\left(\log(n^{c/2}) + \sqrt{\pi \log(n^{c/2})} + 1\right) \sum_{(u,b)\in\mathcal{B}_{\text{nop}}} \Delta_{u,b}^{-2} \right). \tag{1.26}$$

Accordingly, the asymptotic average ECEAR satisfies

$$\lim_{n\to\infty} \frac{E_n}{n} = \mu_{u^\star,b^\star}(\alpha_C, P). \tag{1.27}$$

Theorem 1.2 indicates that compared to the ε-greedy-based BPT solution, the UCB counterpart can achieve better EAR performance. The reason for this is that the UCB policy is an adaptive exploration strategy, which makes better use of collected historical information. The two theorems show that the beam index difference technique can sense the change in environment and adjust the beam sounding strategy intelligently to improve system performance.

1.6 Stochastic Smoothness – Hybrid Data-induced Kalman Filtering

It is a well-known fact in the ordinary differential equation (ODE) theory that a smooth curve can be thought of as the solution of an ODE. Similarly, a family of stochastic smooth curves of some regularity can be regarded as the solution of an SDE. Based on the intuitive fact, we present the HDIKF approach. Another motivation for HDIKF is that Kalman filtering is inference-efficient but modeling-inefficient, while GP is modeling-efficient but inference-inefficient. Hence, it is natural to raise the question, i.e., whether we can incorporate the two methods, which can exploit their advantages while overcoming their drawbacks.

1.6.1 Theoretical Foundation

First and foremost, we shall shed light on the relationship between the two system modeling methods, more specifically, SDE and SP. As mentioned earlier, a system dynamics model can be characterized by a system of SDEs, i.e.,

$$d\boldsymbol{x}(t) = \boldsymbol{f}(\boldsymbol{x}, t)dt + \sigma(t)d\xi(t), \quad \boldsymbol{x}(0) = \boldsymbol{0}. \tag{1.28}$$

It is referred to Section 1.4 for the meaning of $\boldsymbol{x}(t)$, $\boldsymbol{f}(\boldsymbol{x}, t)$, $\sigma(t)$, and $\xi(t)$. To reveal the relationship between SDE and SP, we first investigate a closely related but simpler SDE, where the drift term $\boldsymbol{f}(\boldsymbol{x}, t)$ is absent, i.e.,

$$d\boldsymbol{y}(t) = \sigma(t)d\xi(t), \quad \boldsymbol{y}(0) = \boldsymbol{0}. \tag{1.29}$$

To avoid confusion, we shall explicitly specify the underlying probability space for the SDE in (1.29), which is denoted by $(\Omega, \mathcal{F}, P)$. The notation $\mathbb{E}_P$ is introduced to emphasize that the expectation is taken with respect to the probability measure P. For the SDE in (1.29), we have the following lemma.

Lemma 1.5 The solution of SDE (1.29) is a GP with mean function $\boldsymbol{0}$ and kernel function given by

$$\mathbb{E}_P\left(\boldsymbol{x}(t)\boldsymbol{x}^{\mathrm{T}}(t')\right) = \int_0^{\min(t,t')} \sigma(\tau)\sigma^{\mathrm{T}}(\tau)d\tau. \tag{1.30}$$

The proofs of Lemma 1.5 and the other theorems or corollaries are omitted to save space, which can be found in [31]. Lemma 1.5 indicates that a GP can be derived from an SDE and vice versa. In fact, under very mild conditions, the famous martingale representation theorem [50] asserts that a GP can be represented by an Itô integral with respect to the Brownian motion. Specifically, the SDE can be written as

$$\boldsymbol{y}(t) = \boldsymbol{y}(0) + \int_0^t \theta(\omega, t)d\xi(t), \tag{1.31}$$

where $\theta(\omega, \cdot)$ is an SP adaptive to the Brownian motion, i.e., a function of the Brownian motion. In literature, the SDE in (1.31) is briefly written as

$d\boldsymbol{y}(t) = \Theta(t)d\xi(t)$. The discussion above indicates that, to some extent, SDE and GP are equivalent.

We next consider the general form. The following theorem establishes the relationship between SDEs (1.28) and (1.29).

Theorem 1.3 There exists a Borel measurable transform $\boldsymbol{F}$ and a probability measure Q (still defined on $(\Omega, \mathcal{F})$) such that under the measure Q, $\boldsymbol{x}(t)$ satisfies the following SDE:

$$d\boldsymbol{x}(t) = \sigma(t)d\tilde{\xi}(t), \quad \boldsymbol{x}(0) = \boldsymbol{0}, \tag{1.32}$$

where $\tilde{\xi}(t)$ is a standard Brownian motion under the measure Q. In particular, the Q-law of $\boldsymbol{y}(t)$ is the same as the P-law of $\boldsymbol{x}(t)$, where $\boldsymbol{y}(t)$ is the solution of SDE (1.29).

Remark 1.5 Because the Q-law of $\boldsymbol{y}(t)$ and the P-law of $\boldsymbol{x}(t)$ coincide, they have the same finite-dimensional distributions. Hence, Theorem 1.3, in fact, constitutes the theoretical foundation of the HDIKF approach. First, as a bridge, it links two different branches of system modeling methodologies. In particular, it provides an efficient ML model or structure for a dynamical system, which avoids heuristic (and even) random selection of the prediction model. Moreover, it guarantees that the ML model does not cause a systematic error.

Based on Theorem 1.3, we can present the HDIKF approach, as illustrated in Figure 1.8. To fully exploit the advantages of both Kalman filtering and GP, HDIKF models the system dynamics via GP and performs inference or prediction via Kalman filter, i.e., GP for modeling and Kalman filtering for inference. The key components of HDIKF include historical data collection, prior extraction and implicit representation of system dynamics, model conversion, explicit SDE representation of system dynamics. The roles of these components are as follows.

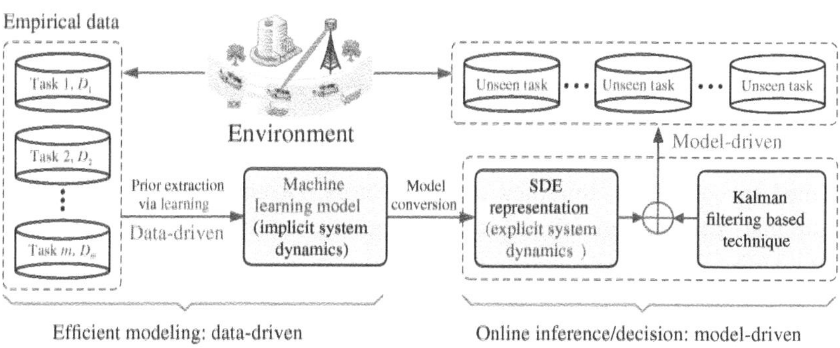

Figure 1.8 The principle of hybrid data-induced Kalman filtering approach.

To train or optimize a learning system or prediction model, we first collect the training samples. Fortunately, the required dataset can be collected online. Then, we choose an appropriate learning model and train the learning model to obtain an implicit representation of system dynamics. With the implicit representation available, we can already make decisions when faced with new tasks, but at the cost of large computational complexity. To tackle this issue and use the Kalman filter, we convert the implicit representation of system dynamics into an explicit one, i.e., an SDE that describes system evolution.

To reach our goal, we still need to solve two problems. The first one is how to choose an appropriate prediction model and, more difficultly, how to train or optimize the prediction model, which makes the learned prior and model applicable to future or new tasks. The second one is how to convert the implicit representation of system dynamics into an explicit one.

1.6.2 Implicit Dynamics Learning via Multitask Learning

In this subsection, we address the first problem. Theorem 1.3 indicates that a GP incorporating the DNN can completely characterize the probabilistic behaviors of a dynamical system (1.28). In fact, the measurable transform or function can be approximated by a DNN parameterized by Θ_T. As for measure Q, we choose a basic GP kernel (i.e., the kernel without incorporating other ML component like neural network, and an example is the squared exponential kernel) denoted by $\mathcal{GP}$ whose parameters are collected into set Θ_B. But if necessary, the basic kernel can be replaced by a complex kernel, which may be enhanced and parameterized by a DNN. The overall prediction model can be denoted by

$$\boldsymbol{x}(t) \sim \mathcal{GP}(t, t' | \Theta_T, \Theta_B). \tag{1.33}$$

Next, we derive an efficient training method for the prediction model in (1.33) in a data-driven manner. To this end, we first investigate single-task GP, which consists of two main steps. The first one is to choose an appropriate model which can encode the prior of interest (e.g., domain knowledge). The prior is updated into a posterior when a dataset is available, which constitutes the second step. The details are as follows. The dataset is denoted by $\boldsymbol{S} = \{(x_1, y_1), \dots, (x_n, y_n)\}$, with x_i and y_i denoting input and output, respectively. Note that x_i and y_i are assumed to be scalars. But they can be vectors. Given $\boldsymbol{S}$, the predictive (or conditional) distribution for an unseen input x_u can be calculated explicitly (See (1.3)–(1.5)).

In practice, we often have multiple datasets $\mathfrak{D} = \boldsymbol{S}_1, \dots, \boldsymbol{S}_m$ from multiple related tasks, where $\boldsymbol{S}_i$ corresponds to task i. As an example, for the BPT problem, $\boldsymbol{S}_i$ takes the form

$$\boldsymbol{S}_i = \{(t_1, a_1), (t_2, a_2), \dots, (t_{n_i}, a_{n_i})\}, \tag{1.34}$$

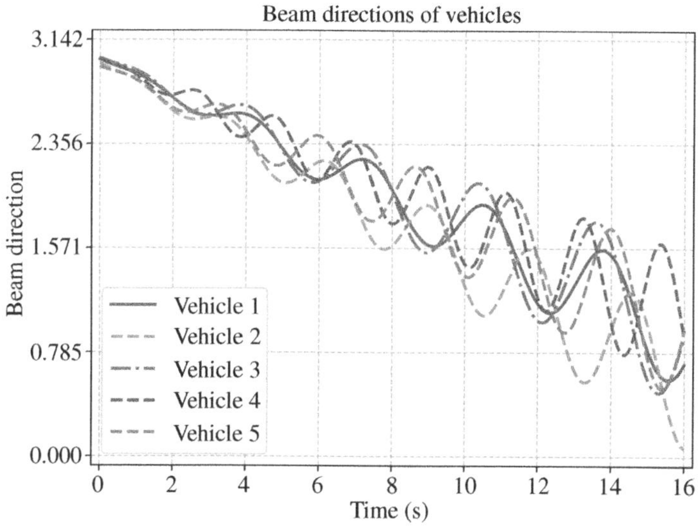

Figure 1.9 An illustration of beam direction trajectories of different vehicles.

where t_i and a_i represent sampling time and beam direction, respectively. For intuitive understanding, an example is shown in Figure 1.9, where each $\boldsymbol{S}_i$ corresponds to the beam trajectory of a vehicle. As shown in Figure 1.9, the distributions of different datasets are different because of different system parameters, i.e., the vehicle speed. Hence, it is problematic to extend the single-task method directly to the multitask case, i.e., to train a single and shared model with $\mathfrak{D}$. But note that the datasets really share many features, e.g., a similar trend, which can be exploited to improve system performance, i.e., to improve the small sample performance of a learning algorithm.

We tackle these issues from the perspective of hierarchical modeling and SP. Specifically, the datasets $\{\boldsymbol{S}_i\}$ are regarded as different realizations of an SP. Mathematically, each $\boldsymbol{S}_i$ is obtained by discretizing a continuous-time function $x_i(t)$, while $x_i(t)$ is sampled from an SP, e.g., $\mathcal{GP}(t, t'|\Theta_T, \Theta_B)$. Let $\Theta = \{\Theta_T, \Theta_B\}$. Then, we shall optimize Θ based on $\mathfrak{D}$. Note that the prior shared across different tasks, which is applicable to future and unseen tasks, should be extracted from $\mathfrak{D}$. To this end, the Bayesian hierarchical modeling is utilized. First, a prior distribution, denoted by $\mathcal{P}(\Theta)$, is imposed on Θ. Without loss of generality, Θ is assumed to be distributed as $\mathcal{N}(\mathbf{0}, \sigma_\Theta^2 \mathbf{I})$, i.e., $\mathcal{P} = \mathcal{N}(\mathbf{0}, \sigma_\Theta^2 \mathbf{I})$. Given $\mathfrak{D}$, it is sufficient to update the prior $\mathcal{P}(\Theta)$ into a posterior $\mathcal{Q}(\Theta)$.

The method to obtain the posterior $\mathcal{Q}(\Theta)$ is closely related to the optimization criterion. Note that the design goal is that the obtained posterior can improve prediction performance on unseen tasks, more specifically, to maximize the generalization performance, e.g., to minimize the transfer error. An effective posterior $\mathcal{Q}^\star$ has been derived in Zhang et al. [31] (see Theorem 1.4).

Theorem 1.4 With the aim of minimizing the generalization or transfer error, an efficient posterior is given by

$$Q^\star(\Theta) = \frac{\mathcal{P}(\Theta)\exp\left(\left(1+\sum_{i=1}^m n_i^{-1}\right)^{-1}\sum_{i=1}^m \frac{1}{n_i} Z(\boldsymbol{S}_i,\Theta)\right)}{\mathbb{E}_{\Theta\sim\mathcal{P}}\left[\exp\left(\left(1+\sum_{i=1}^m n_i^{-1}\right)^{-1}\sum_{i=1}^m \frac{1}{n_i} Z(\boldsymbol{S}_i,\Theta)\right)\right]},$$

where $Z(\boldsymbol{S}_i,\Theta)$ represents the marginal log-likelihood for $\boldsymbol{S}_i$ and Θ. In particular, $Z(\boldsymbol{S}_i,\Theta)$ for GP is calculated as[4]

$$Z(\boldsymbol{S}_i,\Theta) = -\frac{1}{2}\mathbf{a}_i^{\mathrm{T}}(\mathbf{C}_i+\sigma_i^2\mathbf{I})^{-1}\mathbf{a}_i - \frac{1}{2}\log\det(\mathbf{C}_i+\sigma_i^2\mathbf{I}) - \frac{n_i}{2}\log 2\pi. \tag{1.35}$$

To facilitate the use of the Kalman filter and reduce complexity, a point estimate of Θ is preferable. The optimal point estimate is the MAP (maximum a posterior) estimate, i.e.,

$$\Theta^\star = \arg\max_{\Theta} Q(\Theta). \tag{1.36}$$

The optimal point estimate $\Theta^\star$ can be equivalently obtained by maximizing the numerator of $Q^\star(\Theta)$, i.e.,

$$\Theta^\star = \arg\max_{\Theta} \ln\mathcal{P}(\Theta) + \left(1+\sum_{i=1}^m \frac{1}{n_i}\right)^{-1}\sum_{i=1}^m \frac{Z(\boldsymbol{S}_i,\Theta)}{n_i}. \tag{1.37}$$

The complete procedure to obtain the implicit system dynamics is summarized in Algorithm 2.

Algorithm 2 Implicit System Dynamics Modeling

1: **input:** empirical datasets $\mathfrak{D} = \{S_1, S_2, \dots, S_m\}$
2: **construct** and initialize DNN and GP kernel
3: **repeat**
   (a) **sample** a small batch of datasets from $\mathfrak{D}$
   (b) **compute** loss as per (1.37) with sampled data
   (c) **update** parameters of DNN and GP kernel
   until some stopping criterion is met
4: **output:** $\Theta^\star$ - parameters of system dynamics

---

Once $\Theta^\star$ is available, the implicit system dynamics model has been obtained, based on which a Bayesian inference can be made. But the inference is based on GP, which leads to a large computational complexity. Next, we tackle this issue.

4 For a one-dimension dataset, $\boldsymbol{S}_i = \{(t_1,a_1),(t_2,a_2),\dots,(t_{n_i},a_{n_i})\}$, where $a_j = f_i(t_j)+\varepsilon_i$ with $\varepsilon_i \sim \mathcal{N}(0,\sigma_i^2)$. Then, $\mathbf{a}_i$ takes the form $\mathbf{a}_i = (a_1,a_2,\dots,a_{n_i})^{\mathrm{T}}$ and $\mathbf{C}_i$ is calculated as $[\mathbf{C}_i]_{uv} = k(t_u,t_v)$.

1.6.3 SDE Representation and Efficient Inference

To enable efficient inference, the explicit system dynamics, i.e., the SDEs that describe system state evolution and measurement are required. To this end, we shall convert model (1.33) with $\Theta^\star$ obtained based on (1.37) into an SDE representation. Without loss of generality, the fully connected neural network is chosen to parameterize the transform, and the input–output relationship of the neural network can be written as

$$z = h(t, \Theta_T). \tag{1.38}$$

The output of the above neural network is further fed to the basic kernel, which is denoted by $k_B(\cdot, \cdot \mid \Theta_B)$ with parameters Θ_B. Hence, the overall prediction model is given by

$$k(t, t') = k_B\left(h(t, \Theta_T), h(t', \Theta_T) \mid \Theta_B\right). \tag{1.39}$$

In view that many basic kernels are stationary, it is assumed here that the basic kernel $k_B(z, z' | \Theta_B)$ is also stationary. In this case, the basic kernel can be equivalently written as

$$k_B(z, z' \mid \Theta_B) = k_B(z - z' \mid \Theta_B). \tag{1.40}$$

The spectral factorization method can be invoked to obtain an SDE representation [51]. The procedure to convert a stationary kernel into an equivalent SDE representation is as follows:

Step 1: The corresponding spectral density $S(\omega)$ can be obtained by computing the Fourier transform of $k_B(\cdot \mid \Theta_B)$.

Step 2: In many cases, $S(\omega)$ can be written as a rational function taking the form

$$S(\omega) = \frac{1}{p_n(\omega^2)}, \tag{1.41}$$

where $p_n(\cdot)$ is a polynomial of nth order. Otherwise, we can approximate $S(\omega)$ with such a function, e.g., via the Taylor series expansion or Padé approximate.

Step 3: A stable rational transfer function $H(i\omega)$ can be found, which takes the following form

$$H(i\omega) = \frac{\sigma_0}{(i\omega)^n + a_{n-1}(i\omega)^{n-1} + \cdots + a_1(i\omega) + a_0}, \tag{1.42}$$

and meanwhile, the spectral density $S(\omega)$ can be decomposed as $S(\omega) = \sigma_0^2 H(i\omega)H(-i\omega)$.[5]

5 The procedure to find the transfer function is called spectral factorization, which consists of two steps. First, the roots of the denominator are computed. Note that the roots always appear in pairs, where one member of the pair is the complex conjugate of the other one. Then, the denominator polynomial of $H(i\omega)$ can be constructed from the positive-imaginary-part roots only.

With the transfer function $H(j\omega)$ available, we can obtain an SDE of nth order, which can be written as

$$\frac{d^n u(z)}{dz^n} + \cdots + a_1 \frac{du(z)}{dz} + a_0 u(z) = \sigma_0 \frac{d\xi(z)}{dz},$$

where $\xi(z)$ denotes the standard Brownian motion. Let $\boldsymbol{x}(z) = (u(z), du(z)/dz, \dots, d^{n-1}u(z)/dz^{n-1})$ and $\mathbf{V} = \mathrm{diag}(0, \dots, 0, \sigma_0)$. The SDE can be compactly rewritten as

$$d\boldsymbol{x}(z) = \mathbf{C}\boldsymbol{x}(z)dz + \mathbf{V}d\boldsymbol{\xi}(z), \tag{1.43}$$

where $\boldsymbol{\xi}(z)$ denotes the standard Brownian motion of dimension n and matrix $\mathbf{C}$ is given by

$$\mathbf{C} = \begin{pmatrix} 0 & 1 & & \\ & \ddots & \ddots & \\ & & 0 & 1 \\ -a_0 & -a_1 & \cdots & -a_{n-1} \end{pmatrix}.$$

We proceed to derive the measurement equation. We define vector $\mathbf{b} = [1, 0, \dots, 0]^{\mathrm{T}}$, whose dimension coincides with $\boldsymbol{x}(z)$. The measurement equation can be expressed as

$$b(z) = \mathbf{b}^{\mathrm{T}}\boldsymbol{x}(z) + \eta(z), \tag{1.44}$$

where $\eta(z)$ represents the measurement noise. With (1.43) and (1.44) available, the continuous-time Kalman filtering algorithm can be applied, which is, however, computation-unfriendly. It is also problematic to invoke the discrete Kalman filter on the simple discretization model.

To tackle this issue, we propose an equivalent discretization for the SDE in (1.43) by investigating the statistics (e.g., the differential equations that characterize the mean and variance variations) of the corresponding SP. Note that the equivalence means that the probability distributions of the continuous-time SDE and the discrete-time stochastic system coincide at sampling points $\{z_k\}$. The equivalent discretization is formally stated in the following theorem.

Theorem 1.5 The SDE in (1.43) is in distribution equivalent to the following discrete-time stochastic system

$$\boldsymbol{x}(z_{k+1}) = \mathbf{T}_k \boldsymbol{x}(z_k) + \mathbf{q}_k, \qquad \mathbf{q}_k \sim \mathcal{N}(\mathbf{0}, \mathbf{\Pi}_k), \tag{1.45}$$

where $\mathbf{T}_k$ and $\mathbf{\Pi}_k$ are respectively given by

$$\begin{aligned} \mathbf{T}_k &= \exp(\mathbf{C}(z_{k+1} - z_k)) \\ \mathbf{\Pi}_k &= \int_{z_k}^{z_{k+1}} \exp(\mathbf{C}(z_{k+1} - s))\mathbf{V}\mathbf{V}^{\mathrm{T}} \exp(\mathbf{C}(z_{k+1} - s))^{\mathrm{T}} ds. \end{aligned}$$

To invoke the discrete Kalman filter, we still need an initial distribution for $\boldsymbol{x}(t_0)$, which is assumed to be Gaussian. We further assume that the system starts from $t = -\infty$ and it is stable at $t = 0$. Hence, an appropriate choice is the stable distribution of SDE (1.43). Based on these assumptions, an appropriate distribution has been derived in [31], which is given by

$$\text{vec}(\mathbf{P}_0) = -(\mathbf{C} \otimes \mathbf{I} + \mathbf{I} \otimes \mathbf{C})^{-1}\ \text{vec}(\mathbf{V}\mathbf{V}^{\mathrm{T}}). \tag{1.46}$$

Algorithm 3 Hybrid Data-induced Kalman Filtering

1: **input:** historical dataset $\mathfrak{D} = \{S_1, \dots, S_m\}$
2: **construct** machine learning model (as per 1.40)
3: **learn** implicit system dynamics (via Algorithm 3)
4: **convert** implicit model to a SDE representation
5: **compute** initial or prior distribution as per (1.46)
6: **repeat** for each time slot (for a new task)
  (a) **construct** equivalent discrete-time stochastic system (as per Theorem 1.5)
  (b) **predict** beam with discrete Kalman filter
  until current task (e.g., time frame) is terminated

With both the prior distribution and discrete-time dynamical model (or system-state model) available, the discrete Kalman filter can be used. For clarity, the HDIKF algorithm is summarized in Algorithm 3.

1.7 Beam Width Optimization

In this section, we investigate the BWO technique, which is closely related to beam tracking [52]. The motivation of BWO is twofold. First, it can improve system robustness in a challenging rapidly fluctuating wireless environment. Second, system performance, such as effective achievable rate, can also be improved via BWO, because the beam sounding operation can be absent in some cases via just increasing appropriately the beam width or coverage. Similar to the presentation in Sections 1.5 and 1.6, according to the property of the underlying beam process and the time scale of the adjusting beam, two BWO methods are presented.

1.7.1 Stochastic Continuity – Locality Principle of Beam Change and Data Transmission with Multiresolution Beam

The stochastic continuity of the beam process implies an important property, i.e., LPBC, which, in fact, lays the foundation for beam index difference technique. LPBC indicates that although the change of beam direction can be fairly random, beam in the current time slot is more likely to switch to one of its neighbor beams in the next time slot. In contrast to the conventional design philosophy, where the multiresolution beams are used for beam sounding, LPBC implies a novel design philosophy of beam sounding, namely, the narrow beams, which often have large array gains, are advocated for sounding the channel [52].

It should be emphasized that as long as the phenomenon of beam switching occurs, it must be accompanied by the beam drift effect, which is revealed in Zhang and Masouros [52]. Beam drift, caused by nonideal features inherent in practical beams and the rapidly changing environment is referred to as the phenomenon that the center of main lobe of the used beam deviates from the real dominant channel direction, which further deteriorates system performance. Unfortunately, the beam drift effect is inevitable. To mitigate it, the relationship between beam drift and beam width is analyzed in [52] via rigorous mathematical analysis, based on which a novel transmission scheme, referred to as DT with multiresolution beam is proposed.

Based on locality principle of beam change (LPCB) and DT with multiresolution beam, a novel design philosophy of beam sounding and DT is proposed, whose principle is illustrated in Figure 1.10. The key points of the design philosophy are as follows:

- *DT design and DT module*: The mission of DT module is to mitigate the beam drift effect (and allocate transmit power) so as to keep signal to noise ration (SNR) as large as possible and almost all the time in the phase of DT. To this end, we advocate that multiresolution beams are utilized for DT, and beam resolution or width is adjusted as per the rate of change of the environment.
- *BPT design and BPT module*: The BPT module is in charge of finding the optimal beam direction with a minimum cost in terms of sounding overhead. LPBC implies that narrow beams are used to sound the channel, which is in contrast to the conventional practice. To minimize the sounding overhead with a high success rate, the BPT module needs also to adapt the wireless environment.

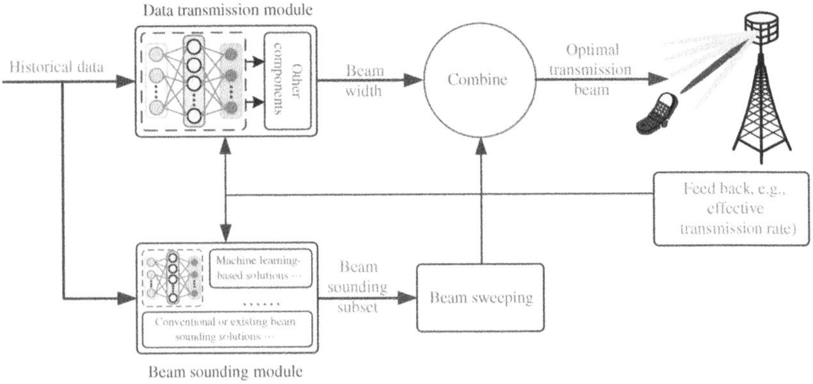

Figure 1.10 The principle of novel data transmission and beam sounding scheme.

An efficient learning-based approach to implement the novel scheme, whose key is to adjust beam width adaptively, is also proposed in [52], which is omitted due to space limitations.

1.7.2 Stochastic Smoothness – Low-frequency Sounding via BWO and Long-term Prediction

The stochastic smoothness property of beam process, whose regularity is often much stronger than continuity, can be used to further improve system performance via the BWO and/or long-term beam prediction techniques. For clarity, the basic idea of long-term beam prediction is presented.

In the classical BPT scheme, beam sounding is performed in each time slot, as shown in Figure 1.11(1). The high frequency of beam sounding inevitably degrades system performance of interest, e.g., the EAR. It is well-known that an SDE can characterize the variation tendency of a physical system. Now, since the SDE that describes the beam variation has been available, we can naturally investigate its long-term variation behavior. As shown in Figure 1.11(2), beam sounding with lower frequency is involved in the long-term scheme, so as to further improve system performance.

In contrast to the conventional BPT scheme, where Δt_k is a constant and is almost always equal to T_S (i.e., $\Delta t_k = T_S$), in the long-term BPT scheme, $\Delta t_k = t_{k+1} - t_k \gg T_S$ often spans across multiple time slots. Hence, we need the predictive distribution of beam direction $y(t)$ at any time t. Without loss of generality, we focus on an arbitrary but fixed time t_k and calculate the distribution of beam direction $b(t)$ for $t > t_k$. Note that since $b(t)$ is a linear function of $\boldsymbol{x}(z)$, Corollary 1.1 which describes the distribution of beam direction $b(t)$ can be obtained immediately from Theorem 1.5.

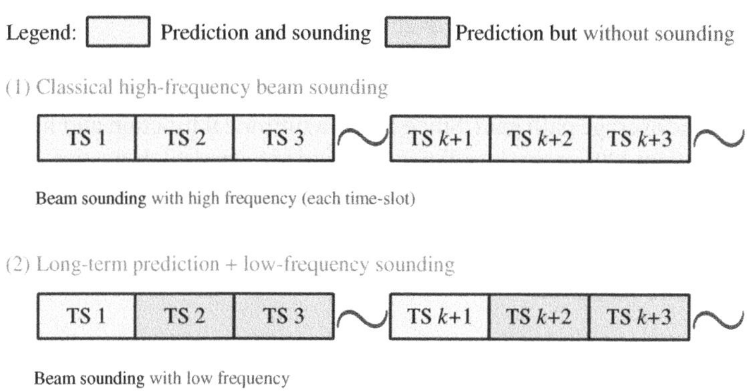

Figure 1.11 The comparison between the classical beam sounding scheme and the proposed novel long-time beam sounding scheme.

Corollary 1.1 The predictive distribution of $b(t)$ is given by

$$b(t) \sim \mathcal{N}(m(t), a(t)), \quad (t > t_k), \tag{1.47}$$

where $m(t) = \mathbf{b}^{\mathrm{T}}\mathbf{T}(h(t_k), h(t))\mathbf{x}(h(t_k))$ and $a(t) = \sigma_m^2 + \mathbf{b}^{\mathrm{T}}\mathbf{\Pi}(h(t_k), h(t))\mathbf{b}$, with $\mathbf{T}(z_k, z)$, and $\mathbf{\Pi}(z_k, z)$ given by

$$\mathbf{T}(z_k, z) = \exp(\mathbf{C}(z - z_k))$$

$$\mathbf{\Pi}(z_k, z) = \int_{z_k}^{z} \exp(\mathbf{C}(z - s))\mathbf{V}\mathbf{V}^{\mathrm{T}} \exp(\mathbf{C}(z - s))^{\mathrm{T}} ds.$$

σ_m^2 denotes the variance of the noise caused by $\eta(t)$ in (1.44).

It is observed that $a(t)$ is monotonously increasing in terms of t, which indicates that the predicted beam becomes more inaccurate as t increases. Apparently, this coincides with our intuition. If $a(t)$ is greater than a predefined threshold value, there are two methods to handle this situation. The first one is to perform beam sounding to find the optimal beam, e.g., via the hierarchical search [9] or adaptive local search [26]. The second one is BWO, i.e., to adjust the beam width.

In general, wide beams incur low array gains. We introduce a function $g(w)$, which characterizes the array gain of a beam with width w. We further introduce an adjustable parameter $c > 0$ to control the beam width. The main lobe of the beam to be chosen is denoted by $\mathcal{I}$, which is given by

$$\mathcal{I} = (m(t) - c\sqrt{a(t)}, m(t) + c\sqrt{a(t)}). \tag{1.48}$$

Under some mild assumptions (similar to [39]), the posterior EAR can be expressed as

$$\begin{aligned} R(c) &= \int_{\mathcal{I}} \frac{e^{-(x-m(t))^2/(2a(t))}}{\sqrt{2\pi a(t)}} \log\left(1 + pHg(2c\sqrt{a(t)})\right) dx \\ &= \int_{-c}^{c} \frac{1}{\sqrt{2\pi}} e^{-x^2/2} \log\left(1 + pHg(2c\sqrt{a(t)})\right) dx, \end{aligned} \tag{1.49}$$

where p denotes the transmit power and H is determined by the channel condition concerned, e.g., the average path gain. Since $R(c)$ is a unimodal function and has a unique maximal point [39], it can be efficiently found via classical derivative-free search methods, e.g., the gold-search method.

1.8 Numerical Results

In this section, simulation results are provided to demonstrate the performance of different intelligent BPT algorithms. The simulation results evaluated under the environments with stochastic continuity and stochastic smoothness are provided in Sections 1.8.1 and 1.8.2, respectively.

1.8.1 Simulation Results for Stochastic Continuity

The simulation setting is as follows. Without loss of generality, the uniform linear array (ULA) and analog discrete Fourier transform (DFT) codebook are considered. The size of the DFT codebook $\mathcal{F}$ is $M = N$. For each MU u, the action space $\mathcal{A}_u$ is given by

$$\mathcal{A}_u = \{(a, b) \,|\, a = -1, 0, 1, 2; b = 2, 4, 6\}.$$

Note that if the environment changes more irregularly, the actions space should be enlarged. The structure of neural network contains one convolution layer, one down-sampling layer and one fully connected layer.

The simulation environment is illustrated in Figure 1.12. The channel between the BS and each MU includes one LOS path and three non-line of sight (NLOS) paths if the LOS path is not blocked. The speed of each MU is stochastic but obeys a probability law. Accordingly, switching to another beam in the next time slot is also stochastic and obeys some probability law. For each MU, if the LOS path is not blocked, the probability that the optimal beam switches to the ith beam of the next S beams is denoted by $p_{S,i}$ $(i = 0, 1, \ldots, S)$, where $p_{S,0}$ represents the probability that the optimal beam in the next time slot is still the current beam. An example of switching probability is given by

$$\text{model 1:} \quad p_{S,i} = e^{-\eta i}\left(\sum_{k=0}^{S} e^{-\eta k}\right)^{-1}. \tag{1.50}$$

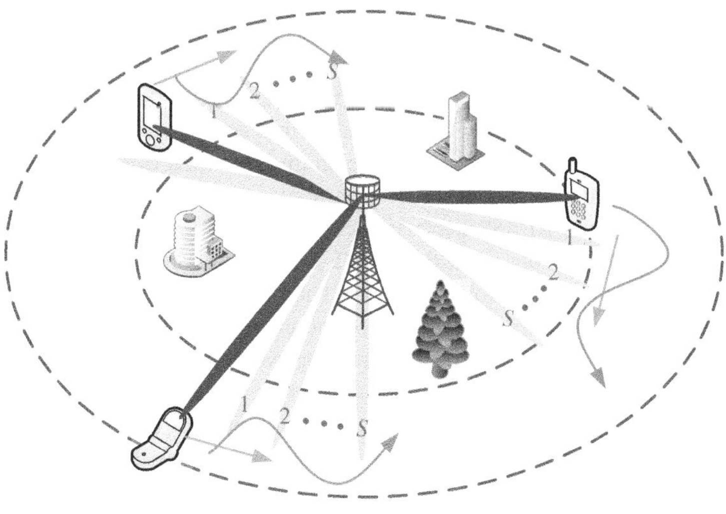

Figure 1.12 U MUs move within an annulus and the BS is located at the center.

$$\text{model 2:} \quad p_{S,i} = e^{-\eta(S-i)} \left(\sum_{k=0}^{S} e^{-\eta k} \right)^{-1}. \tag{1.51}$$

The parameter $\eta > 0$ defines the "decay" rate. The probability model in (1.50) indicates that the MU moves relatively slow, while the probability model in (1.51) implies that the MU moves relatively fast [33].

To demonstrate the superiority of the beam index difference technique, the exhaustive search based BPT (ExSeBT) algorithm, hierarchical search based BPT (HSBT) algorithm [9], and oracle-aided BPT (OABT) algorithm[6] are chosen as benchmarks. For notation simplicity, the stochastic bandit learning based BPT solution (i.e., Algorithm 1) and environment sensing-based BPT solution (incorporating both beam index difference technique [26] and beam image technique [33]) are abbreviated as SBLBT and ESBT, respectively.

First, we demonstrate the crucial role of the beam image technique used to design the BPT algorithm, i.e., ESBT. It is well known that as an important feedback from the external environment to the agent in the RL, the reward is important to decision making. In fact, the reward depends on both beam sounding overhead and the achievable sum rate [33], while the later is further determined by multiple factors, such as transmit power and channel fading, which may affect system performance in terms of robustness and/or stability of the designed algorithm. For example, the reward will change if the transmit power changes, which thus may affect the quality of decisions made by the agent.

Figure 1.13 demonstrates the stability of ESBT. Average effective achievable sum rate (AEASR) is, in fact, the average immediate reward. Note that in addition to the change of beam direction, the transmit power is also changed randomly within $\{0, 3, 6, 9, 12, 15 \text{ dB}\}$ every 200 time slots. We can observe that near-optimal performance can still be achieved near the points where the transmit power has just changed, which shows the robustness of ESBT to the change of transmit power. The reason for this is that the current state is taken into account when making a decision, and the state is designed based on beam image, which can extract and encode important information about the environment, e.g., beam spatial pattern.

We further evaluate the AEASR performance of different BPT algorithms, as shown in Figure 1.14. It is observed that ESBT achieves the best performance among these algorithms and approaches OABT. The reason for this is as follows. First, ESBT can sense the variation pattern of the environment and adaptively adjust the BPT strategy, which reduces the sounding overhead. Second, the beam sounding overhead of HSBT and ExSeBT is too large, although they can find optimal beams and thus achieve large array gains. Similar to ESBT, SBLBT is also designed based on the beam index difference technique, which thus performs

6 The optimal beam is provided by an oracle, causing no sounding overhead. It is an optimal but ideal algorithm, which is often used as a benchmark.

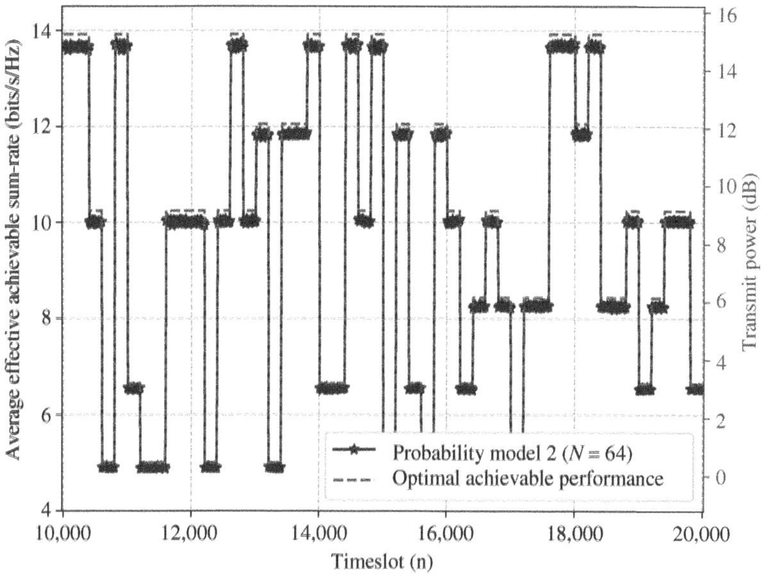

Figure 1.13 The AEASR performance of ESBT (averaged over 300 realizations). The dotted lines represent the optimal performance, i.e., the optimal action is taken each time.

better than ExSeBT and HSBT. However, the change of beamforming gain, e.g., due to random locations, affects the reward, which may induce the agent to take suboptimal actions. In contrast, the beam image technique used in ESBT greatly mitigates and even avoids this adverse effect.

Figure 1.15 demonstrates the AEASR performance of ESBT and ExSeBT for the two probability models. It is not surprising that ExSeBT achieves the same AEASR performance for the two beam switching models because of the fixed and same beam sounding overhead. Nevertheless, it is interesting to observe that AEASR corresponding to model 2 achieved by ESBT is almost the same as that corresponding to model 1, although the environment corresponding to model 2 varies faster than that corresponding to model 1. The reason for this is that ESBT is designed based on beam index difference (which counteracts the rate of change in environment) and the variances of rates of changes in the two models are the same. Hence, the sounding overheads of the two models are the same. This is an important and desirable feature of ESBT.

1.8.2 Simulation Results for Stochastic Smoothness

Without loss of generality, the V2X communication scenario is considered. However, compared to the straight road [15], a more complex road condition,

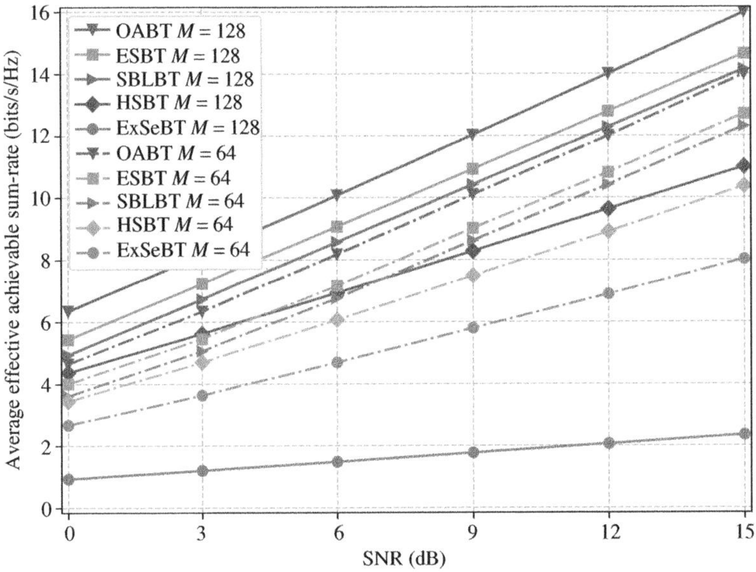

Figure 1.14 The AEASR performance of different BPT algorithms – $S = 4$ and beam switching model 1 in (1.50).

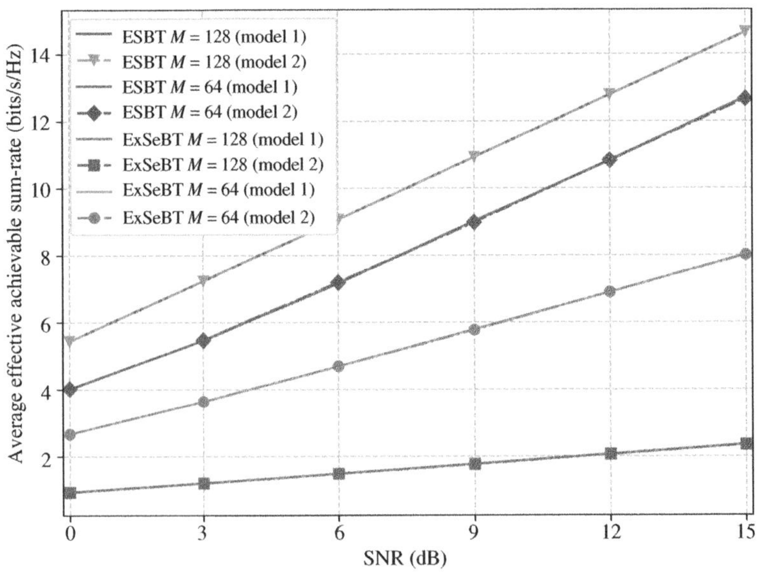

Figure 1.15 The AEASR performance of ESBT and ExSeBT for the two beam switching models – $S = 4$.

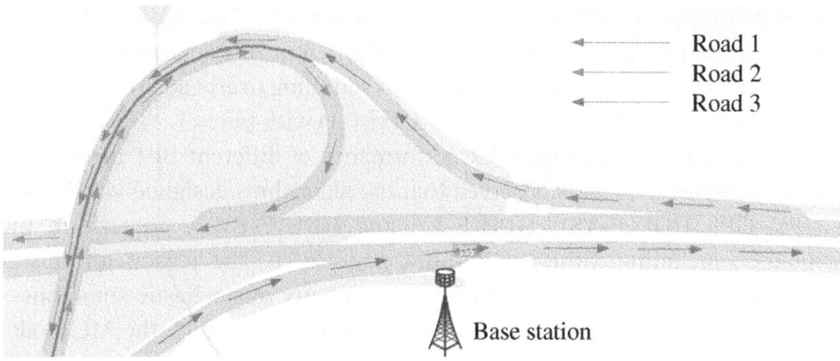

Figure 1.16 The road condition of a typical flyover chosen from the real environment.

i.e., the typical flyover chosen from a real environment, is considered here. As shown in Figure 1.16, three roads are chosen to evaluate different algorithms. The length of each time slot, i.e., transmission time interval, is set to 20 ms. The widths of Roads 1 and 2 are both 12 m, while the width of Road 3 is 20 m. It is allowed to change lane in Road 3, with probability 1/10. The initial velocity, i.e., the velocity that a vehicle enters the coverage area, for Roads 1 and 2 is distributed uniformly as U(54, 80) (km/h), while the initial velocity for Road 3 is distributed uniformly as U(54, 108) (km/h). To simulate the scenario more practically, with probability 0.1, the driver speeds up (or slows down) the velocity to the maximal (or minimal) speed on each road.

To better show the advantages of HDIKF, besides the state-of-the-art benchmarks - HSBT, SBLBT, OABT, and GP regression (GPR-only) based algorithm [37],[7] the following BPT solutions designed (partly) based on HDIKF are evaluated:

- *GPM–GPI (i.e., Gaussian process modeling and Gaussian process inference)*: The implicit system dynamics learning based on Algorithm 3 is incorporated into the BPT solution. However, GPR is still used to perform predictions.
- *HDIKF–TS1*: The BPT solution is designed based on the HDIKF approach, but without the long-term prediction and BWO techniques. As a result, the beam sounding is still operated in each time slot.
- *HDIKF–LT*: Compared to the HDIKF–TS1 algorithm, the long-term prediction technique is incorporated. However, the BWO is not considered.
- *HDIKF–BWO*: In contrast to HDIKF–LT, where the beam sounding is executed if necessary (e.g., wide beam confidence interval), HDIKF–BWO adjusts the beam width.

7 Compared to Algorithm 3 proposed in this paper (named GPM–GPI in this section), the classical GPR method fails to learn the system dynamics.

The probability of successful alignment (PSA), average EAR, and running time are chosen as performance metrics to evaluate different BPT solutions. Note that the EAR also characterizes equivalently beam sounding overhead. The path gain of the LOS path is distributed as $\alpha_{\mathrm{LOS}} \sim \mathcal{CN}(\overline{m}, 0.1)$ with $|\overline{m}| = 1$.

We first evaluate the average EAR performance of different BPT algorithms, as shown in Figure 1.17. It is observed that the algorithms designed via HDIKF, i.e., GPM–GPI, HDIKF–TS1, HDIKF–LT, and HDIKF–BWO, outperform the benchmarks, including SBLBT, GPR-only, and HSBT. The reason for this is that: (i) the underlying environment has the property of stochastic smoothness and (ii) the system dynamics can be efficiently characterized by the ML model proposed in Section 1.6 and also learned by the implicit system dynamics learning method. As a result, the beam sounding overhead is very small. In particular, the BPT algorithms incorporating the long-term prediction scheme approach the ideal oracle-aided algorithm, thanks to the negligible beam sounding overhead.

Figure 1.18 shows the PSA performance of different algorithms. It can also be observed that HDIKF-based solutions achieve better performance than both SBL and HS. In particular, the HDIKF-based algorithms outperform SBL, which is also a learning-based algorithm. The reason for this is twofold. First, stochastic smoothness is much stronger than stochastic continuity. Second, the beam index difference technique involved in SBL predicts the difference of successive beams,

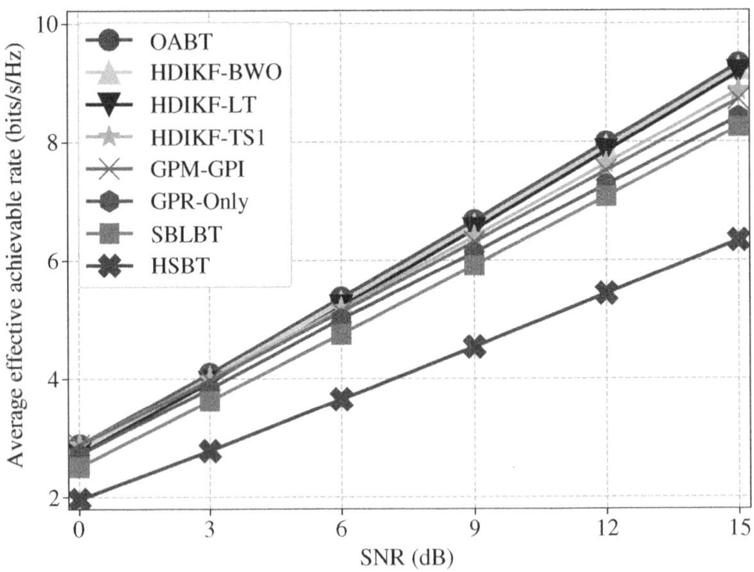

Figure 1.17 The average EAR performance of different BPT algorithms: Road 1 and $N = 128$.

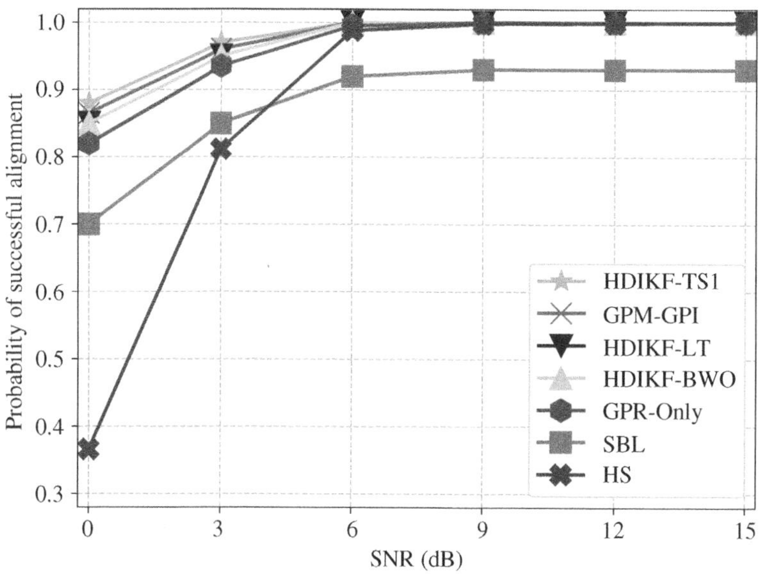

Figure 1.18 The PSA performance of different BPT algorithms: $N = 128$ and Road 1.

which fails to capture and exploit the variation tendency of the beam direction function. In contrast, HDIKF fully exploits this prior information to narrow the beam interval containing the real beam. Due to the noise effect, if less beams are swept, better PSA performance can be achieved [26]. Since the GPR-only method fails to learn the system dynamics, its performance is worse than that of the HDIKF-based BPT solutions.

To intuitively demonstrate the advantage of the long-term prediction scheme, PSA performance varying with respect to the period of beam sounding is shown in Figure 1.19. It can be seen that compared to the conventional beam sounding scheme whose sounding period is one, the BPT algorithms with long-term prediction can achieve a high prediction accuracy even for a relatively large beam sounding period. The reason for this is that the underlying system dynamics, e.g., the beam variation tendency, can be efficiently learned from multiple beam trajectories. When facing a large uncertainty, the BWO method can adjust beam width adaptively and dynamically.

Another appealing advantage of HDIKF is that it inherits the property of low computational complexity from the classical Kalman filtering technique. In particular, thanks to the linear complexity order, the performance curves in terms of accumulative run-time of HDIKF–TS1, HDIKF–LT, and HDIKF–BWO increase very slow, as illustrated in Figure 1.20. In contrast, due to the cubic complexity order, the time resource and computing resource required by both GPR-only

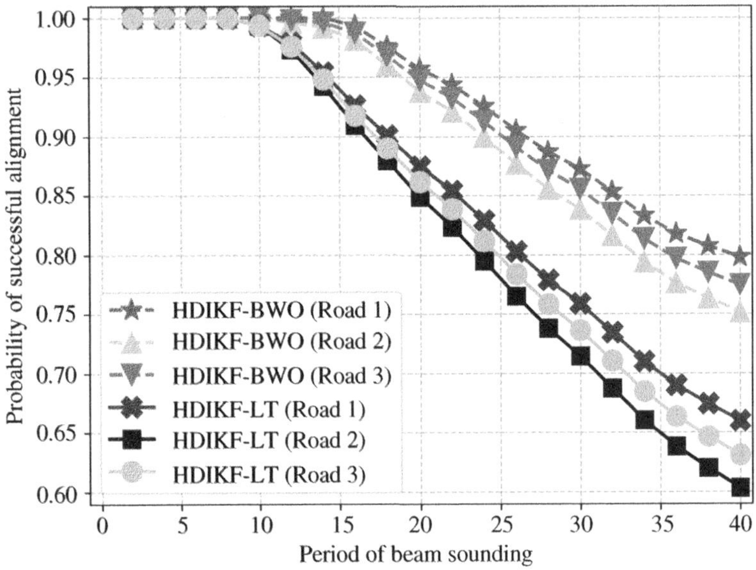

Figure 1.19 The PSA performance varying with the beam sounding period for HDIKF-LT and HDIKF-BWO: $N = 64$ and Road 1.

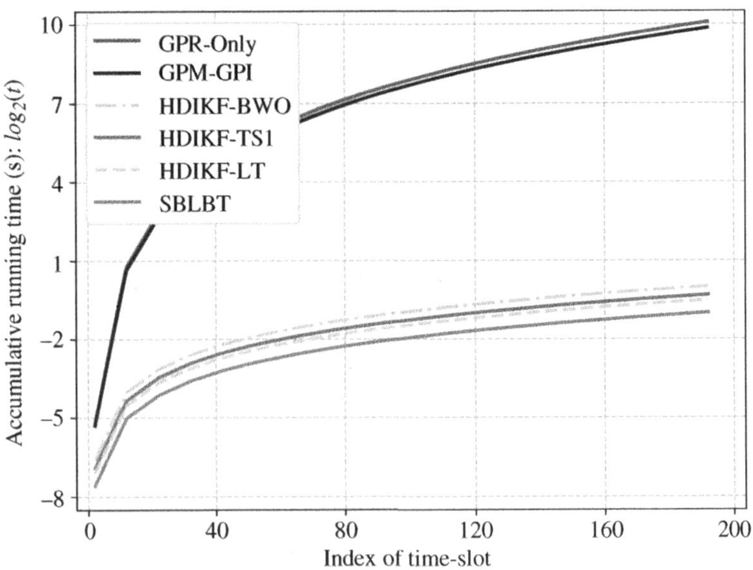

Figure 1.20 The average accumulative run-time of different BPT algorithms: $N = 128$ and Road 1.

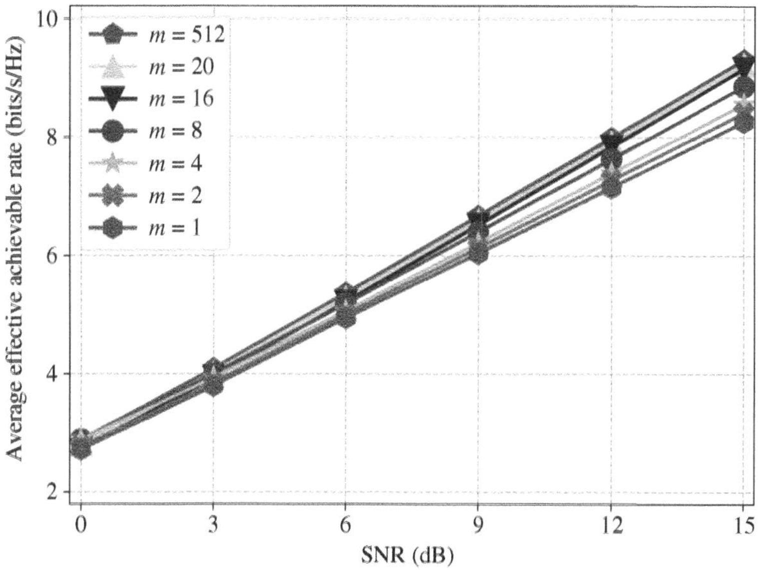

Figure 1.21 The average EAR performance achieved by HDIKF–TS1 models trained with varying amounts of beam trajectories: $N = 128$ and Road 1.

and GPM–GPI increase remarkably. Moreover, thanks to the inherent Kalman filtering characteristics, HDIKF–TS1, HDIKF–LT, and HDIKF–BWO can well adapt to nonstationary environments, which accounts for better performance than GPR-only and GPM–GPI.

In contrast to other ML applications, good small sample performance is desired in wireless communications, which, fortunately, can be achieved by HDIKF. It can be observed from Figure 1.21 that although the number of available training curves is small, e.g., $m \leq 4$, HDIKF can still obtain satisfactory EAR performance. The reason for this is twofold. First, the Bayesian methodology is adopted in HDIKF, which has been shown to achieve good small performance. Second, when deriving the loss function, the transfer error is explicitly minimized to improve the generalization performance.

1.9 Conclusion

BPT plays an important role in high-frequency communications, e.g., millimeter wave and terahertz communications, which is, however, very challenging. In this chapter, we first briefly introduced the concept of BPT, and reviewed different BPT algorithms, including conventional nonlearning methods and recent learning-based solutions. We classified the learning-based BPT methods based on the underlying learning techniques and briefly analyzed their strengths and weaknesses. Then, we analyzed the key challenges of applying ML to BPT. The core of this chapter is to propose a novel design methodology from the point of view of SP by identifying and exploiting the stochastic analytic properties of a beam process. The stochastic continuity and stochastic smoothness were taken as examples to show how to use this methodology to design efficient BPT algorithms. Comprehensive simulation results were provided to demonstrate its effectiveness and superiority.

References

1 Xiao, M., Mumtaz, S., Huang, Y. et al. (2017). Millimeter wave communications for future mobile networks. *IEEE Journal on Selected Areas in Communications* 35 (9): 1909–1935. https://doi.org/10.1109/JSAC.2017.2719924.

2 Adhikary, A., Nam, J., Ahn, J.-Y., and Caire, G. (2013). Joint spatial division and multiplexing–the large-scale array regime. *IEEE Transactions on Information Theory* 59 (10): 6441–6463. https://doi.org/10.1109/TIT.2013.2269476.

3 Björnson, E., Larsson, E.G., and Marzetta, T.L. (2016). Massive MIMO: ten myths and one critical question. *IEEE Communications Magazine* 54 (2): 114–123. https://doi.org/10.1109/MCOM.2016.7402270.

4 Wang, J., Lan, Z., Pyo, C.-w. et al. (2009). Beam codebook based beamforming protocol for multi-Gbps millimeter-wave WPAN systems. *IEEE Journal on Selected Areas in Communications* 27 (8): 1390–1399. https://doi.org/10.1109/JSAC.2009.091009.

5 Hur, S., Kim, T., Love, D.J. et al. (2013). Millimeter wave beamforming for wireless backhaul and access in small cell networks. *IEEE Transactions on Communications* 61 (10): 4391–4403. https://doi.org/10.1109/TCOMM.2013.090513.120848.

6 Biguesh, M. and Gershman, A.B. (2006). Training-based MIMO channel estimation: a study of estimator tradeoffs and optimal training signals. *IEEE Transactions on Signal Processing* 54 (3): 884–893. https://doi.org/10.1109/TSP.2005.863008.

7 Shi, Q., Razaviyayn, M., Luo, Z.-Q., and He, C. (2011). An iteratively weighted MMSE approach to distributed sum-utility maximization for a MIMO interfering broadcast channel. *IEEE Transactions on Signal Processing* 59 (9): 4331–4340. https://doi.org/10.1109/TSP.2011.2147784.

8 Shi, Q., Hong, M., Fu, X., and Chang, T.-H. (2020). Penalty dual decomposition method for nonsmooth nonconvex optimization–Part II: Applications. *IEEE*

Transactions on Signal Processing 68: 4242–4257. https://doi.org/10.1109/TSP.2020.3001397.

9 Zhang, J., Huang, Y., Shi, Q. et al. (2017). Codebook design for beam alignment in millimeter wave communication systems. *IEEE Transactions on Communications* 65 (11): 4980–4995. https://doi.org/10.1109/TCOMM.2017.2730878.

10 Singh, J. and Ramakrishna, S. (2015). On the feasibility of codebook-based beamforming in millimeter wave systems with multiple antenna arrays. *IEEE Transactions on Wireless Communications* 14 (5): 2670–2683. https://doi.org/10.1109/TWC.2015.2390637.

11 Zhang, J., Huang, Y., Wang, J. et al. (2020). Power-efficient beam designs for millimeter wave communication systems. *IEEE Transactions on Wireless Communications* 19 (2): 1265–1279. https://doi.org/10.1109/TWC.2019.2952340.

12 Gao, M., Ai, B., Niu, Y. et al. (2018). Dynamic mmWave beam tracking for high speed railway communications. *2018 IEEE Wireless Communications and Networking Conference Workshops (WCNCW)*, 278–283. IEEE. https://doi.org/10.1109/WCNCW.2018.8368998.

13 Alkhateeb, A., Alex, S., Varkey, P. et al. (2018). Deep learning coordinated beamforming for highly-mobile millimeter wave systems. *IEEE Access* 6: 37328–37348. https://doi.org/10.1109/ACCESS.2018.2850226.

14 Xiao, Z., Xia, P., and Xia, X.-G. (2017). Codebook design for millimeter-wave channel estimation with hybrid precoding structure. *IEEE Transactions on Wireless Communications* 16 (1): 141 153. https://doi.org/10.1109/TWC.2016.2619705.

15 Liu, F., Yuan, W., Masouros, C., and Yuan, J. (2020). Radar-assisted predictive beamforming for vehicular links: communication served by sensing. *IEEE Transactions on Wireless Communications* 19 (11): 7704–7719. https://doi.org/10.1109/TWC.2020.3015735.

16 Larew, S.G. and Love, D.J. (2019). Adaptive beam tracking with the unscented Kalman filter for millimeter wave communication. *IEEE Signal Processing Letters* 26 (11): 1658–1662. https://doi.org/10.1109/LSP.2019.2944255.

17 Liu, F., Zhao, P., and Wang, Z. (2019). EKF-based beam tracking for mmWave MIMO systems. *IEEE Communications Letters* 23 (12): 2390–2393. https://doi.org/10.1109/LCOMM.2019.2940660.

18 Yuan, W., Liu, F., Masouros, C. et al. (2021). Bayesian predictive beamforming for vehicular networks: a low-overhead joint radar-communication approach. *IEEE Transactions on Wireless Communications* 20 (3): 1442–1456. https://doi.org/10.1109/TWC.2020.3033776.

19 Meng, F., Liu, S., Huang, Y., and Lu, Z. (2022). Learning-aided beam prediction in mmWave MU-MIMO systems for high-speed railway. *IEEE*

Transactions on Communications 70 (1): 693–706. https://doi.org/10.1109/TCOMM.2021.3124963.

20 Chen, L., Zhou, S., and Wang, W. (2023). MmWave beam tracking with spatial information based on extended Kalman filter. *IEEE Wireless Communications Letters* 12 (4): 615–619. https://doi.org/10.1109/LWC.2023.3236622.

21 Zhu, P., Lin, H., Bao, J. et al. (2022). Beam tracking for distributed millimeter-wave massive MIMO systems based on the unscented Kalman filter. *IEEE Wireless Communications Letters* 11 (4): 712–716. https://doi.org/10.1109/LWC.2022.3140852.

22 Jayaprakasam, S., Ma, X., Choi, J.W., and Kim, S. (2017). Robust beam-tracking for mmWave mobile communications. *IEEE Communications Letters* 21 (12): 2654–2657. https://doi.org/10.1109/LCOMM.2017.2748938.

23 Va, V., Choi, J., Shimizu, T. et al. (2018). Inverse multipath fingerprinting for millimeter wave V2I beam alignment. *IEEE Transactions on Vehicular Technology* 67 (5): 4042–4058. https://doi.org/10.1109/TVT.2017.2787627.

24 Satyanarayana, K., El-Hajjar, M., Mourad, A.A.M., and Hanzo, L. (2019). Deep learning aided fingerprint-based beam alignment for mmWave vehicular communication. *IEEE Transactions on Vehicular Technology* 68 (11): 10858–10871. https://doi.org/10.1109/TVT.2019.2939400.

25 Va, V., Shimizu, T., Bansal, G., and Heath, R.W. (2019). Online learning for position-aided millimeter wave beam training. *IEEE Access* 7: 30507–30526. https://doi.org/10.1109/ACCESS.2019.2902372.

26 Zhang, J., Huang, Y., Zhou, Y., and You, X. (2020). Beam alignment and tracking for millimeter wave communications via bandit learning. *IEEE Transactions on Communications* 68 (9): 5519–5533. https://doi.org/10.1109/TCOMM.2020.2988256.

27 Cheng, M., Wang, J.-B., Wang, J.-Y. et al. (2019). A fast beam searching scheme in mmWave communications for high-speed trains. *IEEE International Conference on Communications (ICC)*, 1–6. IEEE. https://doi.org/10.1109/ICC.2019.8761773.

28 Wu, W., Cheng, N., Zhang, N. et al. (2019). Fast mmWave beam alignment via correlated bandit learning. *IEEE Transactions on Wireless Communications* 18 (12): 5894–5908. https://doi.org/10.1109/TWC.2019.2940454.

29 Booth, M.B., Suresh, V., Michelusi, N., and Love, D.J. (2019). Multi-armed bandit beam alignment and tracking for mobile millimeter wave communications. *IEEE Communications Letters* 23 (7): 1244–1248. https://doi.org/10.1109/LCOMM.2019.2919016.

30 Zhang, J., Huang, Y., Masouros, C., and You, X. (2023). Data-induced intelligent Kalman filtering for beam prediction and tracking of millimeter wave communications. *IEEE International Conference on Communications (ICC)*, 1–6. IEEE. https://doi.org/10.1109/ICC.2023.8761773.

31 Zhang, J., Huang, Y., Masouros, C. et al. (2024). Hybrid data-induced Kalman filtering approach and application in beam prediction and tracking. *IEEE Transactions on Signal Processing* 72: 1412–1426.

32 Zhang, J., Huang, Y., Wang, J., and You, X. (2019). Intelligent beam training for millimeter-wave communications via deep reinforcement learning. *IEEE Global Communications Conference (GLOBECOM)*, 1–7. IEEE. https://doi.org/10.1109/GLOBECOM38437.2019.9014113.

33 Zhang, J., Huang, Y., Wang, J. et al. (2021). Intelligent interactive beam training for millimeter wave communications. *IEEE Transactions on Wireless Communications* 20 (3): 2034–2048. https://doi.org/10.1109/TWC.2020.3038787.

34 Xu, W., Ke, Y., Lee, C.-H. et al. (2021). Data-driven beam management with angular domain information for mmWave UAV networks. *IEEE Transactions on Wireless Communications* 20 (11): 7040–7056. https://doi.org/10.1109/TWC.2021.3080519.

35 Zhang, J. and Masouros, C. (2021). Learning-based predictive transmitter-receiver beam alignment in millimeter wave fixed wireless access links. *IEEE Transactions on Signal Processing* 69: 3268–3282. https://doi.org/10.1109/TSP.2021.3076899.

36 Zhang, J., Xu, W., Gao, H. et al. (2019). Position-attitude prediction based beam tracking for UAV mmWave communications. *IEEE International Conference on Communications (ICC)*, 1–7. IEEE. https://doi.org/10.1109/ICC.2019.8761981.

37 Zhang, J., Xu, W., Gao, H. et al. (2021). Codebook-based beam tracking for conformal array enabled UAV mmWave networks. *IEEE Internet of Things Journal* 8 (1): 244–261. https://doi.org/10.1109/JIOT.2020.3005394.

38 Song, H.-L. and Ko, Y.-C. (2021). Beam alignment for high-speed UAV via angle prediction and adaptive beam coverage. *IEEE Transactions on Vehicular Technology* 70 (10): 10185–10192. https://doi.org/10.1109/TVT.2021.3103188.

39 Zhang, J., Masouros, C., and Huang, Y. (2023). Beam training and tracking with limited sampling sets: exploiting environment priors. *IEEE Transactions on Communications* 71 (5): 3008–3023. https://doi.org/10.1109/TCOMM.2023.3239927.

40 Wang, C.-X., Ghazal, A., Ai, B. et al. (2016). Channel measurements and models for high-speed train communication systems: a survey. *IEEE Communications Surveys & Tutorials* 18 (2): 974–987. https://doi.org/10.1109/COMST.2015.2508442.

41 Rasmussen, C.E. and Williams, C.K.I. (2006). *Gaussian Processes for Machine Learning*. MIT Press. ISBN: 9780262182539, 9781423769903, 026218253X.

42 Wilson, A.G., Hu, Z., Salakhutdinov, R., and Xing, E.P. (2016). Deep kernel learning. In: *Proceedings of Machine Learning Research*, vol. 51, 370–378. Cadiz, Spain: PMLR.

43 Särkkä, S. and Solin, A. (2019). *Applied Stochastic Differential Equations*. Cambridge University Press. ISBN: 9780262182539, 9781423769903, 026218253X.

44 Kloeden, P.E. and Platen, E. (1995). *Numerical Solution of Stochastic Differential Equations*. Springer Press. ISBN: 9780262182539, 9781423769903, 026218253X.

45 Sutton, R.S. (2018). *Reinforcement Learning: An Introduction*. Cambridge, MA: MIT Press.

46 Auer, P., Cesa-Bianchi, N., and Fischer, P. (2002). Finite-time analysis of the multiarmed bandit problem. *Machine Learning* 47 (2): 235–256.

47 Rappaport, T.S., Ben-Dor, E., Murdock, J.N., and Qiao, Y. (2012). 38 GHz and 60 GHz angle-dependent propagation for cellular & peer-to-peer wireless communications. *IEEE International Conference on Communications (ICC)*, 4568–4573. IEEE. https://doi.org/10.1109/ICC.2012.6363891.

48 Alkhateeb, A., El Ayach, O., Leus, G., and Heath, R.W. (2014). Channel estimation and hybrid precoding for millimeter wave cellular systems. *IEEE Journal of Selected Topics in Signal Processing* 8 (5): 831–846. https://doi.org/10.1109/JSTSP.2014.2334278.

49 Lattimore, T. and Szepesvári, C. (2020). *Bandit Algorithms*. Cambridge: Cambridge University Press.

50 Karatzas, I. and Shreve, S.E. (2021). *Brownian Motion and Stochastic Calculus*. World Publishing Corporation.

51 Sarkka, S., Solin, A., and Hartikainen, J. (2013). Spatiotemporal learning via infinite-dimensional Bayesian filtering and smoothing: a look at Gaussian process regression through Kalman filtering. *IEEE Signal Processing Magazine* 30 (4): 51–61. https://doi.org/10.1109/MSP.2013.2246292.

52 Zhang, J. and Masouros, C. (2021). Beam drift in millimeter wave links: beamwidth tradeoffs and learning based optimization. *IEEE Transactions on Communications* 69 (10): 6661–6674. https://doi.org/10.1109/TCOMM.2021.3088896.

2

Signal Detection with Machine Learning

Jayakrishnan Vijayamohanan, Arjun Gupta, Manel Martínez-Ramón, and Christos Christodoulou

Department of Electrical and Computer Engineering, University of New Mexico, Albuquerque, New Mexico, USA

CHAPTER MENU

2.1 Introduction

Signal detection is a crucial component of the modern digital communication system and has numerous applications across several different areas of signal processing and electrical engineering [1–3]. As a result, it has been widely studied and various statistical methods have been introduced over the better part of the 21st century to detect and estimate various aspects of a signal. The advances and innovations in machine learning (ML) and deep learning (DL) methods have ushered in the age of artificial intelligence (AI). The era of AI has revolutionized signal detection among other areas of signal processing and ML and DL methods have been widely explored and adopted in the context of the detection and estimation of signal parameters [4–8].

Advances in DL in particular have influenced the advances in signal detection tremendously due to the nature of the problem and the ability of DL to provide superior generalization. Advances in fast computation coupled with the availability of huge volumes of data have contributed to larger and deeper models that can be trained and further fine-tuned to perform signal detection and estimation

Artificial Intelligence for Future Networks, First Edition.
Edited by Mohammad A. Matin, Sotirios K. Goudos, and George K. Karagiannidis.

with very high accuracy and precision [5, 9–12]. These data-driven approaches are being widely adopted and largely define the advances in different areas of signal processing for modern communication systems.

We begin the chapter by delving into the signal detection problem and discuss the various strategies for signal detection over a noisy channel. The generic detection framework is then adopted for symbol detection and we discuss the contributions of ML in the advancement of modern symbol detection algorithms [5–7].

Automatic modulation classification (AMC) is an important component of modern communication systems. AMC aims to identify the modulation type of an incoming signal. It is one of the several stages of signal processing in the physical layer to improve spectrum utilization and efficiency. We develop several DL frameworks for AMC and discuss the strategies to train an optimal DL detector [13–15].

We switch topics and dive into source detection, which is crucial for direction and time of arrival estimation. With the advent of mmWave communication, multichannel array signal processing is of particular interest. We introduce the multichannel source detection problem and the conventional techniques in brief and expand on the models and contributions of ML to this sub-domain of signal detection [9, 16].

In this chapter, we explore signal detection from the perspective of symbol detection, modulation detection, and source detection. We provide a brief discussion of the conventional methods used to solve signal detection and the challenges faced by the modern communication system. We elaborate on the data-driven approaches harnessing the power of the DL frameworks and provide intuition on how to build datasets and design models to solve signal detection tasks. We provide insights into training strategies required to successfully train these models to tailor solutions for specific detection tasks.

2.2 Symbol Detection

2.2.1 The Viterbi Algorithm

The most widely used algorithm for symbol detection in communications channels with memory is the well-known Viterbi algorithm [17]. Given an observed sequence of symbols corrupted with noise, the algorithm finds the noiseless sequence with the highest likelihood. If the noise is an additive white Gaussian noise (AWGN), the Viterbi algorithm is the optimal maximum likelihood algorithm for sequence detection with the condition of using it for the detection of an Euclidean distance [18]. The beauty of the algorithm can be seen when compared to the simpler channel equalization algorithms. In these algorithms, an often adaptive finite impulse response (FIR) filter is constructed to approximately cancel the zeros of the channel, which is suboptimal when compared to the maximum likelihood criterion, and it is subject to several limitations, as we will see further. The Viterbi algorithm, instead, learns from the channel rather than fighting it, to model the signal from a Bayesian perspective.

The Viterbi algorithm models the sequence of symbols as a Hidden Markov model with a known set of states S and known transition probabilities between states $p(s_i, s_j)$ and prior probabilities $p(s_i)$. The algorithm also assumes that the observed signal $\mathbf{x}_n$ has a conditional likelihood $p(\mathbf{x}_n|s_i)$. With these elements, a Trellis of a sequence is constructed with the corresponding transition probabilities and the conditional likelihoods. By visiting all possible paths in the Trellis, the joint probability $p(\mathbf{X}, \mathbf{s})$ is maximized, where $\mathbf{X}$ is a sequence of observations and $\mathbf{s}$ is the hypothesis over the symbol sequence.

We explain a simple implementation of this algorithm with an example. Assume that a sequence of binary symbols $z_n \in \{-1, +1\}$ is transmitted through a simple channel whose discrete impulse response is $h_n = \delta[n] + a\delta[-1]$. This simplistic channel assumes that the received signal $x_n = z_n + az_{n-1} + g_n$ contains the symbol transmitted at instant n and an intersymbol interference with delay exactly equal to the duration of one symbol and amplitude a. The received signal is corrupted with AWGN g_n of variance σ_n^2. The receiver constructs N observations $\mathbf{x}_n = [x_n, x_{n-1}]^\top$ with a sliding window of two samples. Therefore, in this case, $\mathbf{X} = [\mathbf{x}_1, \dots, \mathbf{x}_n] \in \mathbb{R}^{2\times N}$ is a matrix of real signals.

Each one of the observations $\mathbf{x}_n$ contains the three symbols z_{n-2}, z_{n-1}, and z_n. Therefore, the hidden state s_n corresponding to the observation is one among the 2^3 possible states, that is $s_n \in \{1, \dots, 8\}$ or, equivalently $\mathbf{z}_n = [s_{n-2}, s_{n-1}, s_n]$. To compute the transition probabilities between states, we assume that the symbols are independent and identically distributed. Given a present state $\mathbf{z}_n = [s_{n-2}, s_{n-1}, s_n]$, the next state can only be $\mathbf{z}_{n+1} = [s_{n-1}, s_n, 1]$ or $\mathbf{z}_{n+1} = [s_{n-1}, s_n, 1]$ with probability 1/2 and the rest of states are not possible. This eases the construction of the Trellis, depicted in Figure 2.1.

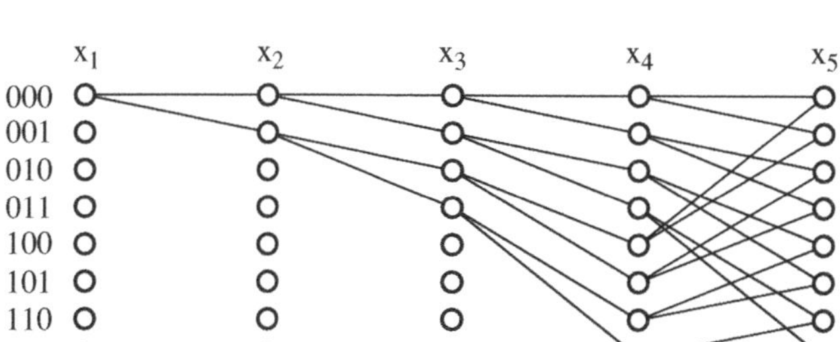

Figure 2.1 Trellis diagram of a sequence of five observations, where the hypothesis for the first state is 000. The first state has two possible transitions to 000 and 001. The possible transitions give 16 possible paths.

To compute the conditional likelihood of observation, since the noise is AWGN, the elements of vector $\mathbf{x}_n$ are conditionally independent assuming the knowledge of the state s_n. The likelihood of sample x_n is

$$p(x_n|s_n) \propto \exp\left(\frac{1}{2\sigma_n^2}|x_n - z_n + az_{n-1}|^2\right) \tag{2.1}$$

and, given the assumption of conditional independence, the likelihood of observation $\mathbf{x}_n$ is

$$p(\mathbf{x}_n|s_n) \propto p(x_n|s_n)p(x_{n-1}|s_n). \tag{2.2}$$

The log-likelihood of this expression is then

$$\log p(\mathbf{x}_n|s_n) = \frac{1}{2\sigma_n^2}\left\|\mathbf{x}_n - \begin{bmatrix} z_n + az_{n-1} \\ z_{n-1} + az_{n-2} \end{bmatrix}\right\|^2 + \text{constant}, \tag{2.3}$$

which is, up to constants, the distance between the observation and the noiseless signal conditional to the hypothesis over the state. A value of the conditional likelihood can be computed for each dot in the Trellis of Figure 2.1. Assuming independence between observations given the states, the log-likelihood of each of the paths is equal to the sum of each one of its likelihoods. The Viterbi algorithm simply chooses the path with the maximum likelihood. The algorithm is not only optimal, but it also has very efficient implementations [19]. In general, one can think of the Viterbi algorithm as a maximum a posteriori (MAP) estimator that, in general, is better than the maximum likelihood criterion. Indeed, a log-posterior probability can be computed just by multiplying the likelihood in Eq. (2.2) by the transition probabilities. In this case, both criteria are equivalent given that the transition probabilities are uniform.

2.2.2 Channel Equalization Through Machine Learning

The Viterbi algorithm as presented above assumes known channel and noise models. While the thermal noise can be straightforwardly estimated in many situations, modeling dispersive channels has a much higher complexity. At any rate, the estimation of the channel is subject to the transmission of sequences of known symbols (usually called training sequences) to estimate the channel, and this has to be done periodically to account for the channel variability.

A suboptimal alternative to the Viterbi algorithm is channel equalization. In this approach, observations are detected one at a time rather than using the entire sequence. The main idea of channel equalization consists of the construction of a linear filter with FIR $h_n = \sum_{i=0}^{D-1} w_i\delta(n-i)$ that is convolved with the incoming signal to approximately cancel the zeros of the channel. The output of the detector is then

$$z_n = x_n * w_n + e_n = \sum_{i=0}^{D-1} w_i x_{n-i} + e_n, \tag{2.4}$$

where e_n is the detection error. If an observation vector is defined as $\mathbf{x}_n = [x_n, \dots, x_{n-D+1}]^\top$, then the filter can be written as the unbiased estimator

$$z_n = \mathbf{w}^\top \mathbf{x}_n + e_n. \tag{2.5}$$

Provided that a sequence of samples is available for training, an ML algorithm can be constructed to optimize the weight vector by applying some optimality criterion, as the minimum mean square error (MMSE) over the error, or a more general criterion as the maximum likelihood used in neural networks, a MAP as in the Gaussian processes (GPs) [20] or in or the maximum margin used in support vector machines (SVMs) [21]. The problem with this approach is that it is limited to channels whose zeros are all inside the unit circle, this is, the FIR filter represented by vector $\mathbf{w}$ is intended to cancel these zeros and, therefore, the impulse response w_n must be a truncated approximation of an infinite impulse response (IIR) filter whose poles are equal to the zeros of the channel. The truncated elements of the impulse response are those elements whose total energy is negligible. Nevertheless, if the channel has zeros outside the unit circle, the IIR impulse response of the equalizer filter becomes unstable. Since it does not vanish, a truncated approximation is not possible, and this is why this approach is suitable only for those channels whose zeros are inside the unit circle.

The above estimator in Eq. (2.5) can be seen as a linear classifier where the hyperplane $\Pi \equiv \mathbf{w}^\top \mathbf{x} = 0$ plays the role of classification boundary. The symbol is decided by just observing the sign of the response. This can be extended to complex algebra to classify quadrature-phase shift keying (QPSK) symbols and, for quadrature amplitude modulation (QAM) symbols, the estimator can be extended to a multilevel classification with several thresholds. It can be easily proven that when some poles of the channel are outside the unit circle, the data is not linearly separable (see, e.g., [22]). In these cases, the linear estimator can be changed by a nonlinear one. A solution based on the kernel trick is used in SVM and GP estimator. The trick implicitly uses a nonlinear transformation $\boldsymbol{\varphi}_n = \varphi(\mathbf{x}_n)$ of the input data to a higher dimensionality Hilbert space. The estimator is represented as $z_n = \mathbf{w}^\top \boldsymbol{\varphi} + e_n$. The transformation leads to a higher computational burden due to the increased dimensionality that, in practical cases, may be infinity. Nevertheless, under nonrestrictive conditions, it is proven through the Generalized Representer Theorem [23] that the weight vector is a linear combination of the training data, this is, assuming that n observations are available for training,

$\mathbf{w} = \sum_{i=1}^{N} \alpha_i \mathbf{x}_i$ and therefore, the estimator can be constructed as

$$z_n = \sum_n \alpha_i \boldsymbol{\varphi}_i^{\mathsf{T}} \boldsymbol{\varphi}_n + e_n, \tag{2.6}$$

where $\boldsymbol{\varphi}_i$ is the set of training observations and $\boldsymbol{\varphi}_n$ is the observation for which the detection is performed.

If the Hilbert space is a Reproducing Kernel Hilbert Space [24], then the dot product is expressable as a function of the input data, this is $\boldsymbol{\varphi}_i^{\mathsf{T}} \boldsymbol{\varphi}_n = k(\mathbf{x}_i, \mathbf{x}_n)$, where $k(\cdot, \cdot)$ is a positive definite function that maps both vectors to a scalar. Such a function is usually called a Mercer's kernel [25]. Coefficients α_i are the so-called dual coefficients, and they can be optimized according to the same criterion used to optimize $\mathbf{w}$. The use of kernel functions allows the generalization of linear algorithms to endow them with nonlinear properties.

Figure 2.2 shows two toy examples of a classifier used as an equalizer. The channel has an impulse response $h_n = \delta_n + a\delta_{n-1}$, and two classifiers have been implemented where the input is a two-sample window $\mathbf{x}_n = [x_n, x_{n-1}]^{\mathsf{T}}$. White dots correspond to observations whose transmitted symbol z_n is positive, while the black ones correspond to the symbol $z_n = -1$. Figure 2.2a shows the case where $a = 0.5$, which produces a channel with one zero inside the unit circle. It can be seen that a linear classifier (acting as a linear filter) can detect the signals. This classifier has been trained using a simple MMSE criterion. Figure 2.2b shows a case where the zero is outside the unit circle ($a = 1.5$). The behavior is nonlinear. The corresponding classifier is an SVM with an exponential kernel.

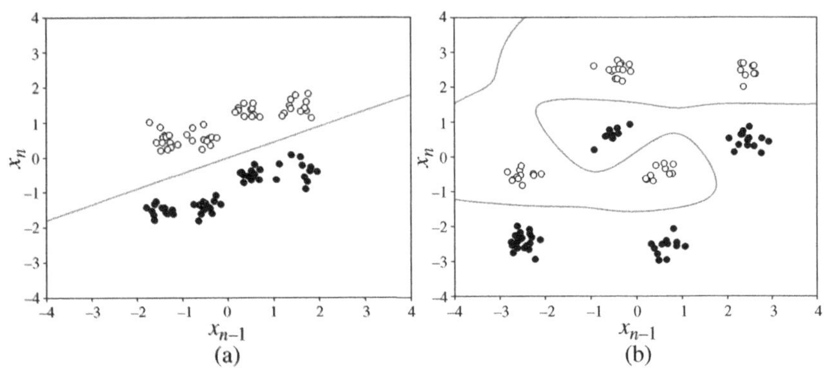

Figure 2.2 Two toy examples of binary symbol detection where the channel has an impulse response $h_n = \delta[n] + a\delta[n-1]$ using simple ML. (a) shows a case where $a = 0.5$, therefore, the only zero of the channel is inside the unit circle. The detection can be performed with a linear classifier. (b) shows an example with $a = 1.5$, this is, the zero is outside the unit circle. The data landscape shows that the symbol detection must be implemented with a nonlinear classifier.

ML approaches for channel equalization are classic and many solutions have been proposed that use virtually all ML structures and optimization criteria. The most classic approaches are based on the multilayer perceptron (MLP). The first adaptive approach was presented in [26] after which complex-valued MLP for QAM signals were introduced (see, e.g., [27]). Recurrent neural networks are known to outperform the performance of the MLP in symbol detection [28].

SVMs and GP have also been proposed for channel equalization. The first complex SVM was presented in [29, 30], where linear and nonlinear versions for symbol detection and antenna array beamforming were proposed (see also [31]).

All these approaches, nevertheless, have been seldom used in practice due to the difficulties of training such structures and their intrinsic suboptimal performance with respect to the Viterbi algorithm in most cases. Besides, these approaches show poor-adaptive performance, or they cannot be trained adaptively, as it is the case of the SVM approaches. Recently, a novel adaptive nonlinear GP for regression [32] has been presented that is trained online and can be suitable for these applications.

2.2.3 Machine Learning Implementations of the Viterbi Algorithm

While the ML presented before is intrinsically suboptimal since they do not use the MAP criterion over the whole transmitted sequences, the Viterbi algorithm may be very complex to implement due to the need to compute the conditional likelihood $p(x_n|s_n)$ of the observation given a hypothesis over the channel state. This probability distribution involves an estimation of the channel, which can be difficult if it is a very complex one. A comprehensive treatment of the different communication channels can be found in [33], where different environments are modeled and simulated. Many communication channels, such as the urban or suburban ones, need to be modeled with a very long impulse response, which translates into a large space state. The computational complexity of the receiver increases exponentially with the length of the channel impulse response [34], and therefore, classically prefiltering techniques are used to shorten this impulse response [35, 36], resulting in a decreased performance of the algorithm. Moreover, these channels change rapidly over time, due, for example, to the movement of the transmitter and receivers, which dramatically changes the propagation paths. Several approaches to DL applied to symbol detection have been presented in [37, 38], or [39] to cite some examples. Nevertheless, improving the Viterbi algorithm seems to be possible, as stated in [40].

An example of the use of DL in symbol detection that implements the Viterbi algorithm was presented in [7], where authors use the deep unfolding technique [41], that is used in the application of deep neural networks (DNN) for tasks as sparse recovery or, more recently, as an alternative to inverse problem solutions in spectroscopic imaging in materials science [42]. The structure does not require the channel impulse response and it has adaptive properties. The structure of the algorithm is based on a DL-based structure that is trained to compute the log-likelihood of each sample with respect to each one of the possible states. This likelihood is passed to the Viterbi algorithm, which otherwise remains unaltered.

The computation of the log-likelihood $p(\mathbf{x}_n|\mathbf{s}_n)$ requires the computation of the probability $p(\mathbf{s}_n|\mathbf{x}_n)$ of the state conditional to the observation, which is carried out by a DL structure, and the marginal probability distribution of the observations $p(\mathbf{x}_n)$. This last probability is estimated by the use of a Gaussian mixture model with the form:

$$p(\mathbf{x}_n) = \sum_{k=1}^{K} \alpha_k p(\mathbf{x}_n|k), \tag{2.7}$$

where $p(\mathbf{x}_n|k)$ is a multivariate Gaussian distribution with mean μ_k and covariance Σ_k and where scalars α_k produce a convex combination $\sum_k \alpha_k = 1$, hence, these coefficients have also probabilistic properties. Each Gaussian describes a cluster of data in the space of observations and coefficients α_k play the role of probability priors of a sample $\mathbf{x}_n$ to belong to cluster k. The means, covariances, and priors are updated online using the well-known Expectation–Maximization algorithm [43].

The product between the conditional probabilities $p(\mathbf{s}_n|\mathbf{x}_n)$ and the marginal probabilities is proportional to the conditional likelihood needed in the Viterbi algorithm (see Figure 2.3).

After a training sequence, assuming that the channel varies sufficiently slowly, the structure can be retrained online using the decision of the Viterbi algorithm as the desired output. The authors of [7] implement a neural network consisting of three layers of fully connected nodes. The first layer has 100 nodes, the second one has 50 nodes. These layers have neurons with rectified linear (ReLU) units used

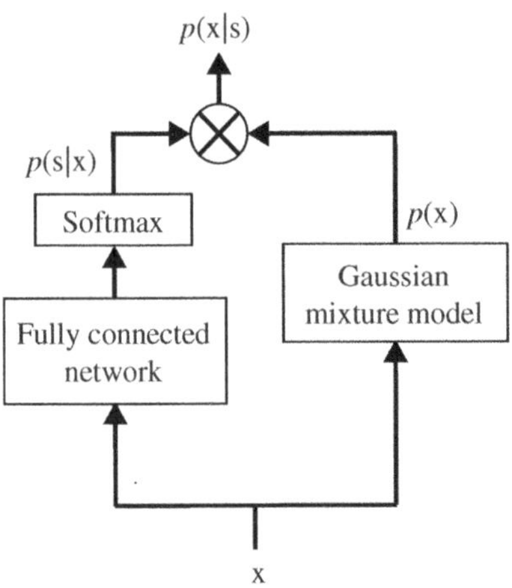

Figure 2.3 Structure of the DNN likelihood estimation.

as nonlinear output. Given an arbitrary input h the ReLU activation is computed as $\max(0, h)$. This activation has been proven to be much more efficient than the classic sigmoidal activations used in neural networks [44]. The output layer has 16 nodes, which assumes that the channel has 16 possible states. The output of this layer is a softmax output. In the softmax output, each neuron uses an exponential activation $\exp(-h_k)$, but these outputs are normalized to have properties of a probability distribution. The structure was trained using 5000 observations to minimize the cross entropy (or equivalently, to maximize the log-likelihood of the network output) [45].

The structure, named ViterbiNet, was compared to the Viterbi algorithm and to the sliding bidirectional recurrent neural network (SBRNN), a decoding algorithm based on a recursive neural network presented in [46] with various channels. Authors compare the performance of the algorithms with perfect knowledge of the channel impulse response or channel state information (CSI) and under CSI uncertainty produced by its estimation. The results can be seen in Figure 2.4.

The channel has a memory of four binary symbols. This produces a total of 16 possible states. The figure shows that the structure, with an estimation of the CSI, is competitive with the rest of the systems under perfect knowledge of the CSI. Nevertheless, the computational burden, the training time, and the needed data increase with the channel length, thus making difficult the scaling of this structure.

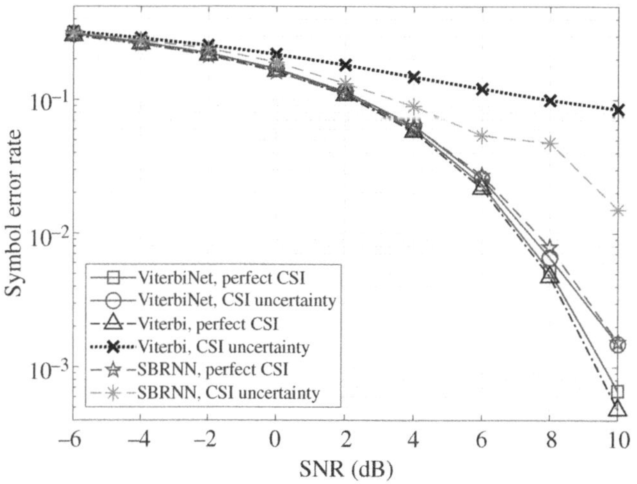

Figure 2.4 Symbol error rate as a function of the SNR in a channel with intersymbol interference and AWGN noise. Source: Reproduced from [47].

2.3 Modulation Detection

Digital modulation is the technique of embedding information onto the carrier wave to facilitate wireless communication [48, 49]. The rapid evolution of communication systems fuelled by the proliferation of handheld devices and faster data rates has given rise to the need for smart and adaptive communication systems capable of optimizing spectral efficiency in densely cluttered radio environments [3, 14, 50]. The congested radio spectrum has given rise to intelligent communication systems capable of transmitting and receiving information over a variety of diverse schemes. Such systems can sense the spectrum and make decisions to transmit information over a particular modulation scheme to minimize interference and maximize throughput [13–15]. This opportunistic spectrum sharing or dynamic spectrum access (DSA) given the sensed radio environment is a crucial component of intelligent radio systems such as cognitive radios (CRs), which can exploit a broad range of diversity schemes. This makes it imperative for the receiver to be capable of identifying the modulation scheme to successfully demodulate the signal and retrieve the information embedded in the carrier signal. As a result, modulation detection has received significant attention over the past few decades and various statistical methods have been introduced to the problem of AMC [15, 51, 52].

AMC is used for sensing and classifying the environment to identify the optimal modulation scheme to transmit as well as to identify the modulation scheme of the received signal for information retrieval. Various modulation techniques have been implemented over the years to pack as much information as possible and squeeze the most out of the available spectrum. Such techniques can be broadly classified into three distinct categories amplitude modulation, frequency modulation, and phase modulation based on the component of the signal that is being modulated to encode the information. The most commonly used modulation schemes are the 16-ary quadrature amplitude modulation (16-QAM), the 64-ary quadrature amplitude modulation (64-QAM), and the 4 pulse amplitude modulation (PAM4). Phase modulation techniques such as phase shift keying have led to the development of binary phase shift keying (BPSK), quadrature phase shift keying (QPSK), and 8-ary phase shift keying (8-PSK), which are most commonly used for communication. Other frequency modulation schemes of use are Gaussian frequency shift keying (GFSK) and continuous phase frequency shift keying (CPFSK), etc. Modern communication systems exploit a variety of modulation schemes to enable fast and reliable communication as mentioned above. In Section 2.3.1, we will develop the signal model, discuss feature extractions, and develop the intuition for data preparation and model development for DL modulation detectors.

2.3.1 Signal Model

Modulation detection or AMC aims to blindly identify the modulation type of an incoming signal from the sampled signal and can be treated as a multiclass classification problem. The in-phase and quadrature components of the analog signal are sampled at discrete time steps and the complex base-band representation of the signal is processed to generate the input features for the classification problem. The transmitted signal at the receiver $\mathbf{x}(t)$ can be written as,

$$\mathbf{x}(t) = c\ \mathbf{s}(t) + \mathbf{n}(t), \tag{2.8}$$

where $\mathbf{s}(t)$ is the signal containing the discrete bits modulated onto the carrier waveform using a variation scheme in amplitude, phase, frequency, trajectory, or a combination of these schemes. c represents the path loss or a constant gain term and $\mathbf{n}(t)$ is the noise term assumed to be i.i.d. Gaussian. The representation in Eq. (2.8) is a good place to start exploring data generation and model building. However, the real-world systems suffer from fluctuations in the gain and noise terms and can be represented as,

$$\mathbf{x}(t) = e^{j*n_O(t)} \int_{\tau=0}^{\tau_1} \mathbf{s}(n_{Ck}(t-\tau))\mathbf{h}(\tau) + \mathbf{n}_{ad}(t), \tag{2.9}$$

where $n_O(t)$ is the modulation done by a residual carrier random walk process, $n_{ck}(t)$ is the sampling done by a residual clock oscillator random walk, $\mathbf{h}(t-\tau)$ is the time-varying rotating nonconstant amplitude channel impulse response, $\mathbf{n}_{ad}(t)$ is the additive noise that may not be white [13]. Each of these variables presents a unique source of time-varying noise resulting in error in detection. Modern communication systems exploit a variety of modulation detection algorithms to enable fast and reliable detection. Methods for modulation detection can be grouped into likelihood-based methods and feature-based empirical learning methods [13, 52]. Likelihood-based methods exploit the probability density function of the different distributions for each of the modulation classes and infer a decision using a hypothesis testing approach. While likelihood-based methods provide optimal performance under the ideal scenario in Eq. (2.8), they suffer severe performance degradation under nonideal conditions since they rely on the knowledge of the ideal signal and channel parameters, which may not be available in real-time systems. Computing multiclass hypothesis testing is also computationally expensive and difficult to implement in real-time systems. Feature-based machine and DL methods exploit empirical learning and are trained on large datasets to achieve superior generalization. Such models and methods take time to train but are rather easy to implement and have a reduced computational footprint in the inference mode.

Several ML and DL methods have been introduced to the problem of AMC over the years exploiting a diverse set of feature selection schemes. Typical feature selection includes first-order statistics such as the power spectral density, standard deviations, and variances of the various attributes of the received signal [13, 15]. DL architectures have been known to extract features from unprocessed data and thus explicit feature extraction is not required. As a result, DL classifiers trained on the raw IQ data have shown promising results in terms of detection capabilities. Image-based preprocessing has also been explored to exploit the capabilities of deep networks such as the convolutional neural networks (CNN). In the Sections 2.3.2–2.3.4, we go over some of these feature extraction and selection techniques employed by the data-driven methods and discuss the pros and cons of the various strategies.

2.3.2 Feature Selection

Various feature extraction techniques have been explored over the years to train classifiers to detect modulation. Maximum power spectral density, deviations in amplitude, phase, frequency, variance of the zero crossing, and spectral correlation function are some of the most widely adopted features in AMC [13, 15]. Simple features such as these are optimal features for high signal-to-interference plus noise ratio (SINR) but suffer severe degradation when the SINR falls below a certain threshold. Spectral correlation function on the other hand has shown to contain enough information for successful detection even under low SINR conditions and has been the feature of choice for ML and shallow neural net algorithms [13, 15]. A time domain representation of the commonly used modulation strategies is plotted in Figure 2.5.

Signal modulations are cyclostationary processes characterized by periodicity in the mean and autocorrelation. Such processes can be sufficiently represented by capturing the cyclic autocorrelation function in the frequency domain [13]. Cyclic autocorrelation is the correlation between the different frequency-shifted versions of the sampled signal. Cyclic autocorrelation is defined as,

$$\mathbf{R}_{xx}[p] = \frac{1}{2N+1} \sum_{n=-N}^{N} \mathbf{x}[n]\mathbf{x}^*[n-p]e^{-j2\pi\alpha n}, \tag{2.10}$$

where $\mathbf{x}$ is the modulated signal and α is the cyclic frequency. The spectral correlation function (SCF) is obtained by computing the Fourier transform of the cyclic autocorrelation given by,

$$\mathbf{S}_{xx}[f] = \sum_{p=-\infty}^{\infty} R_{xx}[p]e^{-j2\pi fp}, \tag{2.11}$$

where $\mathbf{S}_{xx}$ is the SCF and f is the temporal frequency of the modulated signal. The noise component associated with the sampled signal is assumed to be AWGN that

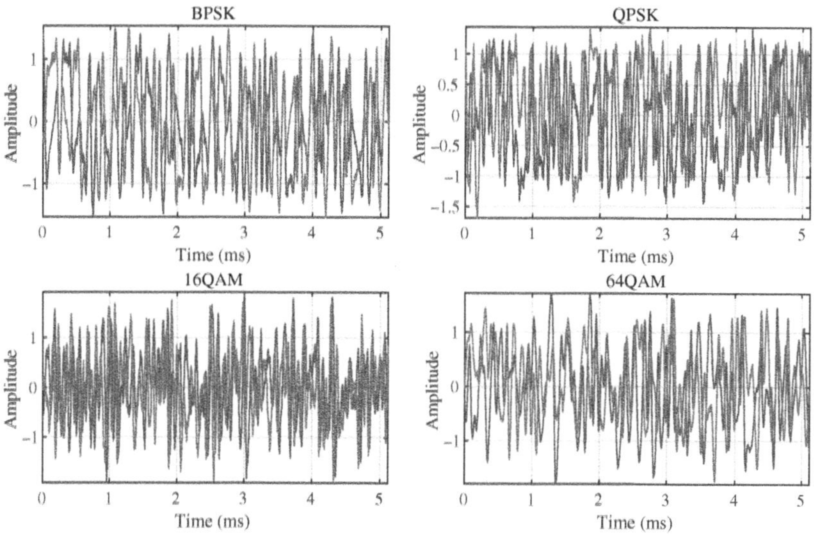

Figure 2.5 Time domain visualization of the modulated signals, starting from top left, BPSK, QPSK, 16-QAM, and 64-QAM.

is a stationary process and thus has no spectral correlation with the cyclostationary process generating the modulated signals.

Image-based preprocessing of the data has been explored to exploit the spatial learning capability of CNNs. CNNs have revolutionized the field of computer vision with their ability to model extreme nonlinearities and the ability to extract increasingly abstract features using very deep networks containing blocks of CNN layers and nonlinear activation functions. Spectral images obtained from computing the short-time Fourier transform (STFT) of the received signal have been used to train CNN classifiers with high accuracy and precision. Spectral images can be further denoised using Gaussian filters improving the detection performance of the CNN-based models under low-SINR conditions [15]. The amplitude spectrum of the bispectrum, cyclic correntropy spectrum, discrete orthonormal Stockwell transform and constellation diagrams have all been successfully used to train CNN-based classifiers. The images are preprocessed and converted into three matrices containing the three different channels representing red, green, and blue (RGB). This preprocessing is a standard procedure for processing image data for feature learning and classifiers with CNN. DL architectures such as CNN do not require explicit feature extraction. Instead, it extracts layers of abstract features using the depth of the model and can be trained on the raw time domain signals, as shown in Figure 2.5. The sampled signals are simply normalized and processed in frames producing the input to the CNN model.

2.3.3 Maximum Likelihood Estimation

Modulation detection can be formally defined as the process of determining the modulation scheme of the noisy sampled signal $\mathbf{x}(t)$ from a set of modulation schemes $C \in \mathcal{R}^{1\times M}$, where M is the number of modulation schemes that can be present in a sampled signal. Maximum likelihood estimation (MLE) provides the optimal solution to this problem from a Bayesian perspective [15, 53]. The presence of multiple modulation schemes among the possibilities makes the MLE a composite hypothesis testing problem that maximizes the probability of the noisy observation given a hypothesis over the transmitted symbol. If the prior probabilities of the symbols are uniform, the maximum likelihood maximizes the posteriori probability $P(C_i|\mathbf{x}(t))$. The MLE criterion can be written as:

$$\hat{C} = \underset{C_i}{\text{argmax}} \ \log \ P(\mathbf{x}(t)|C_i), \tag{2.12}$$

where $\hat{C}$ is the estimated class, $\mathbf{x}(t)$ is the sampled signal from the channel-impaired waveform, a noisy version of the transmitted signal. The likelihood estimation approach relies on the explicit availability of the CSI. In the absence of the CSI, it is either estimated, computed from the preexisting probability density functions (PDFs), or obtained using a combination of different estimation strategies. Average Likelihood Ratio Test (ALRT), Generalized Likelihood Ratio Test (GLRT), Hybrid Likelihood Ratio Test (HLRT), and quasi-HLRT are some of the methods explored to estimate the CSI before a modulation classification is performed using composite hypothesis testing [1, 54].

2.3.4 Neural Modulation Detection

Neural networks have revolutionized entire fields of study such as Computer Vision (CV), natural language processing (NLP), and various domains of signal processing [3, 8, 55, 56]. A plethora of network architectures have been proposed over the last two decades to handle spatial and temporal learning and extract abstract features from larger data sets [44]. Several such neural net architectures have been subsequently adopted to solve the problem of modulation classification with various degrees of success [13, 57, 58]. While the data-driven approaches are suboptimal in a Bayesian sense, they often outperform the MLE when the CSI is not available for the likelihood estimation. They are also easy to deploy and require less computation and memory footprint as opposed to an MLE using composite hypothesis testing. Neural networks are state-of-the-art interpolators and can model extreme nonlinearities in the data. These models are trained on large amounts of data to optimize the network parameters on labeled datasets, otherwise known as supervised learning. A detailed survey of the neural net-based architectures for modulation classification can be found in [52].

2.3.4.1 Convolutional Neural Network

Convolutional Neural Network or CNNs have been tremendously successful in learning increasingly abstract features from structured data such as images [59, 60]. These architectures leverage sparse interactions, parameter sharing, and equivariant representations to provide network learning that is both robust and efficient and provides improved generalization [44, 60]. As a result, they have been adopted to solve various applications dealing with structured data including modulation detection with significant success [52, 57, 58]. Communications signals being time series data, each sampled frame can be treated as a tensor with at least two valid dimensions representing time and one channel for a single channel system, the number of columns increasing as the number of channels increases. These frames are passed through layers of CNNs computing convolutions over space and time and pooling to extract features through successive layers. These features are subsequently classified by one or several fully connected layer of perceptron, providing means of classifying the extracted features using linear or nonlinear activation functions. This ensures no additional feature extraction is required from the input data and the normalized input data can be directly used to classify the modulation of a sampled signal.

As the name suggests, CNNs use layers of linear convolution operations as opposed to matrix multiplications typically used in traditional neural networks such as the MLP [44]. A typical CNN layer consists of three computation steps computed sequentially. The first step performs convolutions in parallel and computes a set of linear outputs. These linear outputs are the input to a non-linear activation function to incorporate nonlinearity into the architecture. The final step involves pooling from a given kernel output and replacing them with summary statistics of the kernel inputs. Convolutions along with pooling layers give CNNs the ability to extract increasingly abstract features from the structured data. The original pictorial description of the CNN architecture is shown in Figure 2.6.

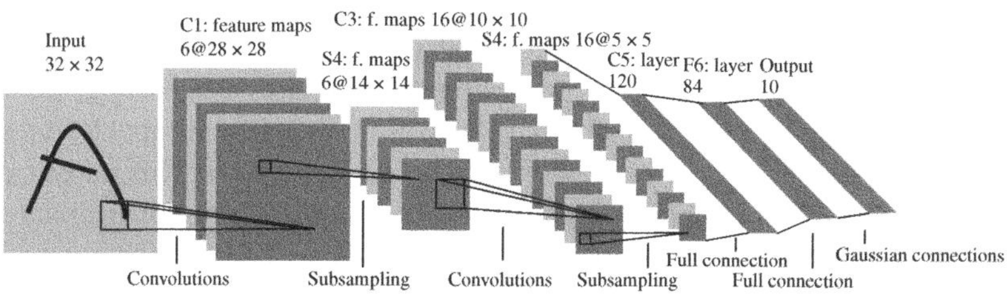

Figure 2.6 CNN architecture showing the convolution and pooling operations for feature extraction and fully connected layers for classification. Source: Adapted from [59].

For a continuous input $\mathbf{a}(t)$, convolution performs a weighted average of the input to obtain a new function $\mathbf{g}(t)$ and is defined as,

$$\mathbf{g}(t) = (\mathbf{a} * \mathbf{w})(t) = \int \mathbf{a}(\tau)\mathbf{w}(t-\tau)d\tau, \tag{2.13}$$

where $(*)$ denotes the convolution operation and $\mathbf{w}$ is the weighting function that is required to be a valid probability density function for the output to be a weighted average [44]. The weighted averaging using convolution puts a higher weight on the more recent observations as opposed to observations in the distant past. The discretized convolution is defined as,

$$\mathbf{g}(t) = \sum_{-\infty}^{\infty} \mathbf{a}(\tau)\mathbf{w}(t-\tau). \tag{2.14}$$

In practice, the summation is taken over a finite set of array elements determining the size of the kernel $\mathbf{w}$ using the assumption that these function values are zero everywhere except the set of points defined by the kernel parameters [44, 59]. For two-dimensional data such as multichannel communication signals and images, a two-dimensional kernel is employed to obtain

$$G(i,j) = (\mathbf{W} * \mathbf{A})(i,j) = \sum_{m}\sum_{n} \mathbf{A}(i+m)(j+m)\mathbf{W}(m,n), \tag{2.15}$$

where (i,j) denotes the position of the two-dimensional output tensor and (m,n) is the size of the two dimensional kernel matrix $\mathbf{W}$. Note that the one-dimensional inputs from Eq. (2.14) have been replaced with their two-dimensional representations in Eq. (2.15).

The convolution kernel $\mathbf{W}$ is usually much smaller than the input to the convolutional layer and slides along the input to compute the linear kernel outputs for each position as showed in Figure 2.7. This layer can be computed in parallel for a known input and kernel dimension. The output is then projected into a nonlinear space using a nonlinear activation function such as a rectified linear activation (ReLU), Sigmoid, or a tangent hyperbolic (tanh) activation function depending on the required output. In the third step, each kernel output is replaced by the summary statistics computed over the kernel outputs obtained from taking the convolutions and nonlinearity is introduced using the activation function. This step is known as pooling and various pooling strategies such as max pooling, mean pooling, and weighted average pooling have been used to modify the input features to the layer to obtain another layer of abstraction on the features. This statistical representation of the neighboring inputs (kernel values) helps provide invariance to local translation, a crucial property of the CNNs making them invariant to small changes and improving the statistical efficiency of the model [44, 60]. These three steps of convolution, activation, and poolings are repeated to create deeper model architectures capable of extracting increasingly abstract features from the data.

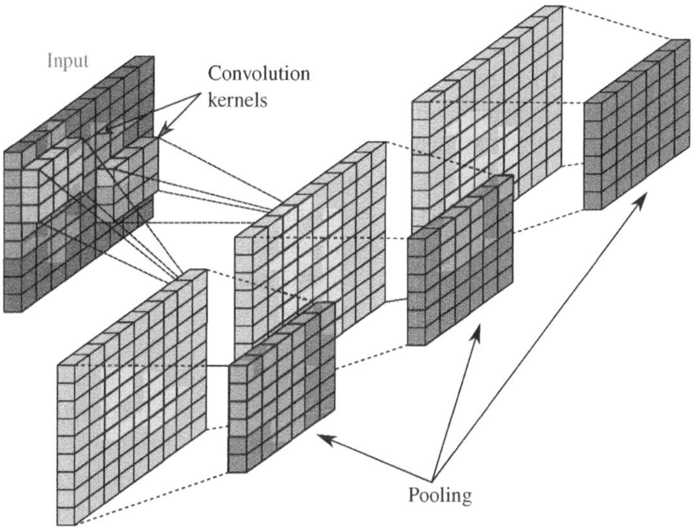

Figure 2.7 Visualization of pooling operation using convolutional kernels.

2.3.4.2 CNN Modulation Detection

Various variants of CNNs have been implemented for modulation classification with significant success [52, 57, 58]. The architectures are relatively simple with layers of CNN cascaded to create sufficient depth to extract the most "relevant" set of features. These features are then classified using one or multiple layers of fully connected perceptrons. These layers help add regularization and generate structured outputs from the extracted features. The number of classes or modulation schemes in this case determines the size of the output of the final layer. The number of convolutional and fully connected layers are treated as hyperparameters of the model and optimized to find the optimal architecture from a DL perspective.

DL employs data-driven empirical learning requiring lots of observations. This makes data generation and/or collection crucial for building robust detectors. Creating a diverse dataset for modulation classification is nontrivial and a majority of the methods in literature use simulated data to show the efficacy of the algorithm. A list of publicly available datasets for modulation detection can be found in [15]. For practical applications, the model can be trained on synthetic data generated from controlled simulations and then fine-tuned on real-world observations collected from experiments. We continue with model building using synthetic data generated using the signal model discussed in Section 2.3.1 [61].

The dataset contains 11 different modulation schemes such as BPSK, QPSK, 8-PSK, 16-QAM, 64-QAM, PAM4, GFSK, CPFSK, broadcast FM (B-FM), double sideband amplitude modulation (DSB–AM), and single sideband amplitude modulation (SSB–AM). To avoid class imbalance, 10,000 frames of each modulation type are generated to create the dataset. For simplicity, we constrain the problem to narrowband signals with a center frequency of 1 GHz for the digital modulation types and 200 MHz for the analog ones. To resemble realistic scenarios, the generated signals are passed through a Rician multipath fading channel, and an AWGN noise is added to the sampled signal resembling the thermal noise at the sensors. The sampled complex-valued baseband signals are then split into their real and imaginary components and concatenated to form the feature vector. Since CNNs are powerful feature estimators, we do not need to compute any explicit feature extraction.

We begin model development with the simplest architecture to identify the scope and model complexity given the data distribution. The base model consists of one convolutional layer with a ReLU activation followed by pooling layers to extract features. These features are then classified using a fully connected layer with a softmax activation. The softmax activation function converts the extracted features into a probability distribution of M possible outcomes. We assign these M outcomes to each of the 11 modulation classes. More formally, the softmax ($\mathcal{S}$) is defined as,

$$\mathcal{S}(\mathbf{y})_i = \frac{e^{y_i}}{\sum_{m=1}^{M} e^{y_m}}, \tag{2.16}$$

where $\mathbf{y}$ is the output vector of the layer and the input to the softmax function $\mathcal{S}$, subscripts i, m denote the ith and mth element of the input vector $\mathbf{y}$ and M is the number of categories or classes. Observe how the right-hand side of Eq. (2.16) is normalized ensuring each element of the output vector $\mathcal{S}(\mathbf{y})_i \in [0, 1]$ and total output sums up to 1. This property of the softmax makes it suitable to predict the probabilities associated with a Multinoulli distribution [44, 62]. The termination of the network with a softmax activation ensures we output a probability for each of our output modulation types instead of margin scores.

Now, we have a network architecture that outputs probabilities for each modulation class and we need a loss function to minimize the loss and train out the network using backpropagation. Conveniently, we can just minimize the negative log-likelihood of obtaining the correct modulation class and backpropagate the error to train the network weights. In other words, we can formulate the problem as,

$$\mathcal{L}_i = -\log\ \mathcal{P}(y_m = \mathbf{M}_i|\mathbf{x}), \tag{2.17}$$

where we are just minimizing the negative log-likelihood of obtaining a class m among the M classes given the observations in $\mathbf{x}$. The likelihood inside the logarithmic is nothing but the probability of each class given an observation. Hence, we can replace it with the output of our final layer since

$$\mathcal{P}(y_i = \mathbf{M}_i|\mathbf{x}) = \frac{e^{y_i}}{\sum_{m=1}^{M} e^{y_m}}. \tag{2.18}$$

Substituting for the probability function from Eq. (2.18) in Eq. (2.17) brings us to the loss function to be minimized to train the network parameters and is defined as,

$$\mathcal{L}_i = -\log\left(\frac{e^{y_i}}{\sum_{m=1}^{M} e^{y_m}}\right). \tag{2.19}$$

$\mathcal{L}_i$ is the cross-entropy loss for softmax classification [63]. Minimizing the cross-entropy loss is analogous to minimizing the distance between the probability distributions of the predicted and the target variables. The network parameters are subsequently trained and optimized using the Stochastic Gradient Descent with Momentum (SGDM) algorithm with **L2** regularization. Alternative stochastic optimization techniques such as adaptive moment estimation (Adam) or root mean square propagation (RMSProp) can also be used for network optimization [3, 44, 60].

To train and validate the model performance the dataset is split into training, validation, and test sets in a [80:10:10] ratio. The validation set is used to benchmark generalization after each epoch and the test set is used to benchmark the model with the best validation score after training is completed. Since the model has seen the validation set, we opt for a completely unseen set of data to benchmark the final model performance. Next, we train the network and optimize the hyperparameters of the architecture to fine-tune and fit the model parameters to the data. The learning rate was set to 0.03 and the model was trained for 50 epochs with a mini-batch size of 1024.

The training and validation losses are plotted as a function of the number of epochs in Figure 2.8a. The loss curves give us an intuition about convergence and the inherent complexity of the model as opposed to the complexity required for an optimal solution. Observe how the gradient of the loss curves grows smaller as the number of epochs grows larger. The change in gradient is almost negligible after the 40th epoch, meaning training further will not improve generalization. The accuracy as obtained on the training and validation sets are plotted in Figure 2.8b. The accuracy curves show the same behavior as the loss curves as expected. The network learns to predict accurately 40% of the time pretty quickly, and then saturates at a maximum training and validation accuracy of 49%. The low accuracy levels were expected since our base model is just a linear convolution followed by a nonlinear activation. In other words, the low complexity of the model is insufficient to represent the complexity of the problem.

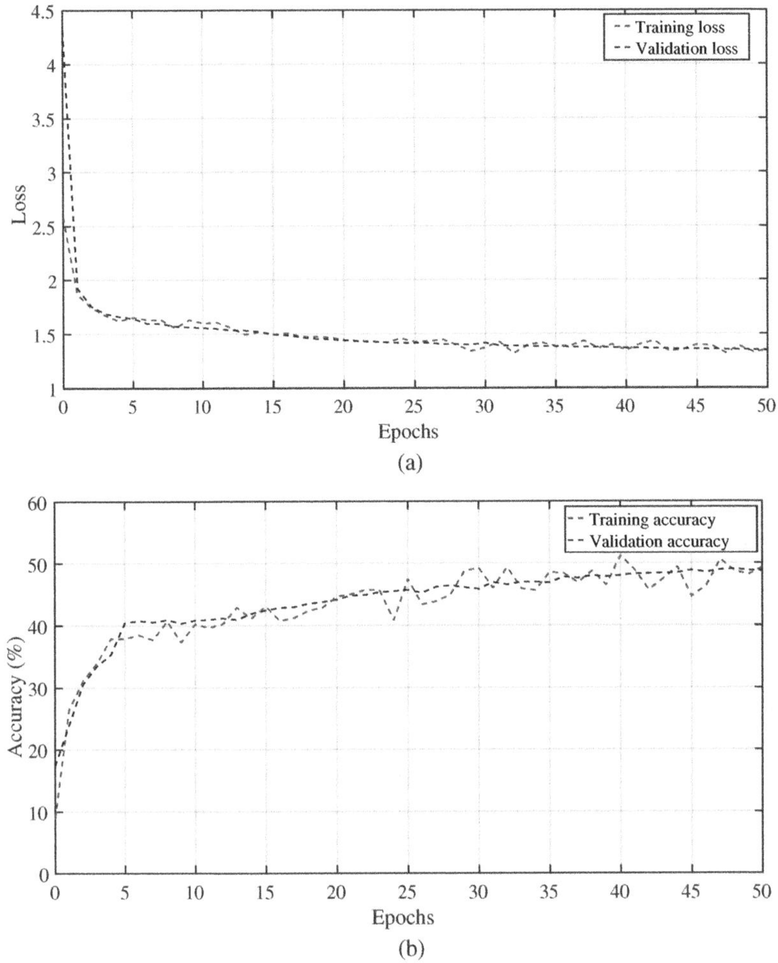

Figure 2.8 (a) Training and validation losses and (b) training and validation accuracies, as obtained from inference on the test set by the base CNN model with one convolution layer.

Accuracy can be a misleading metric, especially when you have a class imbalance in the data set. Consider a detector trying to detect the presence of rare pathogens which occurs every million cases. In such a scenario, the machine will obtain a very high accuracy by simply learning to predict the absence of the pathogen at all times. It would have no incentive to detect the rare occurrences. Furthermore, accuracy as introduced in the plots above is the mean accuracy across all classes and does not present the performance of the detector for every individual class.

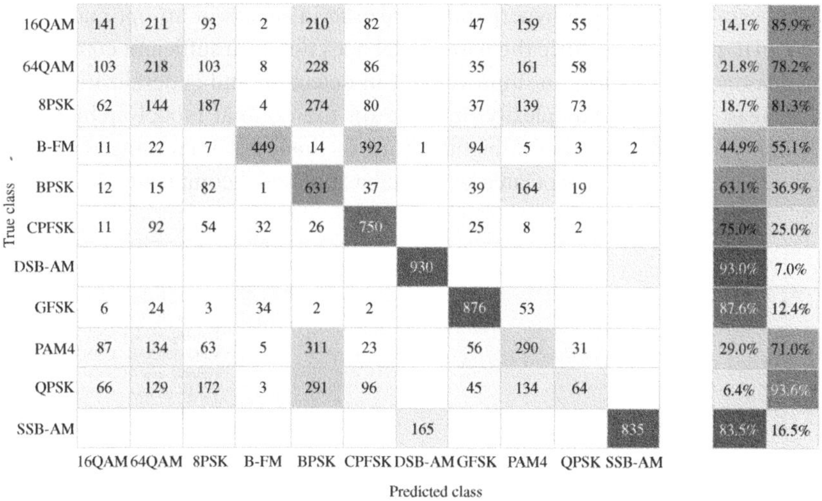

Figure 2.9 Confusion matrix is computed on the test set for the base CNN model with one convolution layer.

A better way to characterize the behavior and performance of classifiers is to compute the confusion matrix given the predicted and true labels of the test data [62]. A confusion matrix is a table where the two orthogonal axes represent the true class and predicted class of a classification model. It is widely used in statistics and ML to benchmark the performance of classifiers. The confusion matrix as obtained given the predicted and true labels on the test set is plotted in Figure 2.9. The diagonal tells us the number of times we predicted a class correctly. Everything outside the diagonal is a misclassification. The table also gives us an understanding of which classes are getting misclassified and what they are being wrongly classified as. It outlines the performance of the model for every modulation type and highlights any confusion the model is facing with respect to distinguishing the features extracted for different classes. The highest individual accuracy was obtained for DSB–AM followed by the GFSK and SSB–AM. More importantly, the confusion matrix tells us which classes are closer in the feature space and which ones are further apart. Observe, how 16QAM is wrongfully predicted as 64QAM and BPSK for nearly 20% of the test cases. Similarly, 64-QAM is wrongfully predicted as 16QAM and BPSK due to the similarity in the features extracted by the convolution layer. The shallow network is unable to extract relevant features to distinguish between these modulation types consistently. The next logical step in model development is to add more complexity to the model so it is able to extract more relevant features from the data.

We begin exploring the hyperparameter space in search of an optimal model from a DL perspective. Since the base model does not have sufficient complexity, it is reasonable to expect an improvement in detection and generalization with an increase in model complexity until we hit the optimal model architecture. Adding complexity is rather straightforward in this case. Increasing the depth of the model is directly associated with increasing model complexity. Increasing

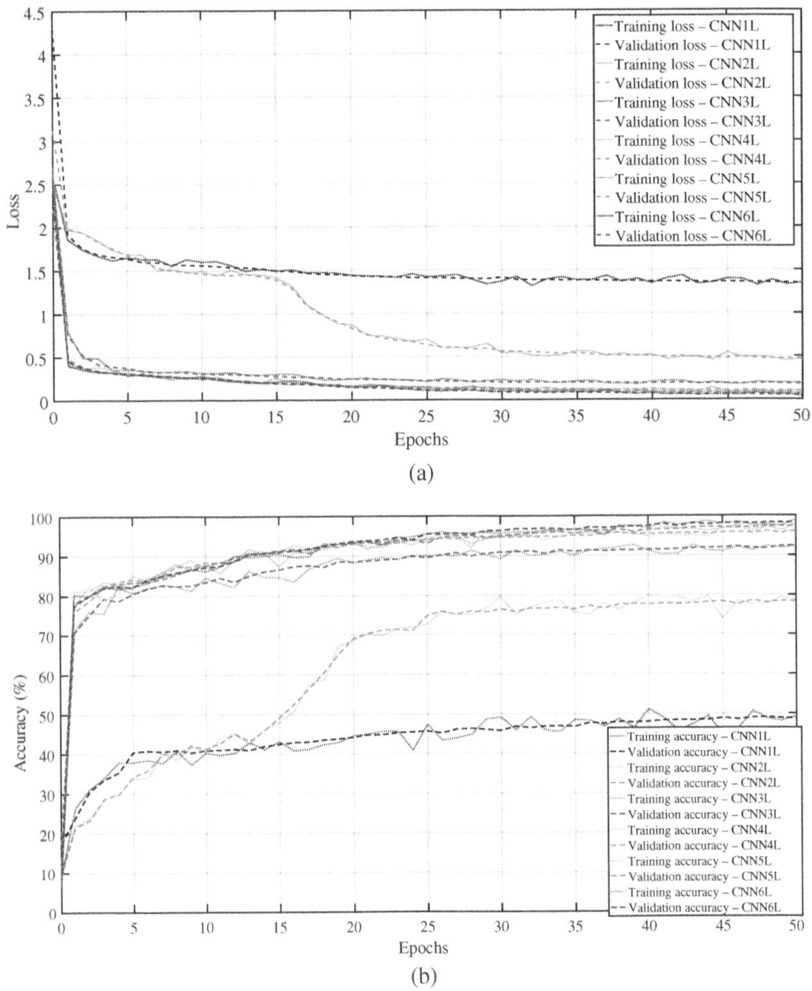

Figure 2.10 (a) Loss as a function of epochs and the number of convolution layers in the six different model architectures denoted by CNNxL, where 'x' refers to the architecture with the as highest number of convolution layers. (b) Accuracy as a function of number of epochs for the six different architectures.

depth would mean stacking several successive convolution and pooling layers. We stack six successive convolution and pooling layers one at a time to obtain incremental performance as a function of the number of convolution layers.

The loss obtained from six different architectures with an increasing number of CNN layers from one to six is plotted in Figure 2.10a. Notice the difference in loss between two successive architectures grows smaller as we increase the number of layers. A similar pattern is witnessed with the accuracy plots for the six different architectures in Figure 2.10b, as expected. The improvement in detection accuracy decreases as the number of layers increases. The architecture with six pairs of convolution and pooling layers resulted in a test accuracy of 98.2%. Adding more layers does not increase accuracy enough to justify the extra computation required, resulting in the optimal architecture from a DL perspective. The confusion matrix as obtained from predicting on the test set is detailed in Figure 2.11. The machine is able to predict every modulation scheme successfully. We observe some confusion between 16QAM and 64QAM exists in the features extracted by the deepest model with six convolution layers, resulting in the lowest accuracy for 64QAM at 93.5%.

This section is meant to serve as an introduction to AMC with ML and DL methods. We developed several AMC models trained with synthetic data generated using simulations closely replicating a real-time communication channel. CNNs were able to extract features that can be classified with a mean accuracy of above 98% across the 11 different modulation classes. Several other model architectures and approaches have been introduced for modulation detection

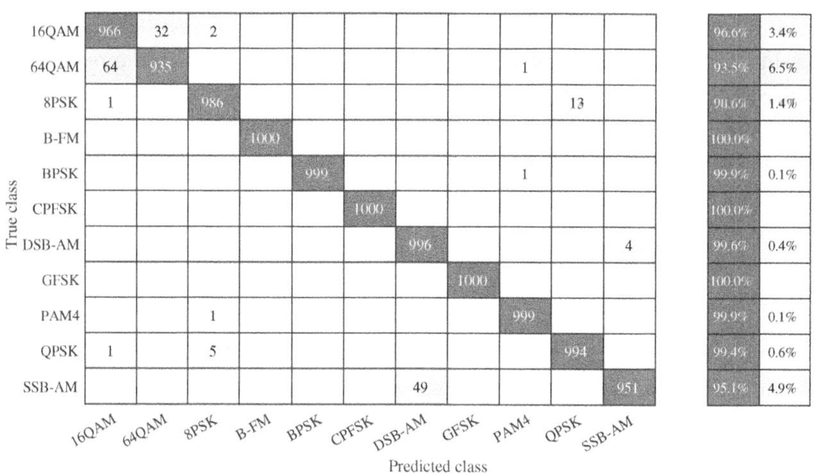

Figure 2.11 Confusion matrix is computed on the test set for the base CNN model with six convolution layers.

with DL, including but not limited to recurrent neural networks, deep belief networks (DBNs), autoencoders, and residual networks (ResNets) [13, 15]. While simulated data gives us a good base model, it is necessary to retrain the model on real data using a transfer learning approach. Fine-tuning the model on actual data should improve performance and generalization by adjusting the weights to account for the noise and channel fading in real-time communication channels.

2.4 Source Detection

Source detection is the method of estimating the number of sources present in a waveform sampled by an array of sensors [3, 9]. The knowledge of the number of sources is essential to localize sources and exploit spatial diversity to maximize throughput [3, 64, 65]. Popular direction finding algorithms such as multiple signal classification (MuSiC), root-MuSiC, estimation of signal parameters via rotational invariance techniques (ESPRITs) and as modern ML and DL algorithms require this knowledge on the number of sources to be known *a priori* [3, 64, 66–69]. As a result, source detection has found applications in radar signal processing, magnetoencephalography (MEG), and electroencephalography (EEG) of brain and biosignals, sonar signal processing, ultrasonography, and modern communication systems among others [3, 70, 71]. Estimation of the number of sources in the sampled spectrum is carried out in the early stages of the array signal processing chain in modern communication systems capable of exploiting spatial diversity to enhance spectral efficiency [3]. This information is then used to understand the radio frequency environment and localize the sources to maximize spectral efficiency by enhancing the signal of interest (SOI) and suppressing the undesired signal sources [3, 65]. The information on the number of sources is contained within the autocorrelation matrix of the sampled signal and eigenspace-based methods have been exploited traditionally to estimate this value [65, 72]. Information theory-based criteria such as the Akaike information criterion (AIC), and the minimum description length (MDL) have been widely used to estimate the number of sources [73–75]. More recently, several DL classifiers have also been proposed to solve the problem of source detection [9, 76–80]. In Sections 2.4.1–2.4.2, we introduce the problem statement along with the signal model and discuss the conventional approaches to source estimation. Finally, we introduce the state-of-the-art DL methods and compare them with the conventional methods under different operating scenarios.

2.4.1 Array Signal Model

The array signal model is different than the single-channel model introduced so far in this chapter. The presence of multiple receivers and interactions between them needs to be accounted for to have an accurate model. The knowledge of the analytical formulation and its properties is crucial for understanding the more advanced topics of source detection and localization. Let us consider a uniform linear array (ULA) of L identical elements, with the center of the array located at the origin of the coordinate system. We assume M far-field sources are emitting simultaneously and being observed by the array, with θ_m being the location of the source, $m = 1, 2, \dots, M$. The received signal $\mathbf{X}$ is a function of time and can be written as,

$$\mathbf{X}(t) = \mathbf{A}(\theta)\mathbf{S}(t) + \boldsymbol{N}(t), \tag{2.20}$$

where $\mathbf{X}(t) \in \mathbb{C}^{L\times N}$ is the received signal, and $\mathbf{S}(t) \in \mathbb{C}^{M\times N}$ is the modulated transmitted signal, with N being the number of snapshots captured by the receiver array for a given frame. The received signal is corrupted by AWGN denoted by $\boldsymbol{N}(t)$ that has a variance of σ_n^2. $\mathbf{S}(t)$ represents the signal transmitted by M sources and is a collection of vectors $[\mathbf{s}_1, \mathbf{s}_2, \dots, \mathbf{s}_M]^{\mathsf{T}}$. $\mathbf{A}(\theta) = [\mathbf{a}(\theta_1), \mathbf{a}(\theta_2), \dots, \mathbf{a}(\theta_M)]$ is the array manifold matrix, consisting of the steering vectors $\mathbf{a}(\theta_m)$ and defined as,

$$\mathbf{a}(\theta_m) = [1, e^{j2\pi\frac{d}{\lambda}\sin(\theta_2)}, \dots, e^{j2\pi\frac{d}{\lambda}\sin(\theta_M)}]^{\mathsf{T}}, \tag{2.21}$$

where d is the inter-elemental distance and λ is the wavelength of the center frequency. For signal processing and numerical computation in general we always use a discretized version of the signal obtained from sampling the function at discrete time steps. The discretized version of the continuous signal in Eq. (2.20) is given by,

$$\mathbf{X}[n] = \mathbf{A}(\theta)\mathbf{S}[n] + \boldsymbol{N}[n]. \tag{2.22}$$

The above signal model assumes all the sources in the sampled signal are uncorrelated. It is a good place to start exploring the problem, but real-life scenarios often contain correlated sources. Correlated sources consider the effects of multipath reflection during propagation, and are modeled as amplitude and phase shifts in the transmitted signal. Thus, if there exists a source $\mathbf{s}_1$, then its k-th correlated counterpart can be written as,

$$\mathbf{s}_k(t) = \rho_k e^{j\phi_k}\mathbf{s}_1(t), \tag{2.23}$$

where ρ_k is the amplitude fading factor, and ϕ_k is the phase shift caused by multipath fading. This way the received signal can be modeled as a collection of both correlated and uncorrelated sources. The presence of correlated sources would result in the matrix being rank-deficient [81]. To avoid this, we employ the forward–backward smoothing technique on the autocorrelation before extracting the feature vectors. To perform smoothing, the array is divided into

K overlapped subarrays with $K = L - L_0 + 1$, L_0 being the dimension of each subarray. The smoothed autocorrelation matrix is given by,

$$\mathbf{R_{fb}} = \frac{1}{2K}\sum_{k=1}^{K}(\mathbf{R_{ff}}^{(k)} + \mathbf{R_{bb}}^{(k)}), \tag{2.24}$$

where $\mathbf{R_{ff}}^{(k)}$ and $\mathbf{R_{bb}}^{(k)}$ are the forward and backward autocorrelation matrices constructed from the kth subarray and $\mathbf{R_{fb}}$ is the spatially smoothed autocorrelation matrix.

In the frequency domain, we can also plot the power spectrum density (PSD) of the received signal as,

$$\mathbf{S}_{xx}(f) = \lim_{T\to\infty}\frac{1}{T}|\tilde{\mathbf{X}}(f)|^2, \tag{2.25}$$

where T is the period of the signal, and $\tilde{\mathbf{X}}(f)$ is the Fourier transform of $\mathbf{X}(t)$ if it exists. This will tell us the distribution of the average power of the signal as a function of frequency. PSD is also a Fourier transform pair with the autocorrelation matrix given by $\mathbf{R}_{xx}$ [82].

$$\mathbf{S}_{xx} \overset{\mathcal{F}}{\longleftrightarrow} \mathbf{R}_{xx}. \tag{2.26}$$

$$\mathbf{R}_{xx} = \frac{1}{N}\mathbf{X}[n]\mathbf{X}^H[n]. \tag{2.27}$$

The eigenvalues corresponding to the M largest eigenvectors of this matrix would correspond to the signal eigenvalues. Since $\mathbf{R}_{xx}$ is a square matrix we can simply perform an eigenvalue decomposition (EVD) to retrieve the eigenvalues and eigenvectors. The EVD of the of the autocorrelation can be write as,

$$\mathbf{R}_{xx} = \mathbf{Q}\boldsymbol{\Delta}\mathbf{Q}^H, \tag{2.28}$$

where $\mathbf{Q}$ is a matrix containing the eigenvectors of the autocorrelation $\mathbf{R}_{xx}$, and $\boldsymbol{\Delta}$ is a matrix containing the corresponding eigenvalues along its diagonal. If we define $\mathbf{Q}_s$ and $\mathbf{Q}_N$ as the eigenvectors of the signal and noise subspaces, respectively, then the eigendecomposition of the autocorrelation matrix can also be written as,

$$\mathbf{R}_{xx} = \mathbf{Q}_s\boldsymbol{\Delta}_s\mathbf{Q}_s^H + \mathbf{Q}_N\boldsymbol{\Delta}_N\mathbf{Q}_N^H, \tag{2.29}$$

where $\boldsymbol{\Delta}_N$ is a diagonal matrix of $(L - M)$ elements containing σ_n^2. The eigenvalues of $\mathbf{R}_{xx}$, denoted by λ, can be sorted as

$$\hat{\lambda}_1 \geq \hat{\lambda}_2 \cdots \geq \hat{\lambda}_M \geq \hat{\lambda}_{M+1} \geq \cdots \hat{\lambda}_L. \tag{2.30}$$

If we have the sufficient statistics on the noise threshold, we can simply determine the number of sources by setting such a threshold on the eigenvalues. This information seldom exists and hence needs to be estimated to determine the number of sources in the sampled waveform.

2.4.2 Conventional Source Detection

The source detection problem can be considered as a model order selection problem [83, 84]. For any such parameter estimation problem, the maximum likelihood solution is the optimal solution from a Bayesian perspective. Two such solutions are the AIC, and the minimum description length (MDL) solution [73–75].

MDL and AIC try to solve this problem by minimizing the negative log-likelihood of the maximum likelihood estimator. If we consider N observations as $\mathbf{X} = [\mathbf{x}(1), \dots, \mathbf{x}(N)]$, and parameterized family of models with known probability densities $f(\mathbf{X}|\Theta)$, then we need to select a distribution from the family of known distributions that best fits the data [72]. The AIC criterion can be simply written as,

$$\text{AIC} = -2\log f(\mathbf{X}|\hat{\Theta}) + 2K, \tag{2.31}$$

where $\hat{\Theta}$ is the maximum likelihood estimate of the parameter Θ, and K is the penalty term. We are estimating the parameters in the eigendecomposition of $\mathbf{R}_{xx}$, and thus the parameter model can be described by $\Theta^{(m)} = [\lambda_1, \dots, \lambda_m, \sigma_n^2, \mathbf{Q}_1^{\mathrm{T}}, \dots, \mathbf{Q}_m^{\mathrm{T}}]$. In a similar approach, Schwartz proposed to select the model that yields the maximum posterior probability [75], which is referred to as the MDL technique [74].

$$\text{MDL} = -\log f(\mathbf{X}|\hat{\Theta}) + \frac{1}{2}K\log N. \tag{2.32}$$

The log-likelihood function was calculated as shown in [72] as

$$\log f(\mathbf{X}|\hat{\Theta}) = N(L-M)\log\frac{\left(\prod_{i=M+1}^{L}\hat{\lambda}_i\right)^{\frac{1}{L-M}}}{\frac{1}{L-M}\sum_{i=M+1}^{L}\hat{\lambda}_i}. \tag{2.33}$$

This expression can be substituted in Eqs. (2.31) and (2.32) and then minimized for different values of M to obtain the correct estimate. The penalty term was also calculated in [65] and found to be

$$K = M(2L - M). \tag{2.34}$$

Thus, Eq. (2.31) can be rewritten as,

$$\text{AIC} = -2\log f(\mathbf{X}|\hat{\Theta}) + 2M(2L - M). \tag{2.35}$$

And the estimate of M can be found as,

$$\tilde{M}_{\text{AIC}} = \underset{m}{\text{argmin}}\ \text{AIC}. \tag{2.36}$$

Similarly, Eq. (2.32) can be rewritten as,

$$\text{MDL} = -\log f(\mathbf{X}|\hat{\Theta}) + \frac{1}{2}[M(2L - M)]\log N, \tag{2.37}$$

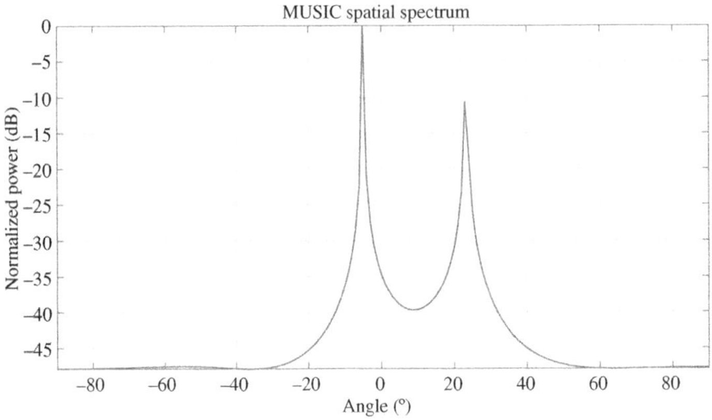

Figure 2.12 Music spectrum generated for an eight element array with two signals transmitted from −5 °, to 34 °.

and,

$$\tilde{M}_{\text{MDL}} = \underset{m}{\text{argmin}}\ \text{MDL}. \tag{2.38}$$

Although both MDL and AIC perform well, they degrade when we consider low signal-to-noise ratio (SNR), less number of snapshots, a high number of signals, and correlated cases [65].

To perform direction of arrival (DoA) estimation, the most popular algorithm is an eigensubspace-based method called as MuSiC [66]. It was first introduced by Schmidt to find the DoA of signals [66]. It makes use of the orthogonality of signal subspace with noise subspace. Once we obtain the eigenvector matrix on decomposing the autocorrelation matrix, if we know the number of sources we are looking for, we can separate the signal eigenvectors from the noise eigenvectors. We use this matrix and compute the following value for all the attainable values in our spectrum,

$$P_{\text{MU}}(\theta) = \frac{1}{\mathbf{a}^H(\theta)\mathbf{Q}_N\mathbf{Q}_N^H\mathbf{a}(\theta)}, \tag{2.39}$$

where, ^{H} denotes the Hermitian operator and $\mathbf{a}(\theta)$ denotes the array steering vector response to a signal arriving from θ as

$$\mathbf{a}(\theta) = [1, e^{j2\pi\frac{d_1}{\lambda}\sin(\theta)}, \dots, e^{j2\pi\frac{d_{L-1}}{\lambda}\sin(\theta)}]^\mathsf{T}. \tag{2.40}$$

Since the steering vectors should coincide with the signal subspace at the angle of arrivals, their dot product with the noise eigenvectors at those angles should be zero due to them being orthogonal. Thus, if we plot the above expression as a function of angle, the peaks should ideally coincide with the angle of arrivals.

The MuSiC spectrum for the scenario of two signals originating from $-5°$ to $34°$ for a ULA of eight elements is shown in Figure 2.12. However, deploying such algorithms in a practical scenario has shown a drop in performance due to issues like multipath propagation, array imperfections, decreased performance with low SNR, and less number of snapshots [85]. One alternative to the standard MuSiC was published in [86]. The approach is rooted in the interpretation that singular value decomposition (SVD)-based algorithms are based on the design of detection filters tuned to the different frequencies of interest. Indeed, the minimum variance distortionless (MVDR) algorithm [87] for DoA detection is obtained by assuming that there exists a spatial filter $y = \mathbf{w}^{\mathsf{T}}\mathbf{x}$ for the signal that is tuned at a given spatial frequency $\omega = \frac{2\pi d}{\lambda}\sin\theta$. The filter is designed so that it has minimum variance and if a unitary amplitude signal contains frequency Ω, the output is unitary. The solution to this problem is a filter with power output

$$P(\theta)_{MVDR} = \frac{1}{\mathbf{a}^{H}(\theta)\mathbf{R}_{xx}^{-1}\mathbf{a}(\theta)}. \tag{2.41}$$

The autocorrelation inverse is decomposed as

$$\mathbf{R}_{xx}^{-1} = \mathbf{Q}_S\mathbf{\Lambda}_S^{-1}\mathbf{Q}_S^H + \mathbf{Q}_N\mathbf{\Lambda}_N^{-1}\mathbf{Q}_N^H.$$

Under the assumption that the SNR is high, this output power can be written as in Eq. (2.39). In work [86], authors change the standard solution of the optimization for the MVDR filter by a support vector regressor, and then they apply the same high SNR approximation as in the MuSiC. The properties of the support vector regression (SVR) make the solution robust against outliers and low sample data, and the machine is able to detect correlated signals where MuSiC cannot, at expense of a higher computational burden. Thus, there is a need for alternate solutions to solve both of these problems. There are several examples of data-driven ML solutions offered in literature to solve these problems.

2.4.3 Neural Source Detection

Recent developments in deep neural network architectures have made a significant contribution to various aspects of signal detection, as we have seen previously in this chapter. Their ability to extract seemingly abstract but relevant features makes them the bedrock of the artificial intelligence revolution [3, 44, 59, 60, 88]. The success of neural networks can be attributed to their ability to learn complex nonlinear patterns from the data and the availability of vast amounts of data available to train these networks [3]. There have been multiple ML-based solutions to perform source detection [9, 76–80]. Here, we have a closer look at the most recent work done in [9]. The authors here sample the incoming waveform and take only the upper triangular elements of the autocorrelation matrix. These elements are normalized and separated to form a one-dimensional feature vector that would be the input to the neural network. The hypothesis of the proposed work is that information on the number of sources should be encoded within this feature vector, and thus a well-trained model should be able to learn this information from the input.

2.4.3.1 CNN Detector

CNNs are excellent feature extractors, as seen in Section 2.34. They have been adopted for various signal detection scenarios, including detecting the number of sources in the sampled waveform [9].

We start with the assumption that CNNs should be able to extract the features relevant to the information on the number of sources from the autocorrelation matrix through layers of convolutions and pooling. The problem of source detection is modeled as a classification problem, with the output labels of the CNN being a one-hot-encoded vector. The upper triangular elements of the autocorrelation matrix are extracted and concatenated to form an unidimensional feature vector. The architecture is relatively simple with five layers of successive convolution and pooling, terminating into a fully connected layer as shown in Figure 2.13. One-dimensional filters with a kernel of size 1×3 are used to perform convolutional operations. Batch normalization and max pooling are subsequently performed after each layer as a form of regularization. To prevent overfitting, the fourth layer is regularized with a dropout rate of 40%. The number of channels in every layer is 128, except the fourth layer has 256 channels. The final layer is terminated to form a vector of size $[1 \times L - 1]$ to provide a structured output for

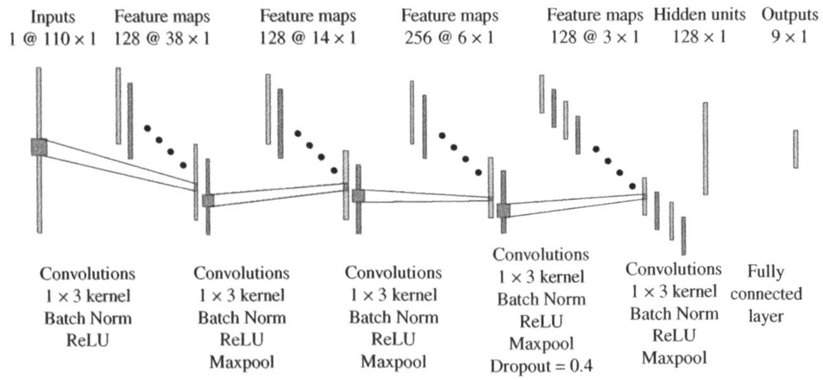

Figure 2.13 Architecture of CNN detector. Source: [9]/IEEE.

the classifier. Finally, the network is trained using the categorical cross-entropy loss, defined as,

$$\mathcal{L} = -\sum_{i=0}^{L-1} y_i \cdot \log(\tilde{y}_i), \tag{2.42}$$

where y_i is the target value of the ith class and $\tilde{y}_i$ is the corresponding prediction of the model. A one hot-encoded target vector is a way to represent the classes numerically by a vector containing a length equal to the number of classes. Each index represents a class and has one in case of a detected class and zero elsewhere. The model parameters were initialized using a Xavier initialization and optimized using Adam optimization with a learning rate of 0.001. This guarantees equal variance across layers and facilitates optimization.

Separate datasets were created for the two different scenarios, one where there are 0–9 uncorrelated sources present, and one where there are anywhere between 0 and 5 correlated sources present in the signal. Two sets of data were created representing two different scenarios to highlight the effect of correlated sources in the sampled waveform. Each dataset was further subdivided into 90, 000 frames for training and 10, 000 frames for validation and testing, respectively.

CNN detector was trained extensively on the two datasets for the two different scenarios. The mean accuracy achieved by the CNN detector for the scenario when there are no correlated sources present is plotted and compared to the conventional source detection methods of AIC and MDL in Figure 2.14a. The test accuracy is plotted as a function of SINR to highlight the effect of a varying SINR on the detection performance. The CNN detector was able to outperform both the conventional methods in terms of mean detection accuracy across classes over all SINR thresholds.

The confusion matrix detailing the classification performance for each class is detailed in Figure 2.15. The confusion matrix provides insights into the performance degradation of the model as the number of sources increases. The model performs better when there are less number of sources to resolve and achieves 99.9% accuracy when there are two to three sources present in the sampled signal. The lowest mean accuracy of 81.5% was observed for the samples containing nine sources.

The mean classification accuracy for the second scenario involving 0–5 correlated sources are plotted in Figure 2.14b. The mean accuracy drops from over 80% in the absence of correlated sources to around 59% with the introduction of the correlated sources, even though there were fewer sources to resolve and hence fewer classes. Increasing the depth of the model further does not provide any benefit hinting we have reached a saturation from a learning perspective. Next, we see how residual learning is introduced to solve this problem and increase detection accuracy in the presence of correlated sources.

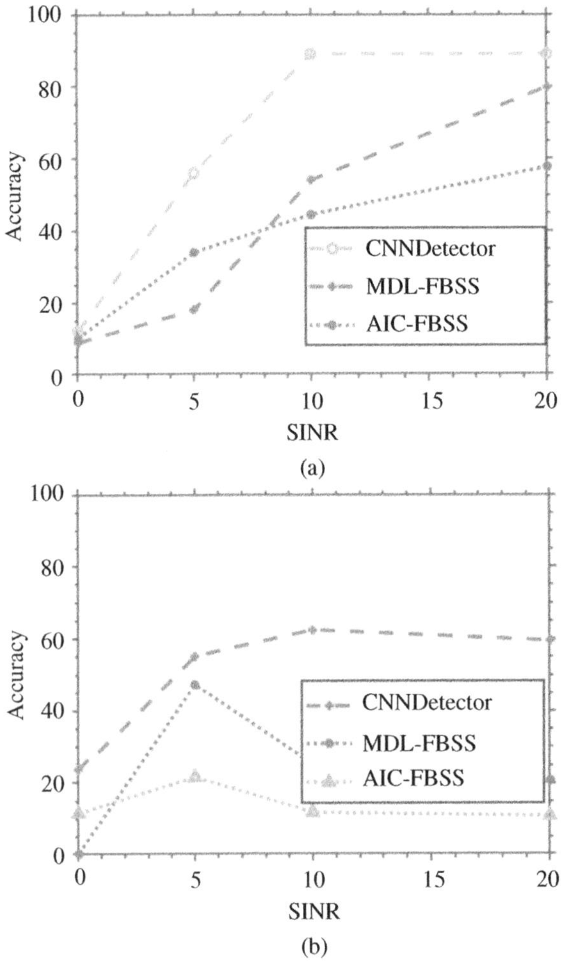

Figure 2.14 (a) Mean accuracy as a function of an increasing SINR for the experiments involving 0–9 uncorrelated sources. (b) Mean accuracy as a function of an increasing SINR for the experiments involving 0–5 correlated sources. Source: [9]/IEEE.

2.4.3.2 RadioNet

A major drawback of very deep CNN architectures is the fact that a large number of convolution layers can possibly result in saturation of learning. Increasing the number of layers after a certain threshold has shown to be detrimental to the learning process [89]. This has been attributed to the increased complexity in the optimization of the weights as we keep increasing the depth.

A possible solution to this problem was proposed in [89], where the authors introduced residual blocks in the architecture to aid the learning process. Residual blocks are trained to learn a residual of the desired mapping. Suppose the desired mapping to be learned is $H(\mathbf{x})$, with $\mathbf{x}$ being the input to the layer, then we let the stacked nonlinear layers train to learn the residual of $H(\mathbf{x})$, i.e., $H(\mathbf{x}) - \mathbf{x}$. At the

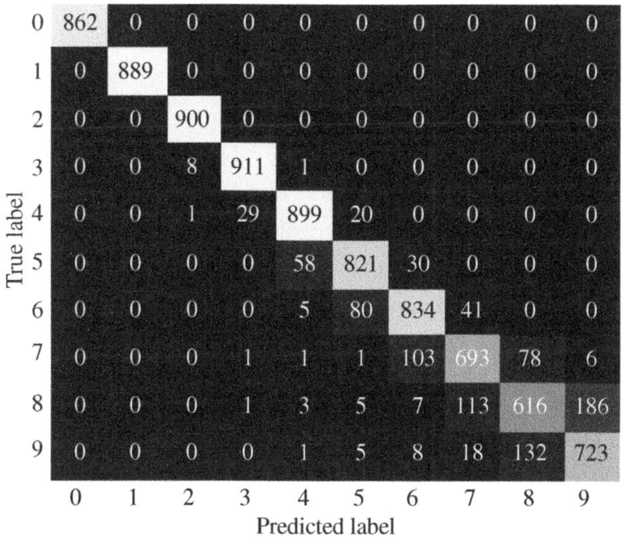

Figure 2.15 Confusion matrix showing the predictions of CNN detector in an experiment involving 0–9 uncorrelated sources at 10 dB SINR. Source: [9]/IEEE.

Figure 2.16 Architecture of a ResNet block demonstrating residual learning.

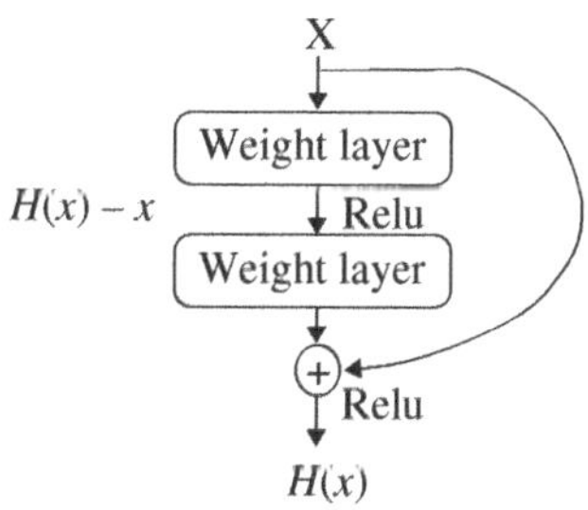

output, we perform identity mapping as the input **x** is added to the output, and thus the desired mapping $H(\mathbf{x})$ is performed as shown in Figure 2.16 [89].

Stacking multiple residual blocks was shown to decrease error and produce improved generalization as compared to a CNN of similar size [9, 89]. A modified unidimensional ResNet architecture referred to as the RadioNet was successfully adopted for source detection [9, 16]. It consists of 16 residual blocks terminating into fully connected layers to provide structured outputs. One-dimensional convolution filters of shape $[1 \times 7]$ and $[1 \times 3]$ are used for different layers. The categorical cross entropy loss described earlier is minimized to train the network parameters, and a grid search approach is used to optimize the network hyper-parameters. The model parameters were optimized using a stochastic gradient descent with an optimized learning rate of 0.0001.

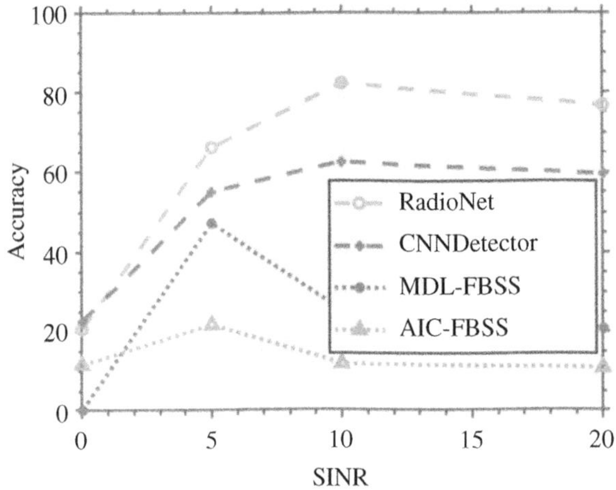

Figure 2.17 Accuracy vs. SINR for the experiment involving 0–5 correlated sources. Source: [9]/IEEE.

The mean accuracy of the RadioNet predictions in the presence of correlated sources in the test set is compared to MDL, AIC, CNN detector in Figure 2.17. RadioNet and residual learning outperforms the vanilla CNNDetector as well as the conventional methods MDL and AIC. RadioNet was able to achieve a mean test accuracy of 81%.

We explored source detection and detailed the intuition and the procedural steps required to develop DL-based methods for radio source detection. Two network architectures, CNN detector composed of stacked vanilla CNN layers and RadioNet, a modified ResNet architecture was introduced and discussed in detail. We demonstrate that if trained correctly the DL methods can outperform conventional methods when it comes to source detection.

2.5 Conclusion

The era of artificial intelligence has revolutionized signal detection among other areas of signal processing. ML-based solutions to signal detection have been widely researched and implemented over the past two decades. The advancements in the methodologies have far outpaced their adaptation into textbooks. We bring together three different signal detection scenarios in a concise manner and discuss the conventional and data-driven strategies to solve these specific signal detection scenarios.

In symbol detection, the literature describes several approaches that are based on ML, but they are suboptimal with respect to the Viterbi algorithm. This technique, on the other side, needs the estimation of the channel impulse response and its state, which can be done with the use of ML techniques with adaptive properties.

We delve into modulation detection and briefly explain the conventional statistical solution in the form of MLE. We discuss the drawbacks of MLE with respect to modulation detection and lay the foundation for the data-driven DL methods for AMC. We discuss model-building strategies to develop the intuition for the problem and solution space. We develop several CNN model architectures and demonstrate how deep CNN models can extract their own features and perform modulation classification with very high accuracy.

Source detection is a classic problem that has been conventionally solved using AIC and MDL algorithms. These algorithms have a reasonable computational burden, but they are not robust against non-Gaussian noise, correlated sources, and others. We introduce two different model architectures in the form of CNN detector and RadioNet both outperforming the conventional methods of AIC and MDL both in the absence and presence of correlated sources.

This chapter is written to serve as a handbook for students and professionals interested in the various signal detection scenarios of the modern communication system and how they are solved using data-driven ML and DL methods. We perused the literature to bring the state-of-the art methodologies, frameworks, and the thought process behind the adaptation of these strategies. The chapter assumes some prior knowledge of signal processing and ML and is suitable for students and professionals with prior knowledge of these two domains.

References

1 Van Trees, H.L. (2002). Detection of signals–estimation of signal parameters. In: *Detection, Estimation, and Modulation Theory: Detection, Estimation, and Linear Modulation Theory*. New York: Wiley.

2 Poor, H.V. (1998). *An Introduction to Signal Detection and Estimation*. Springer Science & Business Media.

3 Martınez-Ramón, M., Gupta, A., Rojo-Álvarez, J.L., and Christodoulou, C.G. (2021). *Machine Learning Applications in Electromagnetics and Antenna Array Processing*. Artech House.

4 Rojo-Álvarez, J.L., Martínez-Ramón, M., Munoz-Mari, J., and Camps-Valls, G. (2018). *Digital Signal Processing with Kernel Methods*. Wiley.

5 Ye, H., Li, G.Y., and Juang, B.-H. (2017). Power of deep learning for channel estimation and signal detection in OFDM systems. *IEEE Wireless Communications Letters* 7 (1): 114–117.

6 Farsad, N. and Goldsmith, A. (2017). Detection algorithms for communication systems using deep learning. *arXiv preprint arXiv:1705.08044*.

7 Shlezinger, N., Farsad, N., Eldar, Y.C., and Goldsmith, A.J. (2020). ViterbiNet: A deep learning based Viterbi algorithm for symbol detection. *IEEE Transactions on Wireless Communications* 19 (5): 3319–3331.

8 Dutta, S.B., Naghibijouybari, H., Gupta, A. et al. (2023). Spy in the GPU-box: covert and side channel attacks on multi-GPU systems. *Proceedings of the 50th Annual International Symposium on Computer Architecture*. Orlando, FL, USA: Association for Computing Machinery. https://doi.org/10.1145/3579371.3589080.

9 Vijayamohanan, J., Gupta, A., Noakoasteen, O. et al. (2023). Source detection with multi-label classification. *IEEE Open Journal of Signal Processing* 4: 336–345. https://doi.org/10.1109/OJSP.2023.3280854.

10 Qin, Z., Ye, H., Li, G.Y., and Juang, B.H.F. (2019). Deep learning in physical layer communications. *IEEE Wireless Communications* 26 (2): 93–99.

11 Kulin, M., Kazaz, T., Moerman, I., and De Poorter, E. (2018). End-to-end learning from spectrum data: a deep learning approach for wireless signal identification in spectrum monitoring applications. *IEEE Access* 6: 18484–18501.

12 Samuel, N., Diskin, T., and Wiesel, A. (2019). Learning to detect. *IEEE Transactions on Signal Processing* 67 (10): 2554–2564.

13 Mendis, G.J., Wei, J., and Madanayake, A. (2016). Deep learning-based automated modulation classification for cognitive radio. *2016 IEEE International Conference on Communication Systems (ICCS)*, 1–6. IEEE.

14 Kim, B., Kim, J., Chae, H. et al. (2016). Deep neural network-based automatic modulation classification technique. *2016 International Conference on Information and Communication Technology Convergence (ICTC)*, 579–582. IEEE.

15 Zhang, F., Luo, C., Xu, J. et al. (2022). Deep learning based automatic modulation recognition: models, datasets, and challenges. *Digital Signal Processing* 129: 103650. https://doi.org/10.1016/j.dsp.2022.103650.

16 Vijayamohanan, J., Gupta, A., Goudos, S., and Christodoulou, C. (2022). Detecting coherent sources with deep learning. *2022 IEEE USNC-URSI Radio Science Meeting (Joint with AP-S Symposium)*, 98–99. https://doi.org/10.23919/USNC-URSI52669.2022.9887433.

17 Viterbi, A. (1967). Error bounds for convolutional codes and an asymptotically optimum decoding algorithm. *IEEE Transactions on Information Theory* 13 (2): 260–269.

18 Forney, G.D.J.R. (1972). Maximum-likelihood sequence estimation of digital sequences in the presence of intersymbol interference. *IEEE Transactions on Information Theory* 18 (3): 363–378.

19 Lou, H.-L. (1995). Implementing the Viterbi algorithm. *IEEE Signal Processing Magazine* 12 (5): 42–52.

20 Rasmussen, C.E. and Williams, C.K.I. (2006). *Gaussian Processes for Machine Learning*, vol. 1. Springer.

21 Cristianini, N. and Scholkopf, B. (2002). Support vector machines and kernel methods: the new generation of learning machines. *Ai Magazine* 23 (3): 31–31.

22 Brandenstein, H. and Unbehauen, R. (2001). Weighted least-squares approximation of FIR by IIR digital filters. *IEEE Transactions on Signal Processing* 49 (3): 558–568.

23 Schölkopf, B., Herbrich, R., and Smola, A.J. (2001). A generalized representer theorem. In: *International Conference on Computational Learning Theory* (ed. Shawe-Taylor, J. and Cristianini, N.), 416–426. Springer-Verlag.

24 Shawe-Taylor, J. and Cristianini, N. (2004). *Kernel Methods for Pattern Analysis*. Cambridge University Press.

25 Aronszajn, N. (1950). Theory of reproducing kernels. *Transactions of the American Mathematical Society* 68 (3): 337–404.

26 Chen, S., Gibson, G.J., Cowan, C.F.N., and Grant, P.M. (1990). Adaptive equalization of finite non-linear channels using multilayer perceptrons. *Signal Processing* 20 (2): 107–119.

27 Pandey, R. (2005). Fast blind equalization using complex-valued MLP. *Neural Processing Letters* 21: 215–225.

28 Kechriotis, G., Zervas, E., and Manolakos, E.S. (1994). Using recurrent neural networks for adaptive communication channel equalization. *IEEE Transactions on Neural Networks* 5 (2): 267–278.

29 Xu, N. (2005). Beamforming using support vector machines. *IEEE Antennas and Wireless Propagation Letters* 4: 439–442.

30 Martınez-Ramón, M., Rojo-Álvarez, J.L., Camps-Valls, G., and Christodoulou, C.G. (2007). Kernel antenna array processing. *IEEE Transactions on Antennas and Propagation* 55 (3): 642–650.

31 Martınez-Ramón, M. and Christodoulou, C. (2022). *Support Vector Machines for Antenna Array Processing and Electromagnetics*. Springer Nature.

32 Gómez-Verdejo, V., Parrado-Hernández, E., and Martınez-Ramón, M. (2023). Adaptive sparse Gaussian process. *IEEE Transactions on Neural Networks and Learning Systems* https://doi.org/10.1109/TNNLS.2023.3294089.

33 Jeruchim, M.C., Balaban, P., and Shanmugan, K.S. (2006). *Simulation of Communication Systems: Modeling, Methodology and Techniques*. Springer Science & Business Media.

34 Falconer, D.D. and Magee, F.R. Jr. (1973). Adaptive channel memory truncation for maximum likelihood sequence estimation. *Bell System Technical Journal* 52 (9): 1541–1562.

35 Herscovici, B. and Gulak, G. (1993). Reduced state Viterbi receivers for digital mobile communications. *Proceedings of MILCOM'93-IEEE Military Communications Conference*, volume 2, 561–567. IEEE.

36 Gerstacker, W.H., Obernosterer, F., Meyer, R., and Huber, J.B. (2002). On prefilter computation for reduced-state equalization. *IEEE Transactions on Wireless Communications* 1 (4): 793–800.

37 Wang, Z.-Y., Yu, H.Y., Shen, Z.X. et al. (2023). Deep learning based symbol detection with natural redundancy for non-uniform memoryless sources. *IEEE Communications Letters* 27 (2): 561–565.

38 Kowshik, A.K., Raghavendra, A.H., and Gurugopinath, S. (2022). Deep learning for signal detection in uplink and downlink rate-splitting multiple access systems. *2022 IEEE 19th India Council International Conference (INDICON)*, 1–6.

39 Chia, K., Baskaran, V.M., Wong, K. et al. (2022). Recurrent network with attention for symbol detection in communication systems. *2022 International Symposium on Intelligent Signal Processing and Communication Systems (ISPACS)*, 1–4. https://doi.org/10.1109/ISPACS57703.2022.10082803.

40 Backurs, A. and Tzamos, C. (2017). Improving Viterbi is hard: better runtimes imply faster clique algorithms *International Conference on Machine Learning*, 311–321. PMLR.

41 Hershey, J.R., Roux, J.L., and Weninger, F. (2014). Deep unfolding: model-based inspiration of novel deep architectures. *arXiv preprint arXiv:1409.2574.*

42 Gan, W., Zhai, Q., McCann, M.T. et al. (2023). PtychoDV: Vision transformer-based deep unrolling network for ptychographic image reconstruction. *arXiv preprint arXiv:2310.07504.*

43 Murphy, K.P. (2012). *Machine Learning: A Probabilistic Perspective*. MIT Press.

44 Goodfellow, I., Bengio, Y., and Courville, A. (2016). *Deep Learning*. MIT Press.

45 Martinez-Ramon, M., Ajith, M., and Kurup, A.R. (2024). *Deep Learning, A Practical Introduction*. Wiley.

46 Farsad, N. and Goldsmith, A. (2018). Neural network detection of data sequences in communication systems. *IEEE Transactions on Signal Processing* 66 (21): 5663–5678.

47 Shlezinger, N., Farsad, N., Eldar, Y.C., and Goldsmith, A.J. (2019). ViterbiNet: A deep learning based Viterbi algorithm for symbol detection. *arXiv preprint arXiv:1905.10750.*

48 Rabiner, L.R. and Gold, B. (1975). *Theory and Application of Digital Signal Processing*. Englewood Cliffs, NJ: Prentice-Hall.

49 Xiong, F. (2006). *Digital Modulation Techniques*, 2e. Artech House Publishers, London.

50 Jayaweera, S.K. (2014). *Signal Processing for Cognitive Radios.* Wiley.

51 Jdid, B., Hassan, K., Dayoub, I. et al. (2021). Machine learning based automatic modulation recognition for wireless communications: a comprehensive survey. *IEEE Access* 9: 57851–57873.

52 Peng, S., Sun, S., and Yao, Y.-D. (2021). A survey of modulation classification using deep learning: signal representation and data preprocessing. *IEEE Transactions on Neural Networks and Learning Systems* 33 (12): 7020–7038.

53 Headley, W.C., Chavali, V.G., and da Silva, C.R.C.M. (2013). Maximum-likelihood modulation classification with incomplete channel information. *2013 Information Theory and Applications Workshop (ITA)*, 1–4. https://doi.org/10.1109/ITA.2013.6503000.

54 Sills, J.A. (1999). Maximum-likelihood modulation classification for PSK/QAM. *MILCOM 1999. IEEE Military Communications. Conference Proceedings (Cat. No. 99CH36341)*, volume 1, 217–220. IEEE.

55 Szeliski, R. (2022). *Computer Vision: Algorithms and Applications.* Springer Nature.

56 Purwins, H., Li, B., Virtanen, T. et al. (2019). Deep learning for audio signal processing. *IEEE Journal of Selected Topics in Signal Processing* 13 (2): 206–219.

57 O'Shea, T.J., Corgan, J., and Clancy, T.C. (2016). Convolutional radio modulation recognition networks. *Engineering Applications of Neural Networks: 17th International Conference, EANN 2016*, Aberdeen, UK (2—5 September 2016), Proceedings 17, 213–226. Springer.

58 Liu, X., Yang, D., and El Gamal, A. (2017). Deep neural network architectures for modulation classification. *2017 51st Asilomar Conference on Signals, Systems, and Computers*, 915–919. IEEE.

59 LeCun, Y. and Bengio, Y. (1995). Convolutional networks for images, speech, and time series. *The Handbook of Brain Theory and Neural Networks* 3361 (10): 1995.

60 LeCun, Y., Bengio, Y., and Hinton, G. (2015). Deep learning. *Nature* 521 (7553): 436–444.

61 The MathWorks Inc. (2022). MATLAB Version: 9.13.0 (R2022b). Natick, MA, USA. https://www.mathworks.com (accessed 1 October 2024).

62 Hastie, T., Tibshirani, R., Friedman, J.H., and Friedman, J.H. (2009). *The Elements of Statistical Learning: Data Mining, Inference, and Prediction*, vol. 2. Springer.

63 Pang, T., Xu, K., Dong, Y. et al. (2020). Rethinking softmax cross-entropy loss for adversarial robustness. *arXiv: 1905.10626 [cs.LG].*

64 Gupta, A., Christodoulou, C.G., Rojo-Álvarez, J.L., and Martínez-Ramón, M. (2019). Gaussian processes for direction-of-arrival estimation with random arrays. *IEEE Antennas and Wireless Propagation Letters* 18 (11): 2297–2300. https://doi.org/10.1109/LAWP.2019.2934944.

65 Van Trees, H.L. (2002). *Optimum Array Processing*, Chapter 7, 710–916. Wiley.

66 Schmidt, R. (1986). Multiple emitter location and signal parameter estimation. *IEEE Transactions on Antennas and Propagation* 34 (3): 276–280. https://doi.org/10.1109/TAP.1986.1143830.

67 Friedlander, B. (1993). The root-MUSIC algorithm for direction finding with interpolated arrays. *Signal Processing* 30 (1): 15–29.

68 Gupta, A., Christodoulou, C., Martínez-Ramón, M., and Rojo-Álvarez, J.L. (2018). Kernel DOA estimation in nonuniform arrays. *2018 IEEE International Symposium on Antennas and Propagation & USNC/URSI National Radio Science Meeting*, 193–194. https://doi.org/10.1109/APUSNCURSINRSM.2018.8608523.

69 Gupta, A., Christodoulou, C.G., Martínez-Ramón, M., and Rojo-Álvarez, J.L. (2020). Deep neural nets for DOA estimation with random arrays. *2020 IEEE International Symposium on Antennas and Propagation and North American Radio Science Meeting*, 2043–2044. https://doi.org/10.1109/IEEECONF35879.2020.9330446.

70 Wax, M. and Adler, A. (2021). Detection of the number of signals by signal subspace matching. *IEEE Transactions on Signal Processing* 69: 973–985. https://doi.org/10.1109/TSP.2021.3053495.

71 Cohen, D., Cuffin, B.N., Yunokuchi, K. et al. (1990). MEG versus EEG localization test using implanted sources in the human brain. *Annals of Neurology* 28 (6): 811–817. https://doi.org/10.1002/ana.410280613. eprint: https://onlinelibrary.wiley.com/doi/pdf/10.1002/ana.410280613.

72 Wax, M. and Kailath, T. (1985). Detection of signals by information theoretic criteria. *IEEE Transactions on Acoustics, Speech, and Signal Processing* 33 (2): 387–392. https://doi.org/10.1109/TASSP.1985.1164557.

73 Akaike, H. (1974). A new look at the statistical model identification. *IEEE Transactions on Automatic Control* 19: 716–723.

74 Rissanen, J. (1978). Modeling by shortest data description. *Automatica* 14 (5): 465–471. https://doi.org/10.1016/0005-1098(78)90005-5.

75 Schwarz, G. (1978). Estimating the dimension of a model. *The Annals of Statistics* 6 (2): 461–464. https://doi.org/10.1214/aos/1176344136.

76 Bialer, O., Garnett, N., and Tirer, T. (2019). Performance advantages of deep neural networks for angle of arrival estimation. *ICASSP 2019 – 2019 IEEE International Conference on Acoustics, Speech and Signal Processing (ICASSP)*, 3907–3911. https://doi.org/10.1109/ICASSP.2019.8682604.

77 Barthelme, A., Wiesmayr, R., and Utschick, W. (2020). Model order selection in DoA scenarios via cross-entropy based machine learning techniques. *ICASSP 2020 - 2020 IEEE International Conference on Acoustics, Speech and*

Signal Processing (ICASSP), 4622–4626. https://doi.org/10.1109/ICASSP40776.2020.9053029.

78 Yang, Y., Gao, F., Qian, C., and Liao, G. (2020). Model-aided deep neural network for source number detection. *IEEE Signal Processing Letters* 27: 91–95. https://doi.org/10.1109/LSP.2019.2957673.

79 Barthelme, A. and Utschick, W. (2021). A machine learning approach to DoA estimation and model order selection for antenna arrays with subarray sampling. *IEEE Transactions on Signal Processing* 69: 3075–3087. https://doi.org/10.1109/TSP.2021.3081047.

80 Hu, W., Liu, R., Lin, X. et al. (2017). A deep learning method to estimate independent source number. *2017 4th International Conference on Systems and Informatics (ICSAI)*, 1055–1059. https://doi.org/10.1109/ICSAI.2017.8248441.

81 Pillai, S.U. and Kwon, B.H. (1989). Forward/backward spatial smoothing techniques for coherent signal identification. *IEEE Transactions on Acoustics, Speech, and Signal Processing* 37 (1): 8–15. https://doi.org/10.1109/29.17496.

82 Hayes, M.H. (1996). *Statistical Digital Signal Processing and Modeling*, 1e. USA: Wiley.

83 Liavas, A.P. and Regalia, P.A. (2001). On the behavior of information theoretic criteria for model order selection. *IEEE Transactions on Signal Processing* 49 (8): 1689–1695. https://doi.org/10.1109/78.934138.

84 Stoica, P. and Selen, Y. (2004). Model-order selection: a review of information criterion rules. *IEEE Signal Processing Magazine* 21 (4): 36–47. https://doi.org/10.1109/MSP.2004.1311138.

85 Krim, H. and Viberg, M. (1996). Two decades of array signal processing research: the parametric approach. *IEEE Signal Processing Magazine* 13 (4): 67–94. https://doi.org/10.1109/79.526899.

86 El Gonnouni, A., Martinez-Ramon, M., Rojo-Alvarez, J.L. et al. (2012). A support vector machine MUSIC algorithm. *IEEE Transactions on Antennas and Propagation* 60 (10): 4901–4910.

87 Capon, J. (1969). High-resolution frequency-wavenumber spectrum analysis. *Proceedings of the IEEE* 57 (8): 1408–1418.

88 Yu, D. and Deng, L. (2011). Deep learning and its applications to signal and information processing [exploratory DSP]. *IEEE Signal Processing Magazine* 28 (1): 145–154. https://doi.org/10.1109/MSP.2010.939038.

89 He, K., Zhang, X., Ren, S., and Sun, J. (2015). Deep residual learning for image recognition. *arXiv: 1512.03385 [cs.CV]*.

3

AI-Aided Channel Prediction

Oscar Stenhammar[1,2], Gábor Fodor[1,2], and Carlo Fischione[1]

[1] *School of Electrical Engineering and Computer Science, KTH Royal Institute of Technology, Stockholm, Sweden*
[2] *Radio Network Algorithms, Ericsson Research, Ericsson AB, Kista, Sweden*

CHAPTER MENU

Acronyms

3GPP	Third Generation Partnership Project
6G	Sixth Generation
AI	artificial intelligence
AR	autoregressive
ARIMA	autoregressive integrated moving average
BER	bit error rate
BS	base station
CNN	convolutional neural network
CSI	channel state information
DL	downlink
D-MIMO	distributed multiple input multiple output
GAN	generative adversarial network
GRU	gated recurrent unit
KF	Kalman filter

Artificial Intelligence for Future Networks, First Edition.
Edited by Mohammad A. Matin, Sotirios K. Goudos, and George K. Karagiannidis.

LS	least squares
LMMSE	linear minimum mean square error
LSTM	long short-term memory
ML	machine learning
MLP	multilayer perceptron
MMSE	minimum mean squared error
MSE	mean squared error
ReLU	rectified linear unit
RNN	recurrent neural network
SNR	signal-to-noise ratio
SVR	support vector machine for regression
TDL	tapped delay line
UE	user equipment
UL	uplink

3.1 Introduction

As 6G wireless communication technologies and services emerge, end users are raising their expectations for mobile broadband, localization, and sensing services [1]. Several fundamental concepts in the development of 5G and the advent of 6G networks, including distributed multiple input multiple output (D-MIMO) systems, ultrareliable low-latency scheduling schemes, and various high-accuracy positioning algorithms, depend crucially on having access to high-quality channel state information (CSI). Unfortunately, acquiring precise CSI at both the transmitter and receiver nodes is exceptionally challenging due to the dynamic nature of the channel [2]. The dynamics of the channel are the results of changes in the scattering environment and the position of the user equipment (UE). The progression of the channel between temporal instances of CSI acquisition is commonly referred to as channel aging, causing a significant hurdle in the design of channel-adaptive algorithms in contemporary wireless systems [3–5].

3.1.1 Channel Aging

In wireless communication, the channel is the link between the transmitter and receiver. The received signal can be modeled as

$$\mathbf{y}[t] = \mathbf{H}[t]\mathbf{x}[t] + \mathbf{n}[t], \tag{3.1}$$

where t is the time instance, $\mathbf{x}[t] \in \mathbb{C}^{N_t}$ is the transmitted signal, $\mathbf{y}[t] \in \mathbb{C}^{N_r}$ is the received signal, $\mathbf{n}[t] \in \mathbb{C}^{N_r}$ is the noise, and $\mathbf{H}[t] \in \mathbb{C}^{N_r \times N_t}$ represents the channel

Figure 3.1 An illustration of how the propagation environment generates the channel that makes the receiver observe a distorted signal $\mathbf{y}[t]$ of the transmitted signal $\mathbf{x}[t]$.

between a transmitter with N_t antennas and a receiver with N_r antennas. The communication model in Eq. (3.1) is illustrated in Figure 3.1. The channel $\mathbf{H}[t]$ is not a measurable physical quantity but can be estimated using channel estimation [6, 7]. In a static environment and assuming a fixed position of the UE, the wireless channel in Eq. (3.1) does not evolve over time. This implies that channel aging in the model of Eq. (3.1) is not present and reliable CSI can be acquired by well-known pilot-based schemes [8]. However, in a practical scenario, the environment changes and the UE moves, which yields the time dependency of the channel $\mathbf{H}[t]$. The mobility of the UE has two consequences: it alters the propagation environment surrounded by the UE and gives rise to Doppler effects. As the channel changes, its estimate becomes outdated, which deteriorates the performance of CSI-adaptive algorithms [9]. To model the temporal correlation of the channel, Jakes' model [10] that assumes Rayleigh fading is commonly considered [11]:

$$r_h[k] = J_0\left(2\pi v f_c T_s |k| \frac{1}{c}\right), \tag{3.2}$$

where $J_0(\cdot)$ is the zeroth-order Bessel function of the first kind, T_s is the channel sampling duration, v is the velocity of the UE, f_c is the carrier frequency, $|k|$ is the delay in terms of the number of symbols and c is the speed of light. As the delay $|k|$ increases, the magnitude of $r_h[k]$ goes to zero, which characterizes the aging of the channel. Alternative channel models to Jake's model exist for other assumptions than Rayleigh fading, such as Rician fading [12]. To meet the challenges of channel aging, channel estimation and channel prediction have emerged as key tools to combat channel aging [1, 4, 5, 11].

3.1.2 Channel Estimation

A fundamental tool to combat the negative effects of channel aging is proper channel estimation updates using properly spaced pilot symbols. It plays a vital role in ensuring the robustness, efficiency, and reliability of wireless communication systems by providing the foundational information to operate adaptive transmission methods to counteract channel conditions, mitigate errors, and optimize resource utilization. It is a fundamental aspect of modern communication system design, contributing to improved performance and user experience. Accurate and regularly updated CSI allows an adaptive transmitter to tune the communication parameters, such as the transmit power, constellation size, coding rate, and multiantenna precoder weights to enhance the performance of the network.

To estimate the channel, established pilot signals are inserted in specific slots within the transmitted signal. The receiver adopts these known symbols to estimate the channel's response by comparing the received pilot signals to the known transmitted ones. Channel estimation has traditionally adopted analytical models, which have recently been challenged by artificial intelligence (AI)-aided schemes.

Although regularly updating the estimate of the channel state is essential to combat the negative effects of channel aging, frequent pilot symbols degrade the spectral efficiency. Recognizing this trade-off, wireless systems can employ adaptive transmission schemes that dynamically adjust pilot spacing and transmission parameters based on the predicted channel conditions. This adaptability helps to maintain reliable communication in the presence of channel aging.

3.1.3 Channel Prediction

To overcome the impact of channel aging, and counteract outdated CSI while keeping the pilot overhead low, channel prediction is a promising scheme. Channel prediction can form the foundation for improving and extending communication solutions [13, 14] allowing adaptive systems to adapt proactively to changing channel conditions by optimizing the performance, reliability, and efficiency of wireless communication systems. This is a crucial part of next-generation networks, leading to improved overall network performance and a better user experience [15].

Once the estimated channel is obtained at the receiver, the future channel state – depending on the autocorrelation function of the channel process – can be predicted. By feedback control, the predicted channel is sent to the transmitter by the receiver, and the adaptive transmitter can proactively react to predicted channel conditions. The architecture of a regular channel prediction scheme is visualized in Figure 3.2.

Because of channel aging, it is probable that for a system experiencing feedback delay τ, the channel becomes outdated as $\mathbf{H}[t] \neq \mathbf{H}[t+\tau]$. It is essential that the

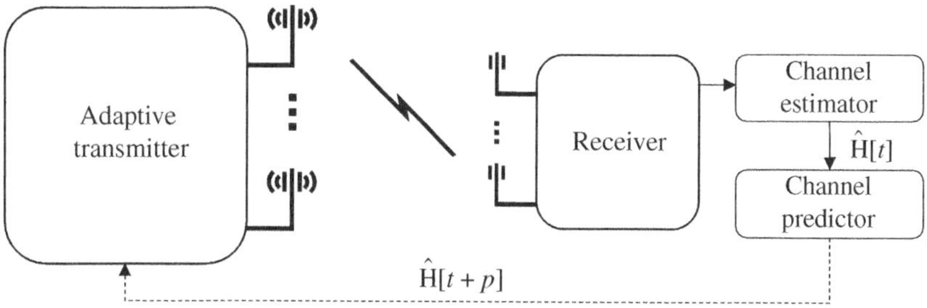

Figure 3.2 Architecture of the channel predictive feedback system to the adaptive transmitter.

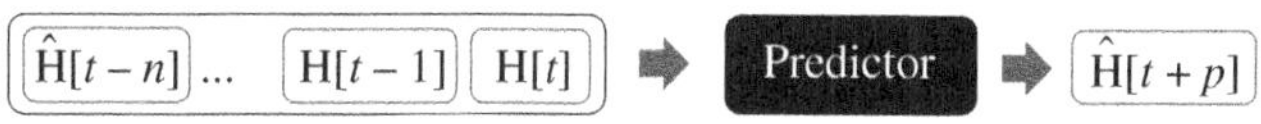

Figure 3.3 A visualization of the mapping from input to output in a channel prediction model.

channel is predicted on time horizon $h \geq \tau$ such that the obtained prediction at the transmitter $\hat{\mathbf{H}}[t+h]$ is useful for the subsequent data transmission.

Many features, such as temporal-, spatial-, network-, and frequency correlation, can be considered as input to a predictive model for channel prediction [16]. A promising, scalable solution is simply to incorporate channel prediction by using several distinct past estimations of the channel and thus treating channel prediction as a time series problem. This prediction procedure that benefits from the temporal correlation is illustrated in Figure 3.3.

Conventional model-based predictive methods have been considered as the foundation of channel prediction. However, recent research using machine learning (ML) predictive methods such as deep learning techniques is becoming increasingly relevant, especially in practical scenarios in that the auto-correlation of the channel is unknown or too costly to obtain [4, 17].

The remainder of this chapter identifies the most promising neural networks for channel estimation and prediction, which have been proven to perform well in previous studies, by presenting a comprehensive survey of the recent work. The selected neural networks are evaluated by comparing the performance results in channel prediction using a dataset obtained by the standardized and widely used realistic 3GPP TDL-A channel model [18]. The ability to predict the channel by neural networks is evaluated with a noisy, respectively, noise-free version of the dataset. Specifically, the performance of predicting fast-fading channels is studied over a large span of prediction horizons, which is suitable for several applications. The advantages and disadvantages of each neural network are discussed to identify the most promising neural network for wireless channel prediction. This can lead to a deeper understanding of state-of-the-art in channel prediction. More accurate and computationally affordable neural networks for channel prediction will result in more reliable communication for a variety of mobile applications, including driving assistance and safety-enhancing automotive applications [15].

3.2 Preliminaries

The growing interest in AI has accelerated the research area and developed several successful AI algorithms. To provide the background of the algorithms used in this chapter, Section 3.2 introduces the theory for neural networks that have shown promising results, focusing specifically on the purpose of channel estimation and prediction based on time series. Both channel estimation and prediction are usually considered as time series problems, which implies that a sequence of historical data is used to estimate the current channel or to forecast the future channel. This section aims to describe how five different neural networks, used in recent research for channel estimation and prediction, work and how they can be beneficial specifically for time series prediction. Section 3.3 will be dedicated to giving an overview of recent research on channel estimation and prediction, while Section 3.4 will perform evaluations connected to these neural networks.

3.2.1 Multilayer Perceptron

The most common type of artificial neural network is the feedforward neural network. It consists of a set of fully connected (FC) layers, meaning that each layer is constructed of multiple neurons connected to all neurons in the adjacent layers. An input vector of the user's desire populates the input layer and the output layer returns the output of the network. Layers between the input and output layers are called hidden layers. A feedforward neural network with one or more hidden layers is called a multilayer perceptron (MLP).

Each connection between neurons in adjacent layers is associated with a weight. The network learns by adjusting these weights during the training process. Biases are additional parameters added to each neuron, contributing to the flexibility and learning capacity of the network. Each neuron in the hidden layers and output layer is associated with an activation function. This function introduces nonlinearity to the model, enabling it to learn complex patterns. Common activation functions include the sigmoid (σ), hyperbolic tangent (tanh), and rectified linear unit (ReLU) functions. To summarize, the value of the ith neuron z_i in layer z is calculated by

$$z_i = \sigma_z \left(b_i + \sum_{n=1}^{N} W_{ni} x_n \right), \tag{3.3}$$

where the scalar W_{ni} is the weight of the nth neuron x_n in the preceding layer onto neuron z_i, b_i is the bias, and σ_z represents some activation function. Eq. (3.3) can be rewritten for layer z, representing a vector of neurons as

$$\mathbf{z} = \sigma_z (W\mathbf{x} + \mathbf{b}), \tag{3.4}$$

where W is the weight matrix, $\mathbf{x}$ is the vector of neurons from the preceding layer, and $\mathbf{b}$ is the bias vector. The architecture of the MLP, described by Eqs. (3.3) and (3.4), is visually explained in Figure 3.4. The illustration only includes one hidden layer and the output vector y with size 1. In general, both the number of hidden layers and the size of the output vector can be manifold.

The task of the MLP, or a neural network in general, is to learn the weights W for each layer. From a dataset populated by input features and related output features, the network learns by inserting input data into the network and calculating the output. A loss function calculates the error between the neural network output and the desired output. Using backpropagation, the error from the loss function is utilized to update the weights in each layer to train the neural network and ultimately learn the relation between input and output data.

The general framework of the MLP has shown promising results. The MLP has a high compatibility with several applications since the input to an MLP must be a one-dimensional vector that can be constructed from any type of data. One problem with this architecture is that the order of the elements in the input vector is disregarded. This might cause a loss of valuable information, possibly with high importance to the problem at hand.

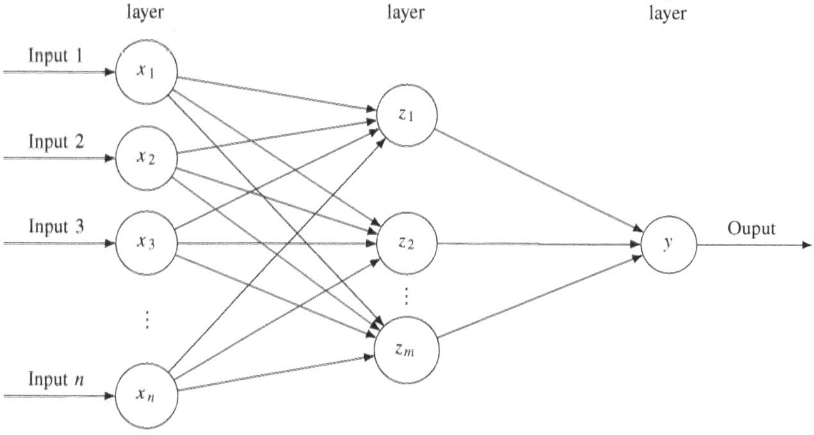

Figure 3.4 Architecture of an MLP, from input to output.

3.2.2 Convolutional Neural Network

A type of neural network specifically designed for processing structured grid data is the convolutional neural network (CNN). Its original idea is to identify patterns in the input matrix, making the position of each element in the input matrix relevant in contrast to the MLP. In a convolutional layer, a kernel strides through the two-dimensional input matrix to calculate a feature map as output. The kernel is a square matrix with trainable weights in each element of the matrix. Nonlinear activation functions and biases are typically applied after convolutional layers to introduce nonlinearity into the network, just as in the MLP. The element at position m, n in the feature map Z of the upcoming layer is calculated by

$$Z_{m,n} = \sigma_z\left(\sum_i \sum_j X_{m+i,n+j} W_{i,j} + B_{m,n}\right), \tag{3.5}$$

where X is the feature map in the previous layer, W is the kernel, i,j are the indices of the kernel, B is the bias matrix, and σ_z is the activation function. A convolutional layer typically has several channels, which represent the number of two-dimensional feature maps in the layer. Convolutional layers are typically followed by pooling layers to reduce the spatial dimensions of the input data while retaining important information. An illustrative example of Eq. (3.5) is depicted in Figure 3.5. This procedure is usually repeated by adding several convolutional, each followed by a pooling layer. After a sequence of convolutional and pooling layers, the CNN is usually finished by being vectorized and inserted in to an FC layer to obtain the desired output. The architecture of the CNN is visualized in Figure 3.6.

CNNs are particularly effective in computer vision tasks, where the goal is to recognize and understand visual patterns in data, and have acquired state-of-the-art status for their ability to detect patterns within image recognition. They are primarily designed for grid-like data, such as images and spectrograms. Adapting them to other data types, like time series, can be less beneficial.

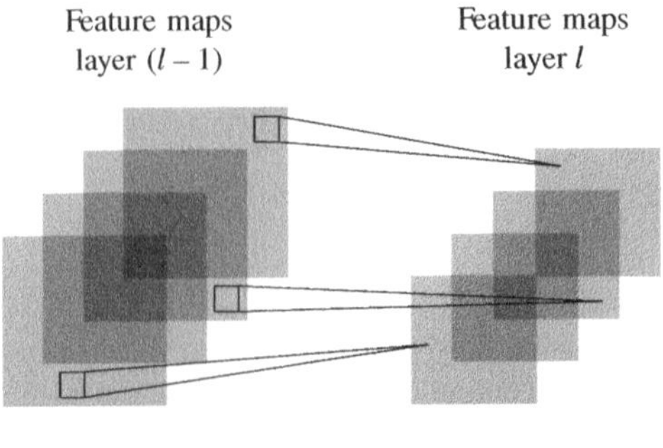

Figure 3.5 Calculation of the elements for each feature map at layer l, using the kernels for each channel at layer l – 1.

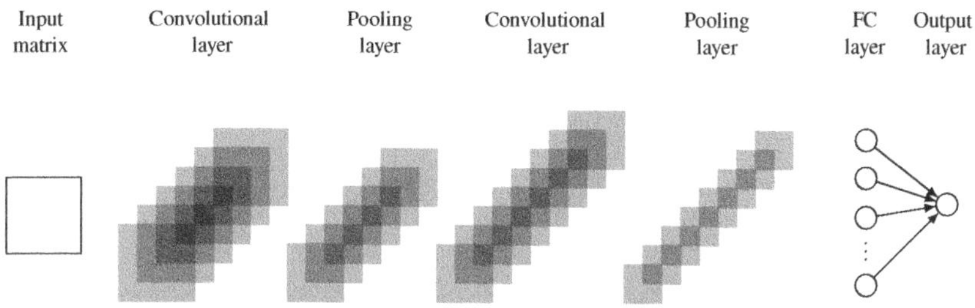

Figure 3.6 Architecture of a complete CNN. In this particular illustrative example, the network has one input matrix, two convolutional layers, two pooling layers, one fully connected (FC) layer, and one output value. The CNN is not limited to these numbers but can be tailored as desired. The calculation of the feature maps in the convolutional and pooling layers is illustrated in Figure 3.5. After the last pooling layer, all feature maps are vectorized to fit into the FC layer.

3.2.3 Recurrent Neural Network

A class of neural networks that is suitable for time series is the recurrent neural network (RNN). Unlike traditional MLPs, which process input data in a single pass, RNNs have connections that form directed cycles, allowing them to maintain a hidden state that captures information about previous instances in the input sequence [19]. Each instance of the input sequence is directed to an FC layer. The output from each sequence instance is provided as input for the upcoming sequence instance, together with the primary input. RNNs are well-suited for tasks where temporal dependencies exist in the data. They can capture and utilize information from previous time steps to make predictions or decisions at the current time step. Recurrent connections enable the network to maintain memory from previous inputs. This looped structure allows the hidden state to capture information across time. However, RNNs can suffer from vanishing or exploding gradient problems [20], which makes it challenging for the network to learn long-term dependencies. Gradients can become extremely small during backpropagation over time, hindering the learning process. To solve these problems, modified networks have been suggested.

3.2.3.1 Long Short-Term Memory

The long short-term memory (LSTM) is an improved RNN algorithm that was created by Hochreiter and Schmidhuber [21] to address the problems of vanishing and exploding gradient problems. As its name suggests, the LSTM cell has one long- and one short-term memory. The LSTM unit at time instance t consists of a cell c_t, an input gate i_t, an output gate o_t, and a forget gate f_t to calculate the hidden state h_t, that is also output vector. The state of all gates and the cell is calculated by individual FC layers, as previously in Eq. (3.4). The formulas for each state and gate in the LSTM are given as

$$f_t = \sigma(W^f x_t + U^f h_{t-1} + b^f), \tag{3.6}$$

$$i_t = \sigma(W^i x_t + U^i h_{t-1} + b^i), \tag{3.7}$$

$$g_t = \tanh(W^g x_t + U^g h_{t-1} + b^g), \tag{3.8}$$

$$o_t = \sigma(W^o x_t + U^o h_{t-1} + b^o), \tag{3.9}$$

$$c_t = \sigma(f_t \odot c_{t-1} + i_t \odot g_t), \tag{3.10}$$

$$h_t = \tanh(c_t) \odot o_t, \tag{3.11}$$

where W and U are weight matrices, and b is the bias, for each gate. To get a graphical overview of the LSTM, see Figure 3.7.

The cell c can remember values over long time intervals, and the three gates regulate the flow of information into and out of the cell. The forget gate decides what information to discard from the previous state, the input gate decides what part of the new information to store in the current state, and the output gate controls

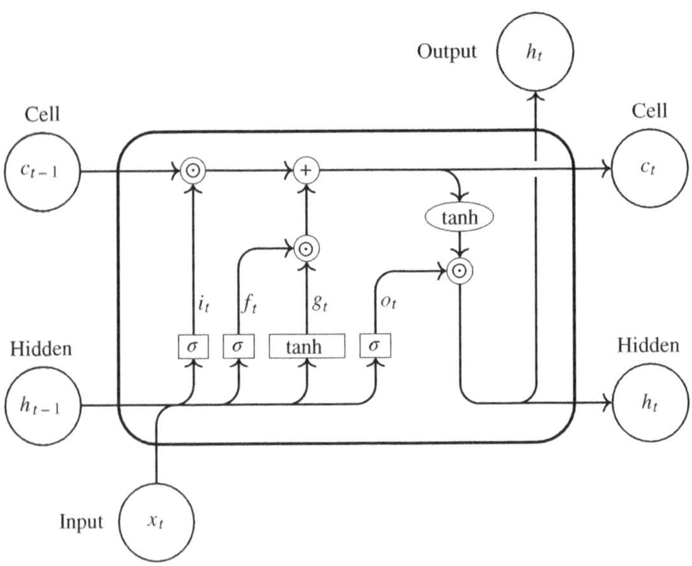

Figure 3.7 An illustration of how the LSTM is designed. The small rectangles containing σ and tanh represent fully connected layers with activation functions $\sigma = \sigma_z$ respectively $\tanh = \sigma_z$. Recall that fully connected layers with activation function σ_z are defined mathematically by Eq. (3.4). The encircled tanh simply represents the hyperbolic tangent function, the encircled $\odot$ represents the Hadamard product (piece-wise multiplication of matrices), and the encircled plus sign simply represents piece-wise addition.

what information is in the current state to output. Selecting the relevant output from the information of the current state allows the LSTM to maintain long- and short-term dependencies.

The LSTM has played an important role in improving the performance of various ML applications that involve sequential data, such as text generation, machine translation, speech recognition, and time series analysis. Since LSTM was proposed, new algorithms based on the LSTM with modifications have been created to try and improve the performance.

3.2.3.2 Gated Recurrent Units

The gated recurrent unit (GRU), proposed by Cho et al. [22], is an efficient and simplified alternative to the LSTM networks while achieving similar performance in many tasks. It has two gates and one hidden state, making it computationally cheaper than the LSTM. The update gate z determines the amount of information from the previous hidden state that should be preserved, and how much should be updated with new information from the input. The reset gate r regulates how much information from the past should be forgotten or reset. The formulas for calculating the gates, or the hidden state, h are given as

$$r_t = \sigma(W^r x_t + U^r h_{t-1} + b_r), \tag{3.12}$$

$$z_t = \sigma(W^z x_t + U^z h_{t-1} + b_z), \tag{3.13}$$

$$\hat{h}_t = \tanh(W^h x_t + U^h (r_t \odot h_{t-1}) + b_h), \tag{3.14}$$

$$h_t = (1 - z_t) \odot h_{t-1} + z_t \odot \hat{h}_t. \tag{3.15}$$

As for the LSTM unit, W and U are weight matrices, and b is the bias, for each gate. The structure represents the neural network from Eq. (3.4). For a more illustrative overview of the GRU, see Figure 3.8.

3.2.4 Transformer

A recently proposed model by Vaswani et al. [23] is the transformer model. It is especially known for its attention mechanism, which enables efficient processing of data sequences. It has grown popular due to its effectiveness and parallelizable structure. It is most commonly used in natural language processing tasks but has also shown significant performance in time series prediction.

The transformer consists of two main blocks: the encoder and the decoder. The main purpose of the encoder is to establish a comprehensive representation of the input sequence, while the primary role of the decoder is to deliver a target sequence based on the representation from the encoder. Transformers

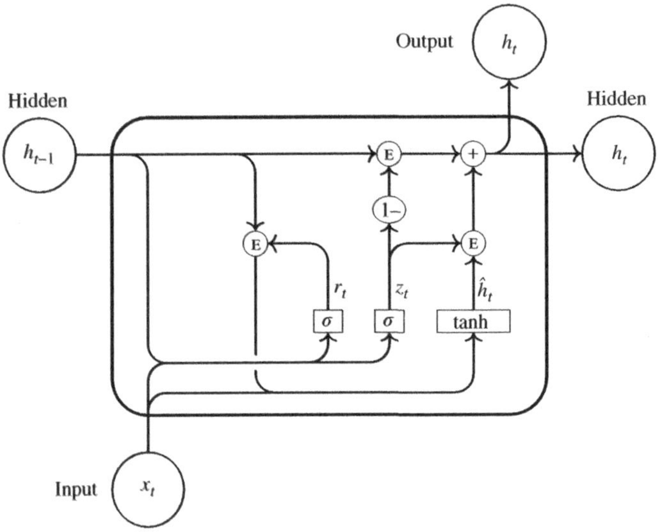

Figure 3.8 An illustration of how the LSTM is designed. The small rectangles containing σ and tanh represent fully connected layers with activation functions $\sigma = \sigma_z$ respectively tanh $= \sigma_z$. Recall that fully connected layers with activation function σ_z are defined mathematically by Eq. (3.4). The encircled 1− represents the action of subtracting 1 by the input vector, the encircled ⊙ represents the Hadamard product (piecewise multiplication of matrices), and the encircled plus sign simply represents piece wise addition.

are not provided with inherent notions of the order of elements in a sequence. Instead, positional encodings are added to the input embeddings to provide the model with the positional information of the elements in the sequence. The core of the transformer is the self-attention mechanism, which allows it to weigh the importance of different elements in the input sequence when making predictions. The self-attention mechanism in the transformer is often multihead, meaning it is computed multiple times with different learned projections of the input. This allows the model to focus on different aspects of the input sequence.

After the self-attention layers, the block called *Add & Norm* adds and normalizes the input and output of the self-attention layer. The FC layers are used to further process the information. Before the output is complete, the output from the decoder is linearly transformed and processed by the Softmax activation function. A flowchart of the transformer model is portrayed in Figure 3.9.

The transformer's versatility and effectiveness have made it a central component in the field of deep learning and artificial intelligence (AI). With its ability to capture long-range dependencies and parallel processing, the transformer has significant performance in time series prediction.

Figure 3.9 A flowchart of the transformer model.

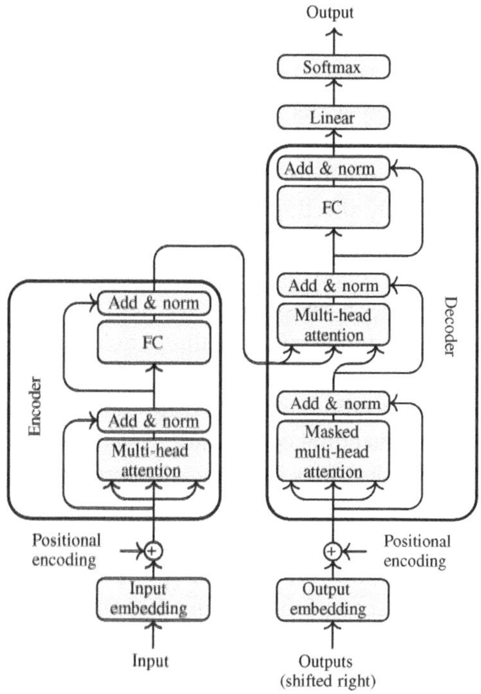

3.3 Previous Work

Channel estimation and channel prediction are two research areas where a considerable amount of work is conducted constantly. This section aims to provide a comprehensive survey of the previous work, with a focus on work toward AI-aided methods.

3.3.1 Previous Work in Channel Estimation

Traditionally, channel estimation has primarily relied on model-based approaches [24]. Among the conventional channel estimation methods, the least squares (LS) estimation stands out as a prevalent choice. LS estimation is widely adopted in various domains, including channel estimation [25], mainly due to its simplicity. Nevertheless, it should be noted that LS estimation fails to leverage valuable side information from noise characteristics and statistical properties of the channel during estimation. Consequently, when applied in intricate propagation environments, LS estimation may result in substantial channel estimation errors.

To address the limitations of the LS approach, a viable alternative is the use of linear minimum mean square error (LMMSE) estimation [24], a method that minimizes the mean square error. LMMSE estimation holds the advantage of incorporating the influence of noise, resulting in enhanced channel estimation accuracy. However, it is important to note that LMMSE estimation requires prior knowledge of channel statistical properties, leading to increased computational complexity compared with LS estimation. Moreover, obtaining the exact distribution of channel impulse responses can be challenging in many cases [26], potentially affecting the reliability of LMMSE estimation performance.

To overcome the limitations associated with LS and LMMSE estimations, researchers have introduced AI-aided channel estimation methods [27, 28]. These approaches harness the power of ML to improve the efficiency and adaptability of the estimation process within wireless communication systems. ML's strengths lie in its adaptability to complex environments and its ability to learn nonlinear and time-varying characteristics. ML models utilize their trainable parameters and architectural configurations to transform input data into meaningful predictions or decisions. Training a supervised ML model involves tuning parameters and weights to yield satisfactory outputs, often achieved by optimizing a loss function. A common approach in the context of channel estimation is to feed historical time series as input to produce an estimate of the current state. This can be accomplished by the data-driven neural network approach. In contrast to model-based methods, the data-driven approach eliminates the need to make assumptions about an underlying model, making the predictive model more flexible and less dependent on assumptions. Neural networks can learn from realistic data, which is advantageous when the available channel is based solely on gathered datasets. Over the past few years, research has explored channel estimation, employing various techniques, with neural networks emerging as a prominent tool.

Because of the scalable nature of the MLP, as explained in Section 3.2.1, and its comparably low computational complexity to reach good results, the MLP is the most frequently appearing channel estimation model in recent research. The most obvious way to implement an MLP to estimate the channel is to use the input variables for the model-based approach as input vector $\mathbf{x}$ in Eq. (3.4) to the MLP. This approach has been applied by Hu et al. [29], Chun et al. [30], and Zheng and Lau [31].

Another prevalent approach to incorporate the MLP into channel estimation and overcome the drawbacks of LS and LMMSE estimations is to use the model-based channel estimations as the input vector $\mathbf{x}$ in Eq. (3.4) and use simulated channel data as the labels to train the model. This procedure has been evaluated by Mei et al. [26] and Le Ha et al. [27]. The evaluations of these works of the MLP show better performance than the model-based methods in a mean squared error (MSE) sense. Le Ha et al. [27] show that the bit error rate (BER) using the MLP estimates is the same as for the model-based methods. However, the authors declare that according to the universal approximation theorem, there exist deep neural networks with the ability to obtain similar or better performance with a limited data volume than the MLP. They leave it as an open research question to identify the deep neural network with the optimal qualities for the purpose of channel estimation.

The state-of-the-art neural network for image recognition, the CNN, thoroughly disclosed in Section 3.2.2, has been applied in the area of channel estimation by Kang [32] and Fan et al. [33]. These works are inclined to the favor of CNNs, especially at high noise levels and long pilot lengths. Furthermore, the recent work by Wang et al. [34] shows that with the LS estimate of the channel as input to a CNN, it is possible to outperform the achievable BER by the model-based LS estimate. This result is obtained by using the extended CNN from [35]. However, the minimum mean squared error (MMSE) estimator actually obtains better BER than the CNN. Notably, the experimental evaluations in the overviewed works so far have only been comparing the results with the model-based channel estimators.

In the works by Helmy et al. [36] and Bai et al. [37], the properties of channel estimation were evaluated among several neural networks. Helmy et al. [36] conducted a thorough comparison of the MLP, CNN, generative adversarial network (GAN), and the proposed combined scheme of the LSTM and GRU. The combination of the two subclasses of RNN, introduced in Section 3.2.3, yields the best result for channel estimation among neural networks according to this evaluation. Bai et al. [37] show that the recurring memory from previous states h_{t-1} and c_{t-1} of the LSTM in Eqs. (3.6)–(3.11) and h_{t-1} in Eqs. (3.12)–(3.15) for the GRU, brings high achieving properties in channel estimation compared to the MLP.

An attempt to incorporate the transformer model from Figure 3.9 into channel estimation has been carried out in the recent work by Sun et al. [38], including a comparison of the obtainable BER for different neural network estimation approaches. Results show that the proposed transformer model is able to achieve a lower BER than other neural networks, indicating that it may be the new state-of-the-art in channel estimation.

3.3.2 Conventional Channel Prediction

A frequently adopted technique for studying the evolution of the wireless channel is the utilization of autoregressive (AR) models, as discussed by Truong and Heath [11]. This method is referred to as model-based because it relies on analytical models that describe how the wireless channel varies over time. The AR model conceptualizes the state of the wireless channel at a specific moment as a result of a linear combination of prior channel states coupled with an additional component of random noise. The parameters governing these linear coefficients are determined by assessing the autocorrelation structure of the channel over a specific time interval. This analytical approach enables a deeper understanding of how the wireless channel evolves over time, making it a valuable tool in the study of channel dynamics. Assuming that the channel evolves according to an AR process, model-based methods such as Kalman and Wiener filtering have conventionally been used for channel prediction.

In the context of channel evolution modeled by an AR process with Gaussian noise and under the condition that the second-order statistical properties of the AR model are either known or attainable, the Kalman filter stands out as an optimal choice for channel prediction in terms of MSE [39]. However, as the bandwidth and the number of antennas increase, the computational complexity of Kalman filters experiences rapid growth. In addition, the effectiveness of the Kalman filter in this setting is directly tied to the quality of the system identification step, which is responsible for determining the parameters of the AR model. It is worth noting that the computational complexity of this system identification step scales proportionally to the square of the number of previous channel measurements incorporated into the model, as discussed by Truong and Heath [11].

To achieve the desired performance level, higher-order models with a substantial number of parameters are frequently required. As a result, the intrinsic trade-off between model complexity and the associated computational demands imposes limitations on the effectiveness of traditional model-based approaches. Moreover, a prevalent issue with conventional channel prediction methods lies in the real-world behavior of channels. In practical scenarios, the channel's dynamics often deviate from the characteristics of an AR model. For instance, in urban environments, the validity of these models tends to diminish due to the high prevalence of obstructions and interference, leading to rapid and unpredictable fluctuations of the channel.

A recently introduced strategy to address the challenges posed by channel aging involves the deployment of predictor antennas on moving vehicles [40]. These predictor antennas are designed for high-speed vehicle movement and are positioned on the vehicle's exterior, ahead of the primary antenna. This placement allows the predictor antenna to anticipate and estimate the channel conditions at the location the main antenna is about to reach in the near future, which is especially valuable for rapidly moving vehicles. However, it is important to note that installing predictor antennas on registered vehicles may be practically problematic. An alternative, more cost-effective solution involves software development of a prediction scheme that can deliver satisfactory performance without necessitating hardware modifications to existing vehicles, but instead enhancing the predictive capabilities through software adjustments. Given these considerations, the increasing popularity and advancements in AI and especially neural networks over recent years offer a promising avenue for addressing wireless channel prediction challenges.

3.3.3 Previous Work in AI-Aided Channel Prediction

Recent progress in the field of AI has accelerated advancements across various research domains, including channel prediction, indicating that AI has the potential to outperform conventional model-based methods. One key tool in AI is neural networks that can learn from realistic data. This is advantageous when predicting future channels based solely on past channel estimations. In essence, forecasting future channels based on previous channel estimates transforms the channel prediction task into a time series learning problem. The model-based Kalman filter and the data-driven MLP, introduced in Section 3.2.1, have been compared through experimental evaluations and a complexity analysis by Kim et al. [4]. The historical instances from the channel data were used as input vector $\mathbf{x}$ in Eq. (3.4) to the MLP. In an attempt to replicate the training process with the fidelity of real channel data, Kim et al. [4] opted for training the MLP, assuming that noise had been introduced to the simulated channel data. Upon examining the performance of both the Kalman filter and the MLP, it was observed that the Kalman filter held a slight advantage. This finding points toward the necessity for more advanced neural network architectures. It's worth noting that all other papers included in this overview that utilize simulated data make the assumption of having complete and precise knowledge of the channel during the neural network training phase. This is not practically feasible.

Numerous studies have introduced more sophisticated architectures aimed at improving prediction accuracy. As described in Eq. (3.5), CNNs have the capacity to process input data in the form of matrices, unlike MLPs that operate on vectors as input. It is frequently employed in image recognition and learns by identifying patterns within smaller segments of the input matrix. For instance, Yuan et al. [41] proposed a CNN model to forecast AR coefficients for channel evolution by constructing a matrix with dimensions determined by the time steps and the number of antennas.

To capitalize on the temporal relationships within sequential data, studies have exploited the memory properties of RNNs exposed in Eqs. (3.6)–(3.11) and Eqs. (3.12)–(3.15) as an alternative to CNNs for channel prediction. In a specific case described by Jiang and Schotten [42], a subset of frequency subcarriers was individually forecasted using an RNN, followed by an interpolation step to predict the entire frequency domain used by the antenna. The channel has also been predicted for several communicating frequencies in parallel by Silva et al. [43] using a CNN.

Other research has explored hybrid models combining CNNs and RNNs for channel prediction. Combinations of CNN and LSTM networks have been employed by Wang et al. [44], Zhou et al. [45], and Wu et al. [46]. The LSTM is superior to the RNN because of its ability to address issues related to training convergence, as discussed in Section 3.2.3.1, which are common challenges in standard RNNs. A study featured by Liu et al. [47] performed a comparative analysis between LSTM and traditional model-based methods, exploring the impact of varying velocities on channel prediction. In a publication by Xu et al. [48], the authors leveraged LSTM to model the channel between a UE and a base station (BS) through a reconfigurable intelligent surface, assuming stationarity between the BS and the reconfigurable intelligent surface. Moreover, the work by Pasic et al. [49] performed an extensive measurement campaign and implemented an LSTM network as a predictor. Furthermore, a GAN was implemented to synthetically generate channel data, based on the measured channel data, as an alternative to expensive measurement campaigns.

Another promising option for enhancing the sequential RNN's performance, in addition to the LSTM, is the GRU, known for its computational efficiency relative to LSTM as mentioned in Section 3.2.3.2. In the domain of channel prediction, GRU has encountered empirical testing by several researchers. For instance, Kulkarni et al. [50] conducted a comparative analysis of LSTM and GRU within a prediction model that employed an encoder–decoder framework. This scheme integrated either LSTM or GRU layers on both the encoder and decoder sides. The evaluation, based on various datasets, including one with 4G measurements, indicated a slight advantage in favor of LSTM. Furthermore, a comprehensive examination of channel prediction has been presented by Jiang and Schotten [17], where LSTM and GRU were thoroughly investigated across different prediction horizons, with the results favoring the utilization of GRU. The GRU framework has also been evaluated for real measured data by Liu et al. [51], where the GRU was combined with Experience Replay to reach the best result.

The recent neural network approach for channel prediction, as outlined by Jiang et al. [52], has recently emerged, incorporating the transformer model in Figure 3.9. The transformer model has the ability to predict multiple future time steps concurrently, thanks to its capacity to recognize and focus on crucial patterns in sequential data, which is comprehensively described in Section 3.2.4. Further work by Zheng et al. [53] has pursued to improve the attention-based transformer model by proposing a combination of the attention mechanism and the LSTM.

In another instance of a transformer-based model, presented by Xiao et al. [54], the focus is on predicting the channel impulse response. This model relies on the UE's location as a crucial input, diverging from the previous papers that utilized historical channel measurements. Despite this distinction, the transformer-based model demonstrates commendable performance compared to LSTM.

Table 3.1 systematically categorizes and summarizes the contributions from each paper, where each row is populated by a paper. The first column, *Model*, declares the main neural network model that has been investigated for the

Table 3.1 Summary of contributions for the surveyed papers.

Model	Performance	Prediction horizon	Noisy label	Mobility	Data generation	Prediction procedure	Paper
LSTM	—	1 – 80 μs	Y	Medium & high	M/G	Time series	[49]
LSTM	Superior to ARIMA and SVR	0.1–1 ms	N	Medium high	S	Time series	[47]
LSTM/CNN	Superior to LSTM	0.5 ms	Y	High	M	Time series/Spatial	[45]
CNN	—	0.5 – 5 ms	Y	Low	S	Time series	[43]
Transformers	Superior to LSTM	0.625–3.125 ms	N	Medium	S	Time series	[52]
Transformer/ LSTM	Superior to MLP	1 ms	N	High	S	Time series	[53]
LSTM & GRU	Superior to RNN	1–5 ms	N	High	S	Time series	[17]
RNN	Inferior to KF	1.28 ms	N	Medium & high	S	Time series	[42]
MLP	Inferior to KF	40 ms	Y	Low	S	Time series	[4]
LSTM & GRU	Superior to ARIMA	1–20 s	N	Low & high	S	Encoder–decoder	[50]
GRU	Superior to LSTM	—	Y	—	M	Time series	[51]
CNN/RNN	Superior to KF	—	N	Low & medium	S	Time series	[41]
LSTM/CNN	Inferior to LSTM and CNN	—	N	Medium	S	UL—DL & subcarriers	[44]
LSTM	Superior to minimum variance unbiased	—	N	Low	S	Time series	[48]
C-GRBFnet	Superior to LSTM	None	N	None	S	Spatial prediction	[54]

Y means yes, N means no, S means simulated, M means measured, and G means generated.

purpose of channel prediction. If a cell is populated by two prediction methods, the "&" reveals that two models were evaluated, and the "/" indicates that a combination of the two models was proposed. The second column, *Performance*, presents the performance of the predictive model stated in the first column compared by some benchmarks from the experimental evaluation of each paper. The rows are sorted by the size of the investigated prediction horizon, marked by the column *Prediction horizon*.

A notable observation within the column labeled *Prediction procedure* is that in all the papers listed in Table 3.1, channel prediction predominantly relies fully or partially on time series data. The sole exception to this trend is found in the work by Xiao et al. [54]. While some papers introduce slight variations in their prediction procedures, such as using encoding and decoding of the time series [50] or a series of the uplink (UL) subcarrier channels to predict the downlink (DL) subcarrier channels [44], time series data consistently forms the fundamental basis, augmenting the efficiency and scalability of the prediction process.

Another notable observation stemming from the findings presented in Table 3.1 is that the majority of the papers assess the performance of channel prediction methods using simulated data. This practice is pragmatic, as collecting simulated data proves to be a more cost-effective and time-efficient endeavor. However, when channel prediction is to be applied in real-world scenarios, it necessitates the use of realistic data, which reflects reality and inherently carries noise. The process of conducting measurements and estimating the channel is inevitably influenced by various sources of noise. In recognition of this, some papers have incorporated noise deliberately into the training process. For example, noise was introduced during training by Kim et al. [4], alongside the authentic data. The presence of noise in the training process is reflected in the column labeled "Noisy label," indicating whether a paper considers the introduction of noisy labels for model training. When data are generated through measurements, the labels inherently carry noise. Remarkably, there is only one paper that accounts for noisy labels while utilizing a simulator to generate the dataset for training the channel prediction model. In the upcoming study by this chapter, the substantial impact of this factor is demonstrated by the evaluation of prediction methods.

The prediction horizons, as observed in Table 3.1, are predominantly associated with the mobility of the UE. Papers tend to define the prediction horizon based on the level of mobility: higher mobility leads to shorter horizons, and conversely, lower mobility implies longer horizons. This relationship arises from the inherent challenges in forecasting the channel over extended periods when dealing with a dynamic channel environment. When the prediction horizon surpasses the coherence time, the temporal correlation of the channel dissipates, rendering it unfeasible to infer the channel state solely from current or past measurements. Consequently, the prediction horizons in the papers featured in Table 3.1 tends to be relatively short. Notably, about half of the papers do not explicitly specify the time horizon for channel prediction. It is also worth mentioning that none of the papers have conducted experimental evaluations with prediction horizons extensive enough to witness a failure in the performance of the prediction models. Section 3.4 aims to address this research void by performing an extensive investigation of the influence of the prediction horizon.

The examination of prior research presented in Table 3.1 reveals several notable research gaps. First, there is a lack of clarity regarding which specific neural network is the most suitable choice for channel prediction. Second, while various data-driven models demonstrate promising results individually, they have not been systematically compared against each other. In the existing literature, data-driven models are typically evaluated by comparing the performance to traditional model-based methods. Unfortunately, most of the reviewed papers do not conduct extensive comparisons among different data-driven models, and if they do, the comparisons are often limited in scope. For instance, in the work by Jiang and Schotten [17], a comparison is made between LSTM and a deep GRU.

Previous work in channel estimation provides a wide comparison of the behavior of various neural networks. Such a comparison is not provided in channel prediction. From this consideration, the remaining part of this chapter aims to provide a comprehensive and thorough comparative analysis among the most prominent data-driven approaches, addressing these research gaps.

3.4 Experimental Evaluations

To evaluate the neural networks discussed in Section 3.2 for the purpose of channel prediction, and to address the identified research gaps in Section 3.3, an empirical evaluation is conducted. In order to carry out the evaluation, fast-fading channel data is simulated using the standardized 3GPP TDL-A model [18]. This specific channel model is based on Rayleigh fading and is useful when simulating the channel for cellular systems in a non-line-of-sight scenario.

3.4.1 Simulation Setup

To simulate the channel using the TDL-A model, simulation parameters have to be set. In this simulation scenario, both the BS and UE are equipped with 2 antennas and operate at a frequency of 2 GHz. The mobility of the UE is set to 20 km/h, resulting in a maximum Doppler shift of approximately 37 Hz. With a total of 52 resource blocks, the system features 624 subcarriers. The sampling frequency of the channel data is 14 kHz and the signal to noise ratio (SNR) is set to 0 dB. To maintain consistency with the literature earlier discussed, the simulation focuses

Table 3.2 Selected parameters for the simulator.

Parameter	Input
Channel model	TDL-A
UE speed	20 km/h
Duration	3 s
Carrier frequency	2 GHz
Sampling rate	14 kHz
Subcarriers	624
Antennas	2×2
SNR	0 dB

on one subcarrier at a time for both input and output data in the prediction process. This configuration generates an extensive dataset comprising around 26 million data points. The simulation parameters are summarized in Table 3.2.

The distribution of the original channel dataset is zero mean circularly symmetric complex Gaussian, which is a consequence of the TDL-A model assuming Rayleigh fading. From the original dataset, a training dataset consisting of 90,000 randomly selected samples and a test dataset with 10,000 samples were collected for subsequent analysis.

In addition to evaluating channel prediction for the simulated noise-free channel data, the element of noise is introduced to the dataset to reflect a more realistic scenario and to address the research gap identified in Section 3.3. The real-world scenario that accounts for uncertainties arising from channel estimation and measurement noise is obtained by adding complex Gaussian noise sampled from $\mathbf{H}' \sim \mathcal{CN}(0, \sigma^2\mathbf{J})$ to the simulated channel. The size of the additive noise matrix is determined by $\mathbf{J}$, which is the all-ones matrix with dimensions determined by the number of transmit antennas m and receiving antennas n.

The variance σ^2 of the Gaussian noise is determined in relation to the variance of the simulated channel σ_H^2 to bring an SNR of 20 dB to the channel data as

$$\text{SNR} = 10 \cdot \log \frac{\sigma_H^2}{\sigma^2} = 20 \text{ dB}. \tag{3.16}$$

This resulted in a noise variance of $\sigma^2 \approx 5\text{e} - 3$, incorporated into the inputs and outputs of the training data. Consequently, separate models are trained using real and noisy channel data. In the test data, however, only the inputs are subjected to noise distortion to facilitate a precise evaluation comparing the predicted data

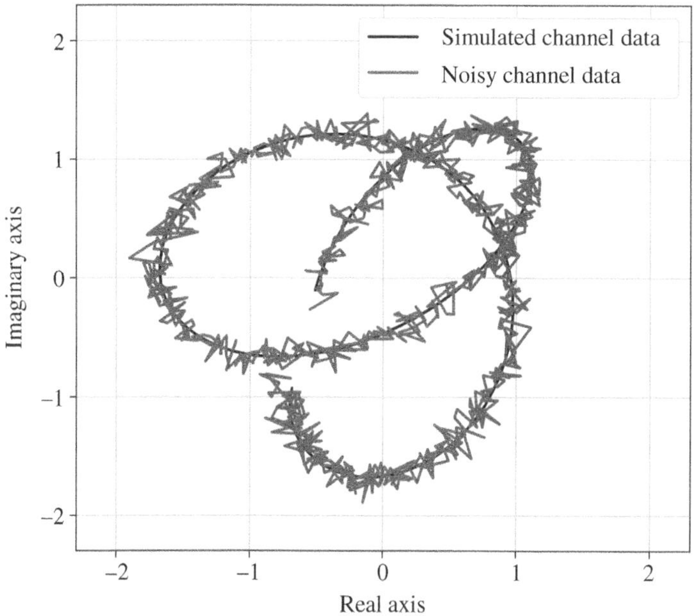

Figure 3.10 A sequence of the simulated channel in the complex plane for the duration of 53 ms, including the clean and distorted version. This sample is collected from the channel between the first transmitting antenna and the first receiving antenna. The signal with additive noise is obtained by adding Gaussian noise with an SNR of 20 dB.

with true simulated channel data. Figure 3.10 provides visual representations of small samples depicting both noisy and true channel data.

To investigate the adaptability of the predictive models in the presence of abrupt changes in the additive noise distribution, the additive noise $\mathbf{H}'$ is randomly sampled in yet another experiment. The additive noise is sampled as $\mathbf{H}'_n \sim \mathcal{CN}(0, \sigma_n^2\mathbf{J})$, where the variance is sampled from $\sigma_n^2 \sim \mathcal{U}(1\text{e}-4, 1\text{e}-3)$, and the time duration of each additive noise matrix $\mathbf{H}'_n$ is sampled from a uniform distribution $t_{\sigma_n} \sim \mathcal{U}(10 \text{ ms}, 100 \text{ ms})$. For a visual explanation, see Figure 3.11.

3.4.2 Neural Network Setup

To forecast the future channel $\mathbf{H}[t+p]$ on time horizon p, n historical channel instances serve as input vector

$$\mathbf{x}_t = \left[\mathbf{H}[t-k\cdot n], \dots, \mathbf{H}[t-k], \mathbf{H}[t]\right]$$

to the predictive model, where the time interval between each sample is determined by k. In this study, k is consequently set to $k = 14$ throughout the work so

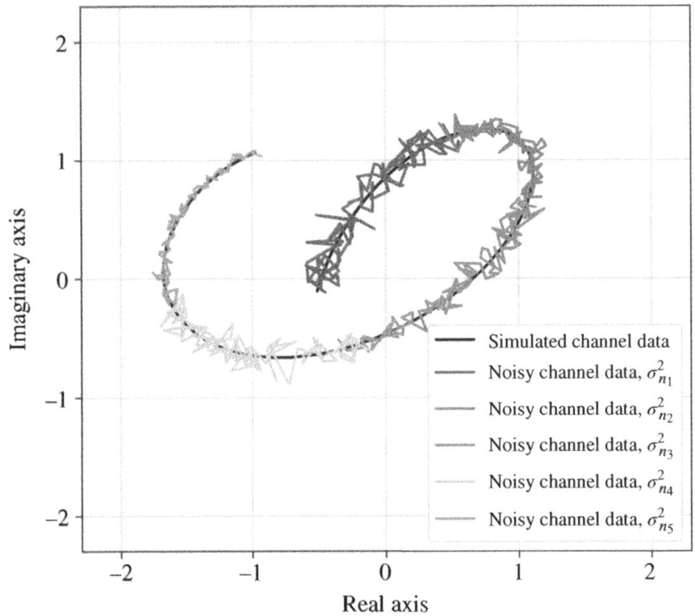

Figure 3.11 An example of how the abrupt changes of the noise variance were implemented.

that the interval between historical samples in the input is 1 ms. The number of historical instances n of the channel, also referred to as the lookback, is evaluated in the range $1 - 10$ to choose the optimal number. The goal is to use neural networks introduced in Section 3.2 to learn the objective function f in

$$f(\mathbf{x_t}) = \hat{\mathbf{H}}[t+p], \tag{3.17}$$

that minimizes the MSE loss function

$$\text{MSE} = \frac{1}{I}\sum_{i=0}^{I} \|\hat{\mathbf{H}}_i[t+p] - \mathbf{H}_i[t+p]\|^2, \tag{3.18}$$

where $\hat{\mathbf{H}}[t+p]$ is the predicted channel, $\mathbf{H}[t+p]$ is the true channel and I is the number of samples. The input channel matrix $\mathbf{H}[t]$ real and imaginary components are separated since neural networks can't handle complex values. To reconstruct the predicted channel after prediction, the output vectors from the neural networks are reformatted into complex-valued channel vectors at each time instance. The neural networks are trained to predict horizon p spanning from 1 to 40 ms, with the objective of exploring extended prediction horizons, an aspect that has received limited attention in prior research overviewed in Section 3.3.3.

Table 3.3 System parameters for the training procedure.

Parameter	Selection
Dataset size	10^5
Batch size	64
Optimizer	Adam
Loss function	MSE
Learning rate	8e−5
Learning rate decay γ	0.6
Epochs	200
Prediction horizon	1–20 ms
Lookback	5 samples

The neural networks are trained to enhance accuracy for each distinct prediction horizon. Using the Adam optimizer, the objective is to minimize the MSE over a span of 200 epochs. The learning rate is set to $8e^{-5}$ at first and decays with the multiplicative factor $\gamma = 0.6$ for every 10th epoch. The parameters for the learning algorithms are summarized in Table 3.3. The MLP is configured with six layers. For the CNN, it employs four convolutional layers, each succeeded by a pooling layer, culminating in two FC layers equipped with ReLU activation functions, mirroring the setup of the MLP. Meanwhile, the RNNs incorporate three stacked layers. A consistent choice of 150 hidden states is maintained for the RNNs. The parameter of the transformer is tuned as by Jiang et al. [52].

An empirical assessment of the sizes of all models employed in this study is conveniently summarized in Table 3.4, reporting the number of trainable parameters

Table 3.4 An empirical evaluation of the complexity for each prediction model.

Model	Trainable parameters	Elapsed time per prediction
MLP	1,990,402	27 μs
CNN	4338	33 μs
LSTM	455,102	34 μs
GRU	341,402	28 μs
Transformer	463,874	97 μs

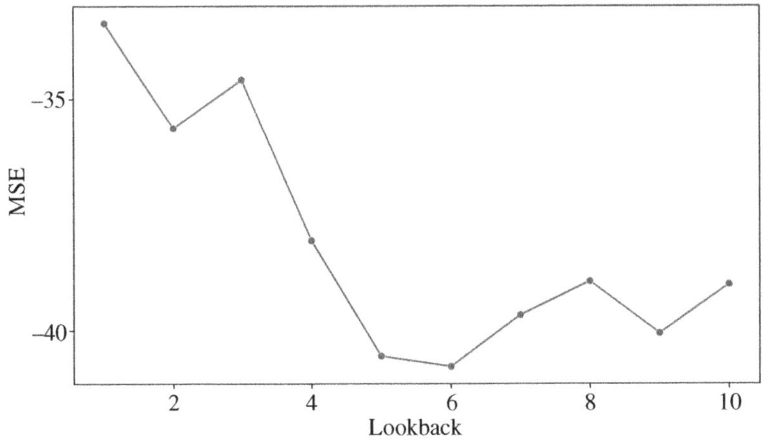

Figure 3.12 The MSE of the GRU predictor as a function of the previous number of channel samples as input.

in each neural network predictor and the elapsed time per prediction during inference. Comprehensive complexity assessments for the various models can be found in [4, 17, 52].

3.4.3 Experimental Results

To begin the empirical experiments, the optimal number for the lookback was investigated. The aim was to find the value of the lookback that results in a low MSE while keeping the computational complexity low. In other words, a short lookback is desirable. The result of the predicted channel data for the GRU model for prediction horizon $p = 1$ ms with lookbacks n in the range of 1–10 is plotted in Figure 3.12. It shows that using five previous values as input gives a low prediction error. According to the plot, using the lookback $n = 6$ gives a slightly smaller error. However, this also gives a slightly larger complexity of the model. In this case, the computational complexity is prioritized and the lookback $n = 5$ is chosen to be used throughout the remaining evaluations to be consistent among the predictive models.

The naive predictor that assumes that the future channel remains identical to the outdated (current) channel, is included as a benchmark for the comparison among predictive models. To further assess the effectiveness of neural networks in channel prediction, the study also incorporates the use of Kalman filtering. For extended prediction horizons longer than 1 ms, where predictions span multiple sampling times ahead, the predicted channel at 1 ms in the future is employed as historical measurements when using the Kalman filter. This process is iterated

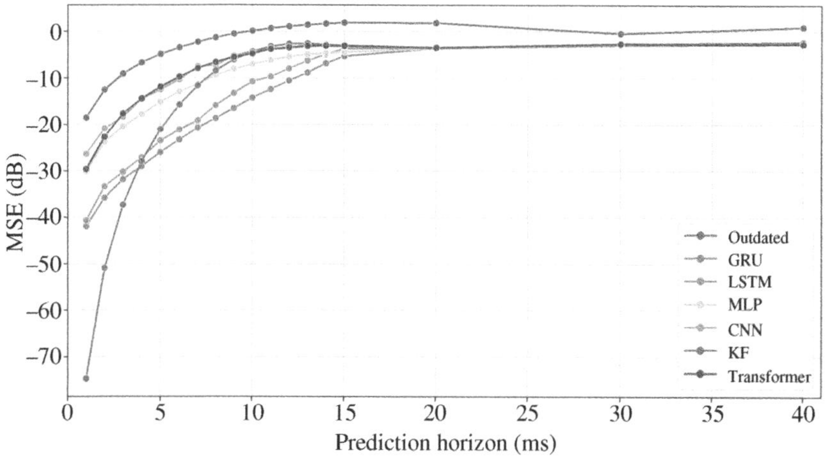

Figure 3.13 The prediction error of the simulated noise-free channel, as a function of the prediction horizon. The MSE is evaluated for all considered models, and compared to the Kalman Filter and the outdated channel.

until the desired prediction horizon is achieved, following a methodology similar to [4]. To facilitate a fair performance comparison between the Kalman filter and the neural networks, the MSE of the Kalman filter is calculated once it has achieved convergence.

The MSE for the test dataset is presented in decibels of the MSE as a function of the prediction horizon in Figure 3.13. The performance of the two RNNs appears quite similar, with a slight advantage observed for the GRU. This similarity in performance can be attributed to their closely related architectures. The MLP, CNN, and transformer exhibit slightly higher MSE but still demonstrate comparable performance. As the prediction horizon increases, the MSE exhibits consistent and gradual growth. With the expansion of the prediction horizon, the temporal correlation of the channel weakens, making channel prediction progressively more challenging. Notably, at a prediction horizon of 15 ms, the performance of the neural networks stabilizes at a relatively constant level of error. This observation suggests that the coherence time of the channel is approximately 15 ms, which is an indication of the predictive time horizon where there is no temporal correlation to the current channel. The coherence time, the rate at which the channel ages, is intricately linked to the mobility of the UE. Lower UE mobility in the simulation would lead to improved neural network performance for longer prediction horizons.

The same pattern repeats in the performance evaluation of neural networks trained using the channel dataset with additive Gaussian noise of constant variance, as depicted in Figure 3.14. Notably, the performance of neural networks

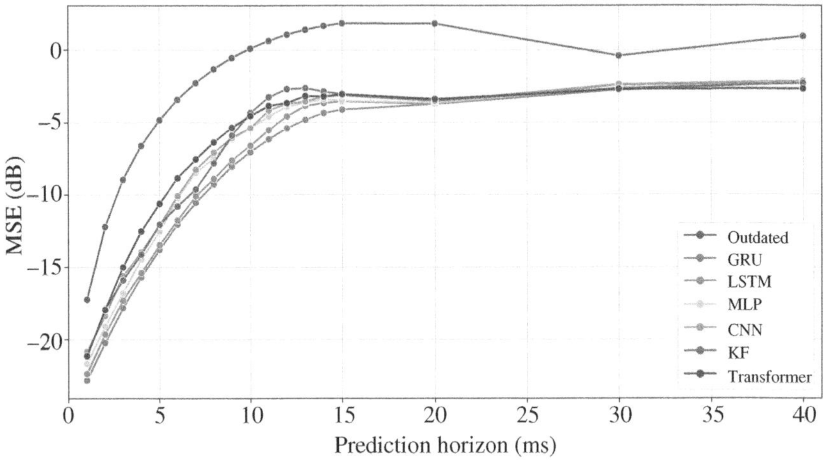

Figure 3.14 Prediction error of the channel is distorted by additive Gaussian noise with an SNR of 20 dB as a function of the prediction horizon. The prediction error is evaluated for all considered models and compared to the Kalman Filter and the outdated channel.

remains relatively stable for prediction horizons exceeding 15 ms. When prediction horizons are shorter, the performance of all neural networks maintains the same relative order, whether in the presence of noise or noise-free conditions. It is important to note that the MSE of the test data significantly increases when noise is introduced. However, when noise is introduced and the prediction horizon is short, there is no substantial performance difference among the RNNs, MLP, CNN, and transformer. To get an intuition of how the prediction of the noisy channel compares to the true simulated channel for one antenna element, see Figure 3.15.

In a noise-free setting with short prediction horizons, the Kalman filter demonstrates strong performance. However, for prediction horizons exceeding 5 ms, both the LSTM and the GRU outperform the Kalman filter. In a noisy environment, the Kalman filter's performance aligns more closely with that of the neural networks. As illustrated in Figures 3.13 and 3.14, it becomes evident that the RNNs outperform the Kalman filter when dealing with extended prediction horizons.

In the presence of abrupt changes in the additive noise variance, it is clear from Figure 3.16 that the two RNNs perform very similarly to each other and better than the other predictive neural networks. However, the MLP is close to the prediction accuracy of the two RNNs. The worst predictor in the case of abrupt changes in the additive noise variance is by far the Kalman filter. It is clear that the neural networks have a strong advantage over the model-based predictor in noisy scenarios.

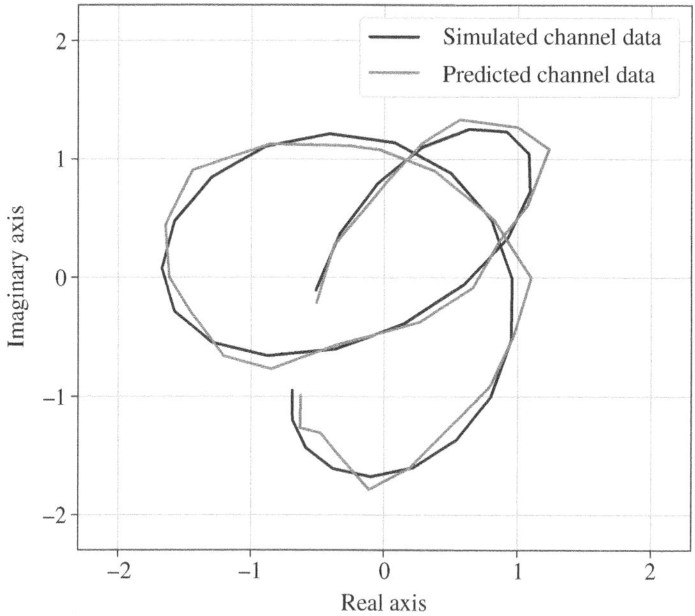

Figure 3.15 An example of how the predicted channel relates to the true simulated channel. The sequence is the same as in Figure 3.10, with a duration of 53 ms. The prediction model used for the predicted channel in this figure is the GRU and the prediction horizon is $p = 2$ ms.

3.5 Discussion

As illustrated in Figure 3.13, it becomes evident that the recurrent memory cell in the RNNs yields a substantial advantage in predicting wireless channel time series. This advantage results in consistently robust performance for both the GRU and LSTM models, outperforming the MLP, CNN, and transformer across various prediction horizons. Based on these findings, it can be inferred that the GRU stands as the state-of-the-art in channel prediction. This can intuitively be attributed to the GRU's inherent capacity to identify and leverage temporal correlations in sequential data. Moreover, the GRU's leaner parameter configuration, in comparison to the similar LSTM network, contributes to its ability to learn more effectively and swiftly.

When assessing practical feasibility within real base stations, limitations in computational power and energy consumption should be acknowledged. In this context, it may be more prudent to favor the MLP among the neural network options due to its relatively lower complexity, especially during the training phase. The computational time and power required to execute the prediction model play pivotal roles in real-world implementations. Subsequent research endeavors could explore ways to adapt these identified methods for practical, real-world use, including aspects such as quantization, continual learning, and one-shot learning.

In the scenario involving a distorted channel, reflecting the uncertainties of channel estimation, the disparity in error between the GRU and the MLP remains relatively constant at approximately 1 dB. The MLP's advantage in terms of lower computational complexity while obtaining a satisfactory prediction accuracy is indicated by the computational time in Table 3.4. This result positions the MLP as a potentially more suitable choice for real-world implementations, especially when computational resources are constrained by factors like time and energy efficiency. This alignment is reinforced by findings in Kim et al. [4], which indicate that the MLP exhibits lower computational complexity compared to the Kalman filter. Furthermore, real-world channels may not consistently exhibit the smooth patterns seen in Figure 3.10. The data-driven nature of the MLP makes it better equipped to adapt in such situations compared to the Kalman filter, enhancing its suitability for practical implementation.

By introducing randomness in the variance of the additive Gaussian noise, the predictor's resistance to abrupt changes was explored. By comparing Figure 3.14 with Figure 3.16, it becomes apparent that the model-based Kalman filter suffers from the randomness in the variance of the noise. The error of the Kalman filter becomes larger than the outdated channel almost exclusively. One explanation for this is the Kalman filter's stationary dynamics. Once it has learned a process with a

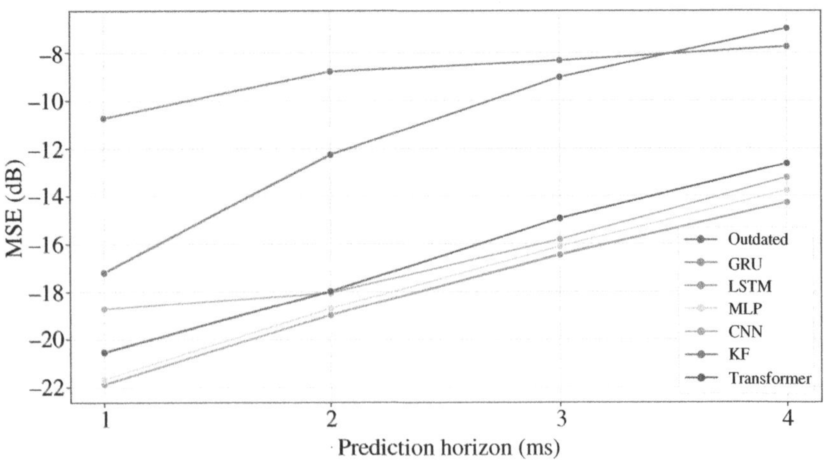

Figure 3.16 The performance of the predictive models was measured by the MSE in dB for random abrupt changes in the noise distribution over the prediction horizon. The performance of the GRU is almost not visible due to its practically identical performance to that of the LSTM.

certain noise level, the Kalman filter has an inhabitant inertia to adapt to new scenarios with different properties. However, for the predictive neural networks, the prediction error does not change significantly. The error for the networks is fairly constant when introducing random variance for the additive noise. It is interesting to note that the transformer does not perform well under these abrupt changes. One conclusion is that the attention mechanism misleads the predictions during abrupt changes.

An important point to note regarding the distorted channel data is that the results that were obtained investigated the ability of channel prediction during poor channel estimation conditions. One way to improve the predictions in this work would have been to preprocess the distorted channel data with a filter, such as the moving average. However, such processing would belong to the channel estimation algorithm, which was not the purpose of this investigation. Again, the channel data was distorted to simulate channel prediction difficulties under channel estimation uncertainties.

This chapter has addressed the non-line-of-sight communication scenario using the standardized 3GPP TDL-A channel model. The dataset employed for simulation comprises 26 million data points, ensuring statistical sufficiency to represent the TDL-A scenario. Consequently, the results presented in this chapter are widely applicable due to the comprehensiveness of the standardized TDL-A channel model. It would be valuable for future research to explore the resilience of neural networks in wireless channel prediction when faced with sudden changes in the communication environment, such as from the emergence or disappearance of dominant transmission paths. In instances where such changes occur, it is inclined that the neural networks trained in this study require only five consecutive historical samples after the abrupt transition to regain accurate predictive capabilities. This is primarily because the neural network input is based on five historical samples. Moreover, line-of-sight communication tends to exhibit more stability and slower fading compared with non-line-of-sight communication. Consequently, the neural network models established in this research are expected to perform effectively in scenarios characterized by smaller variations, even beyond the scope of the training dataset. In contrast, the Kalman filter typically demands more than five historical samples to achieve convergence toward reliable outcomes, as previously observed by Kim et al. [4].

3.6 Summary

This chapter has elaborated on how to combat channel aging using channel estimation and channel prediction. An overview of key findings in AI-aided channel prediction and channel estimation from the existing literature has been provided. Especially the findings in models using neural networks were the focus of the survey. The primary advantage of using AI lies in its ability to adapt without relying on predefined models, making it flexible and capable of autonomously learning from available data. From the comprehensive overview of previous works in AI-aided channel estimation, both the GRU and the transformer model show promising estimating methods. This conclusion from previous work was repeated for channel prediction as well.

Through simulations conducted in a non-line-of-sight scenario using the 3GPP standardized TDL-A model, this study involved the training and validation of neural networks for the purpose of channel prediction. Experiments were conducted for different versions of the dataset, the original simulated channel data, and two distorted versions of the channel by additive Gaussian noise. The first version of the distorted dataset contained additive Gaussian noise with constant variance, while random abrupt changes were introduced to the variance of the second version of the channel data containing additive Gaussian noise.

In a noise-free channel scenario, the numerical experiments presented in this chapter revealed that for prediction horizons up to 15 ms, the GRU and LSTM outperformed the MLP, CNN, and transformer models. However, the Kalman filter demonstrated superior performance compared to all neural networks up to a prediction horizon of 4 ms. When dealing with channel measurements affected by noisy data, the performance gap between the neural networks was less pronounced. Nevertheless, the relative performance order of the neural networks remained consistent across both noise-free and noisy data scenarios. In the presence of noise, the Kalman filter's performance aligned more closely to the MLP, CNN, and transformer models. However, when introducing random abrupt changes in the variance of the additive noise to the channel data, the performance of the Kalman filter predictor deteriorates significantly. The neural networks, especially the MLP, LSTM, and GRU, are barely affected by the abrupt changes to the noise, compared to the case with constant variance of the additive noise.

In conclusion, this survey indicates that the GRU is the most convenient choice for channel prediction that holds great promise. The obtained results are based on the TDL-A channel model that was used to generate the data. However, the comprehensiveness of the standardized TDL-A model gives the results a wider applicability for future real-world applications. In future research, data-efficient ML models should be investigated, particularly when making estimations or predictions on resource-limited wireless devices.

References

1 Wang, C.-X., You, X., Gao, X. et al. (2023). On the road to 6G: visions, requirements, key technologies, and testbeds. *IEEE Communications Surveys & Tutorials* 25 (2): 905–974. https://doi.org/10.1109/COMST.2023.3249835.

2 Schwarz, S. and Rupp, M. (2016). Society in motion: challenges for LTE and beyond mobile communications. *IEEE Communications Magazine* 54 (5): 76–83. https://doi.org/10.1109/MCOM.2016.7470939.
3 Chiu, L.-K. and Wu, S.-H. (2015). An effective approach to evaluate the training and modeling efficacy in MIMO time-varying fading channels. *IEEE Transactions on Communications* 63 (1): 140–155. https://doi.org/10.1109/TCOMM.2014.2382103.
4 Kim, H., Kim, S., Lee, H. et al. (2021). Massive MIMO channel prediction: Kalman filtering vs. machine learning. *IEEE Transactions on Communications* 69 (1): 518–528. https://doi.org/10.1109/TCOMM.2020.3027882.
5 Fodor, S., Fodor, G., Gürgünoğlu, D., and Telek, M. (2023). Optimizing pilot spacing in MU-MIMO systems operating over aging channels. *IEEE Transactions on Communications* 71 (6): 3708–3720. https://doi.org/10.1109/TCOMM.2023.3261384.
6 Kermoal, J.-P., Schumacher, L., Pedersen, K.I. et al. (2002). A stochastic MIMO radio channel model with experimental validation. *IEEE Journal on Selected Areas in Communications* 20 (6): 1211–1226.
7 Vu, M. and Paulraj, A. (2007). On the capacity of MIMO wireless channels with dynamic CSIT. *IEEE Journal on Selected Areas in Communications* 25 (7): 1269–1283.
8 Fodor, G., Di Marco, P., and Telek, M. (2014). Performance analysis of block and comb type channel estimation for massive MIMO systems. *1st International Conference on 5G for Ubiquitous Connectivity*, 62–69. https://doi.org/10.4108/icst.5gu.2014.258076.
9 Weichselberger, W., Herdin, M., Ozcelik, H., and Bonek, E. (2006). A stochastic MIMO channel model with joint correlation of both link ends. *IEEE Transactions on Wireless Communications* 5 (1): 90–100.
10 Jakes, W.C. (1974). *Microwave Mobile Communications.* Wiley-IEEE Press.
11 Truong, K.T. and Heath, R.W. (2013). Effects of channel aging in massive MIMO systems. *Journal of Communications and Networks* 15 (4): 338–351. https://doi.org/10.1109/JCN.2013.000065.
12 Baddour, K.E. and Beaulieu, N.C. (2004). Accurate simulation of multiple cross-correlated Rician fading channels. *IEEE Transactions on Communications* 52 (11): 1980–1987. https://doi.org/10.1109/TCOMM.2004.836559.
13 Wu, Q., Zhang, Y., Yang, Z., and Shikh-Bahaei, M. (2023). Deep channel prediction-based energy-efficient intelligent reflecting surface-aided terahertz communications. *IEEE Transactions on Wireless Communications* 23 (4): 2946–2960. https://doi.org/10.1109/TWC.2023.3304597.
14 Ying, M. and Chen, X. (2023). Channel prediction-based robust multibeam precoding design for LEO satellite Internet of Things. *2023 IEEE/CIC International Conference on Communications in China (ICCC)*, 1–6. IEEE.

15 Rydén, H., Farhadi, H., Palaios, A. et al. (2023). Next generation mobile networks' enablers: machine learning-assisted mobility, traffic, and radio channel prediction. *IEEE Communications Magazine* 61 (10): 94–98. https://doi.org/10.1109/MCOM.001.2200592.

16 Becvar, Z., Gesbert, D., Mach, P., and Najla, M. (2023). Machine learning-based channel quality prediction in 6G mobile networks. *IEEE Communications Magazine* 61 (7): 106–112. https://doi.org/10.1109/MCOM.001.2200305.

17 Jiang, W. and Schotten, H.D. (2020). Deep learning for fading channel prediction. *IEEE Open Journal of the Communications Society* 1: 320–332.

18 3GPP (2020). Study on channel model for frequencies from 0.5 to 100 GHz (3GPP TR 38.901 version 16.1.0 Release 16). ETSI.

19 Rumelhart, D.E., Hinton, G.E., and Williams, R.J. (1986). Learning internal representations by error propagation, parallel distributed processing, explorations in the microstructure of cognition, ed. De Rumelhart and J. McClelland. vol. 1. 1986. *Biometrika* 71 (599–607): 6.

20 Ribeiro, A.H., Tiels, K., Aguirre, L.A., and Schön, T. (2020). Beyond exploding and vanishing gradients: analysing RNN training using attractors and smoothness. *International Conference on Artificial Intelligence and Statistics*, 2370–2380. PMLR.

21 Hochreiter, S. and Schmidhuber, J. (1997). Long short-term memory. *Neural Computation* 9 (8): 1735–1780.

22 Cho, K., van Merrienboer, B., Gülçehre, Ç. et al. (2014). Learning phrase representations using RNN encoder-decoder for statistical machine translation. *CoRR*, abs/1406.1078. http://arxiv.org/abs/1406.1078.

23 Vaswani, A., Shazeer, N., Parmar, N. et al. (2017). Attention is all you need. *Advances in Neural Information Processing Systems 30.*

24 van de Beek, J.-J., Edfors, O., Sandell, M. et al. (1995). On channel estimation in OFDM systems. *1995 IEEE 45th Vehicular Technology Conference. Countdown to the Wireless Twenty-First Century*, volume 2, 815–819. https://doi.org/10.1109/VETEC.1995.504981.

25 Verdu, S. (1998). *Multiuser Detection.* Cambridge University Press.

26 Mei, K., Liu, J., Zhang, X. et al. (2021). Performance analysis on machine learning-based channel estimation. *IEEE Transactions on Communications* 69 (8): 5183–5193. https://doi.org/10.1109/tcomm.2021.3083597.

27 Le Ha, A., Van Chien, T., Nguyen, T.H. et al. (2021). Deep learning-aided 5G channel estimation. *2021 15th International Conference on Ubiquitous Information Management and Communication (IMCOM)*, 1–7. https://doi.org/10.1109/IMCOM51814.2021.9377351.

28 Liu, Y., Tan, Z., Hu, H. et al. (2014). Channel estimation for OFDM. *IEEE Communications Surveys & Tutorials* 16 (4): 1891–1908.

29 Hu, Q., Gao, F., Zhang, H. et al. (2021). Deep learning for channel estimation: interpretation, performance, and comparison. *IEEE Transactions on Wireless Communications* 20 (4): 2398–2412. https://doi.org/10.1109/TWC.2020.3042074.

30 Chun, C.-J., Kang, J.-M., and Kim, I.-M. (2019). Deep learning-based channel estimation for massive MIMO systems. *IEEE Wireless Communications Letters* 8 (4): 1228–1231. https://doi.org/10.1109/LWC.2019.2912378.

31 Zheng, X. and Lau, V.K.N. (2022). Simultaneous learning and inferencing of DNN-based mmWave massive MIMO channel estimation in IoT systems with unknown nonlinear distortion. *IEEE Internet of Things Journal* 9 (1): 783–799. https://doi.org/10.1109/JIOT.2021.3085659.

32 Kang, J.-M. (2023). Deep learning-based robust channel estimation for MIMO IoT systems. *IEEE Internet of Things Journal* 11 (6): 9882–9895. https://doi.org/10.1109/JIOT.2023.3324667.

33 Fan, X., Zou, Y., and Zhai, L. (2023). Spatial attention-based channel estimation in IRS-assisted mmWave MU-MISO systems. *IEEE Internet of Things Journal* 11 (6): 9801–9813. https://doi.org/10.1109/JIOT.2023.3326174.

34 Wang, M., Wang, A., Liu, Z., and Chai, J. (2023). Deep learning based channel estimation method for mine OFDM system. *Scientific Reports* 13 (1): 17105.

35 Dong, C., Loy, C.C., and Tang, X. (2016). Accelerating the super-resolution convolutional neural network. *Computer Vision—ECCV 2016: 14th European Conference*, Amsterdam, The Netherlands (11—14 October 2016), Proceedings, Part II 14, 391–407. Springer International Publishing.

36 Helmy, I., Tarafder, P., and Choi, W. (2023). LSTM-GRU model-based channel prediction for one-bit massive MIMO system. *IEEE Transactions on Vehicular Technology* 72 (8): 11053–11057. https://doi.org/10.1109/TVT.2023.3262951.

37 Bai, Q., Wang, J., Zhang, Y., and Song, J. (2020). Deep learning-based channel estimation algorithm over time selective fading channels. *IEEE Transactions on Cognitive Communications and Networking* 6 (1): 125–134. https://doi.org/10.1109/TCCN.2019.2943455.

38 Sun, T., Lv, J., and Zhou, T. (2023). A transformer-based channel estimation method for OTFS systems. *Entropy* 25 (10): https://doi.org/10.3390/e25101423.

39 Aronsson, D. and Sternad, M. (2007). Kalman predictor design for frequency-adaptive scheduling of FDD OFDMA uplinks. *2007 IEEE 18th International Symposium on Personal, Indoor and Mobile Radio Communications*, 1–5. IEEE. https://doi.org/10.1109/PIMRC.2007.4394065.

40 Björsell, J. (2022). Predictor antennas: enabling channel prediction for fast-moving vehicles in wireless broadband systems. PhD thesis. Uppsala University, Signals and Systems.

41 Yuan, J., Ngo, H.Q., and Matthaiou, M. (2020). Machine learning-based channel prediction in massive MIMO with channel aging. *IEEE Transactions on Wireless Communications* 19 (5): 2960–2973.

42 Jiang, W. and Schotten, H.D. (2019). Recurrent neural network-based frequency-domain channel prediction for wideband communications. *2019 IEEE 89th Vehicular Technology Conference (VTC2019-Spring)*, 1–6. IEEE. https://doi.org/10.1109/VTCSpring.2019.8746352.

43 Silva, P.E.G., Moualeu, J.M., Nardelli, P.H., and de Souza, R.A.A. (2023). An efficient machine learning-based channel prediction technique for OFDM sub-bands. *arXiv preprint arXiv:2305.19696.*

44 Wang, J., Ding, Y., Bian, S. et al. (2019). UL-CSI data driven deep learning for predicting DL-CSI in cellular FDD systems. *IEEE Access* 7: 96105–96112. https://doi.org/10.1109/ACCESS.2019.2929091.

45 Zhou, T., Zhang, H., Ai, B. et al. (2022). Deep-learning-based spatial—temporal channel prediction for smart high-speed railway communication networks. *IEEE Transactions on Wireless Communications* 21 (7): 5333–5345. https://doi.org/10.1109/TWC.2021.3139384.

46 Wu, R., Liu, X., Huang, W., and Jiang, H. (2022). ConvLSTM based channel prediction for 5G network. *2022 IEEE 5th Advanced Information Management, Communicates, Electronic and Automation Control Conference (IMCEC)*, volume 5, 644–647. https://doi.org/10.1109/IMCEC55388.2022.10020010.

47 Liu, G., Xu, Y., He, Z. et al. (2019). Deep learning-based channel prediction for edge computing networks toward intelligent connected vehicles. *IEEE Access* 7: 114487–114495.

48 Xu, W., An, J., Xu, Y. et al. (2022). Time-varying channel prediction for RIS-assisted MU-MISO networks via deep learning. *IEEE Transactions on Cognitive Communications and Networking* 8 (4): 1802–1815. https://doi.org/10.1109/TCCN.2022.3188153.

49 Pasic, F., Di Cicco, N., Skocaj, M. et al. (2023). Multi-band measurements for deep learning-based dynamic channel prediction and simulation. *IEEE Communications Magazine* 61 (9): 98–104. https://doi.org/10.1109/MCOM.003.2200718.

50 Kulkarni, A., Seetharam, A., Ramesh, A., and Herath, J.D. (2019). DeepChannel: Wireless channel quality prediction using deep learning. *IEEE Transactions on Vehicular Technology* 69 (1): 443–456.

51 Liu, Q., Wang, P., Sun, J. et al. (2023). Wireless channel prediction of GRU based on experience replay and snake optimizer. *Sensors* 23 (14): https://doi.org/10.3390/s23146270.

52 Jiang, H., Cui, M., Ng, D.W.K., and Dai, L. (2022). Accurate channel prediction based on transformer: making mobility negligible. *IEEE Journal on Selected Areas in Communications* 40 (9): 2717–2732. https://doi.org/10.1109/JSAC.2022.3191334.

53 Zheng, W., Liu, Z., Yuan, Y. et al. (2023). Fading channel prediction based on attention mechanism. *2023 5th International Conference on Electronic Engineering and Informatics (EEI)*, 400–404. https://doi.org/10.1109/EEI59236.2023.10212879.

54 Xiao, Z., Zhang, Z., Huang, C. et al. (2022). C-GRBFnet: A physics-inspired generative deep neural network for channel representation and prediction. *IEEE Journal on Selected Areas in Communications* 40 (8): 2282–2299. https://doi.org/10.1109/JSAC.2022.

4

Semantic Communications

Qiyang Zhao[1], Hang Zou[1], Mehdi Bennis[2], and Merouane Debbah[3]

[1] *Technology Innovation Institute, Abu Dhabi, United Arab Emirates*
[2] *Centre for Wireless Communications, University of Oulu, Oulu, Finland*
[3] *KU 6G Research Center, Khalifa University of Science and Technology, Abu Dhabi, United Arab Emirates*

CHAPTER MENU

4.1 Introduction

6G is envisioned to be a network of intelligent agents where artificial intelligence (AI) will power massive wireless devices. Conventional communication paradigms dated back to Shannon aiming at transmitting bits efficiently regardless of the meaning of the message [1]. The following three-level structure of communication system is proposed by Shannon and Weaver [2]: (i) level A: "How accurately can the symbols of communication be transmitted;" (ii) level B: "How precisely do the transmitted symbols convey the desired meaning;" (iii) level C: "How effectively does the received meaning affect conduct in the desired way." Shannon's approach, i.e., reconstructs a digital sequence generated at the transmitter, asymptotically error-free at receiver side roots obviously in level A. Level B concentrates on the "meaning" of transmitted message. Finally, Level C generally refers to communication at the effectiveness level, i.e., goal- or task-oriented communication [3, 4]. Up to now, the boundary of semantic communication and goal-oriented communication is still blurry. Therefore, the topics in this chapter will not only be limited to semantic communications in the narrow sense but also include ideas, approaches, and algorithms attempting to

Artificial Intelligence for Future Networks, First Edition.
Edited by Mohammad A. Matin, Sotirios K. Goudos, and George K. Karagiannidis.

clarify the meaning of transmitted symbols and minimize the information sent to the receiver.

Research in semantic communications has been growing rapidly in recent years. However, there are different perspectives or definitions under the terminology of goal-oriented and emergent communications [5]. On the contrary, the role of semantic communication in the future communication system raises several debates. For example, some works see it as an advanced joint source-channel coding [6], in which only the relevant information for the underlying task is communicated to the receiver. While others suggest that semantic communications learn concepts to perform contextual reasoning or pragmatic reasoning [7]. Such semantic representation can enable system a two-level communication system, from physical layer coding to multiple access control (MAC) layer protocols [8]. Following this perspective, a knowledge and reasoning driven semantic communication network is proposed [9], whose paradigm goes beyond data compression, application-aware and goal-oriented communications, toward path to knowledge-driven AI nativeness, reasoning based and symmetric communication system.

A typical paradigm of semantic communications is illustrated in Figure 4.1. The source information is encoded by a semantic encoder to produce semantic representation, which is then converted into bits by conventional source and channel encoder for transmission over noisy channels. The decoding process at the channel and semantic level applies at the receiver end. A virtual semantic channel connects the semantic encoder and semantic decoder. Different from the physical channel, where the gain or loss depends on the accuracy of reconstructed bits, the semantic channel captures the meaning of information effectively conveyed rather than the specific format or medium of communication. For example, it can be the effectiveness of information used to perform specific tasks or the resiliency of information being used over time. According to [9], semantic communication transforms the bit-driven transmitter in to a knowledge-driven teacher, which is capable of extracting semantic content elements from source data and finding a minimal representation of the meaning. Similarly, the bit-driven receiver is transformed into a knowledge-driven student, which is capable of understanding the

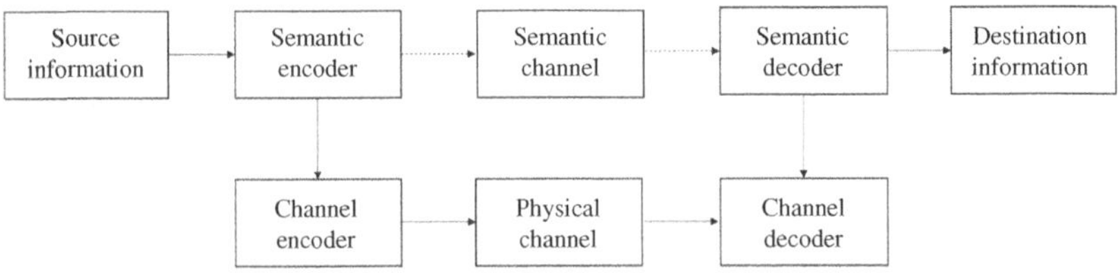

Figure 4.1 Semantic communication paradigm.

minimal semantic representation from the teacher and generating the semantic content elements. In doing so, the communication channel is transformed from a bit-pipeline to a semantic language, which represents the smallest distinct meaningful element from the information.

The transformation from physical communication to semantic communication can be summarized as follows.

- From compression to minimalism. While physical communication utilizes source coding to compress different data modalities, semantic communication characterizes the structure of information with reduced message exchanges.
- From specialization to generalization. Conventional communication system aims to develop optimal coding schemes and protocols in each specific scenario, such as turbo, LDPC, convolutional codes optimized for eMBB, and URLLC on data and control channels. On the contrary, semantic communication aims to develop a generalized semantic representation that is invariant in changes of data distribution, domain, and context, while describing a particular underlying structure.
- From accuracy to effectiveness. The error rates at bit, frame, and symbol levels are commonly used to evaluate the performance of physical communications. In semantic communication, the ability of regenerating information at high fidelity in the least time is the key requirement. This goes beyond reconstructing information sent from transmitter, but more importantly to generate content from representation effectively to accomplish the targeted task or goal jointly between transmitter and receiver, or even with multiple agents.

Building a minimal, generalized, and effective semantic communication system has significant challenges. There are many works addressing semantic communication problems in different data modalities (language, image, video, sound, etc.) for different applications (generation, classification, detection, recognition, etc.), as illustrated in Figure 4.2. However, reaching full semantic-native communication with the above-mentioned objectives still requires significant research efforts.

In the remaining part of this chapter, we will first take the angle of "AI for network (AI4NET)" to discuss the role of machine learning technologies in enabling semantic communication. We discuss the theories and candidate technologies modeling semantic information and then review the design of a semantic-native communication system from physical, MAC to application layers at system 1 and system 2 levels. Thereafter, we will take the angle of "network for AI (NET4AI)" to discuss the role of semantic communication in enabling a collective intelligent network. We discuss the trend of distributed AI and the capabilities of large language model (LLM) in understanding, extracting, and generating from semantic information, and then provide a perspective of semantic communication as an enabler for efficient distributed and collaborated

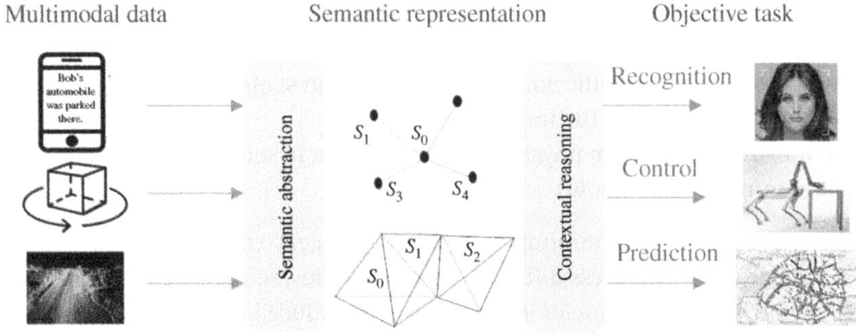

Figure 4.2 Application of semantic communication. Source: (a) LizardTech; (b) An optimized solution for face recognition by Jennifer Michalowski. Massachusetts Institute of Technology (MIT). (c) Boston Dynamics. (d) raphaelauv/Paris-Traffic-Prediction/GitHub, Inc.

LLM in wireless networks. Finally, we summarize this chapter by discussing the challenges and research directions in semantic communications.

4.2 Semantic Information and Semantic-Native Communication

In this chapter, we first introduce the semantic information theory with a mathematical analysis of entropy, rate distortion, loss, etc. Thereafter, we will discuss candidate semantic representations for source and channel semantics on topological spaces. Finally, we will explain the differences between system 1 and system 2 level semantic communications and semantic protocol.

4.2.1 Semantic Information Theory

Arising from the meaningful aspect of the natural language, there is no universal or widely accepted definition for semantics. Here, we define semantics as the meaningful part of any informative representation. According to [9], semantics should satisfy the principles of being minimal, generalizable, and efficient at the same time. To the best knowledge of the authors, there is no such definition in the community satisfying all these principles. The attempt to formalize semantic information and the corresponding theory started from the work of Carnap and Bar-Hillel, which defines the semantic entropy of a sentence given a language system [10]. The semantic entropy of the sentence s given the evidence e is defined as:

$$H(s,e) \triangleq -\log c(s,e), \tag{4.1}$$

where $c(s, e)$ is the degree of confirmation of sentence s on the evidence e that is defined as

$$c(s, e) \triangleq \frac{m(s, e)}{m(e)}, \tag{4.2}$$

where $m(s, e)$ and $m(e)$ denote the logical entropy of s on e and e.

Later, a different definition based on the language comprehension model is proposed in [11], which can be expressed as

$$H(\boldsymbol{v}_t) = -\sum_{\boldsymbol{v}_M \in \mathcal{V}_M} \mathbb{P}(\boldsymbol{v}_M | \boldsymbol{v}_t) \log \mathbb{P}(\boldsymbol{v}_M | \boldsymbol{v}_t), \tag{4.3}$$

where $\mathcal{V}_M = \{\boldsymbol{v}_M | \boldsymbol{v}_t = 1 \text{ iff } M_i = M \text{ and } M \text{ is a unique model in } \mathcal{M}\}$ is the set of meaning vectors that identify unique models in $\mathcal{M}$.

Besides, the information-theoretic method is also applied to translation tasks in [12] by using translational distributions of words. The semantic entropy of each word w is defined as

$$H(w) \triangleq -\sum_{t \in T} \mathbb{P}(t|w) \log \mathbb{P}(t|w) + \mathbb{P}(\text{NULL}|w) \log F(w), \tag{4.4}$$

where T represents the set of target words and $F(w)$ is the frequency of w.

Rate-distortion theory is one of the most important results of Shannon's classical approach to lossy source transmission. The authors in [13] adapt this principle to the source model modeled by a joint pair of random variables (RVs) (S, X) correlated with joint probability distribution $p(s, x)$ in a memoryless channel. RV X and S represent the extrinsic observations and intrinsic semantic features of the source, respectively. This semantic source model is illustrated in Figure 4.3. The decoder has two tasks to achieve: (i) reconstruct the extrinsic observation blocks under the observation distortion measure d_o; (ii) replicate the intrinsic state block under the semantic distortion measure d_s. The observation distortion measure $d_o : \mathcal{X} \times \hat{\mathcal{X}} \to \mathbb{R}_+$ and semantic distortion measure $d_s : \mathcal{S} \times \hat{\mathcal{S}} \to \mathbb{R}_+$ are two

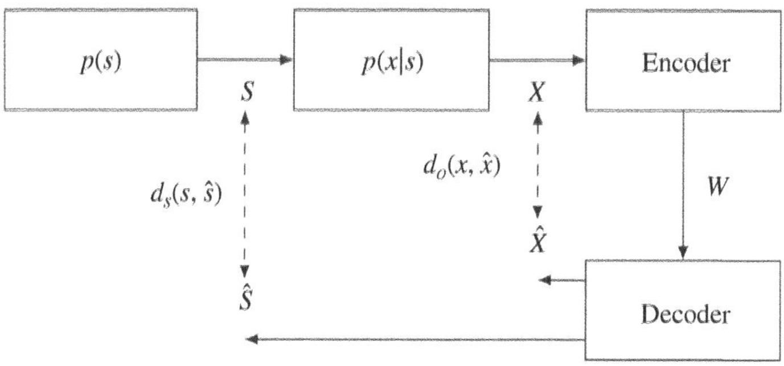

Figure 4.3 An illustrating schematic of a semantic source and its loss compression. Source: [[13], Figure 1].

distortion measures defined over source product alphabet $\mathcal{S} \times \mathcal{X}$ and the reproduction product alphabet $\hat{\mathcal{S}} \times \hat{\mathcal{X}}$, respectively. A tuple $\left(R, D_s, D_o\right)$ is said to be achievable if one can find such a code for any $\varepsilon > 0$ and all sufficiently large lengths of the code such that $\mathbb{E}\left[d_s\left(S, \hat{S}\right)\right] \leq D_s + \varepsilon$ and $\mathbb{E}\left[d_o\left(X, \hat{X}\right)\right] \leq D_o + \varepsilon$. For a given semantic source (S, X) with joint distribution $p(s, x)$ over the source product alphabet $\mathcal{S} \times \mathcal{X}$ and the reproduction product alphabet $\hat{\mathcal{S}} \times \hat{\mathcal{X}}$ equipped with observation distortion d_o and semantic distortion d_s, the rate distortion function $R\left(D_s, D_o\right)$ is as follows:

$$
\begin{aligned}
R\left(D_s, D_o\right) = & \min_{p(\hat{s}, \hat{x}|x)} I\left(X; \hat{S}, \hat{X}\right) \\
& \text{s.t. } \mathbb{E}\left[d_o\left(X, \hat{X}\right)\right] \leq D_o + \varepsilon, \\
& \mathbb{E}\left[\hat{d}_s\left(X, \hat{S}\right)\right] \leq D_s + \varepsilon,
\end{aligned} \tag{4.5}
$$

where $I(\cdot)$ is the mutual information function, $\hat{d}_s(x, \hat{s}) = \mathbb{E}\left[\hat{d}_s(S, \hat{s})\,|x\right]$ and $S, X, \hat{S}, \hat{X}$ constitute a Markov chain $S \longleftrightarrow X \longleftrightarrow \left(\hat{S}, \hat{X}\right)$. Apart from standard rate distortion proof, the key to obtain this rate-distortion function is to convert the original semantic distortion between S and $\hat{S}$ into a new distortion between X and $\hat{S}$ and treat this problem as an indirect rate distortion problem. The authors then apply this theorem to the case with a Gaussian observation with linear state–observation relationship under quadratic distortions. In the meanwhile, given $p(x)$ and $p(z|x)$, authors in [14] find optimal parametric solution by using the Lagrange duality theorem and proposed a generalized Blahut–Arimoto algorithm for finite-alphabet scenarios. Authors in [15] formulate two different fundamental problems, namely, language utilization and language design for semantic communication. The language utilization is to minimize the misinterpretations on the transmitter side (semantic encoding) or the receiver side (semantic decoding) given a agreed semantic and technical language. The language design problem is to such efficient semantic and technical language before the transmission allowing transmitter and receiver to negotiate. We briefly summarize their idea due to the limit of pages here. The authors characterize a semantic language by four ingredients: word, syntax, expression, and interpretation. Words are the small elements of a message, and syntax is a set of rules on arranging words; messages are sets of words and syntax conveying meanings; the expression is a mapping from the set of meanings to the set of messages, while the interpretation is an inverse mapping from the set of messages to the set of meanings. Based on these definitions, the authors then introduce semantic distortions and costs for semantic encoding/decoding, respectively. Then, the authors characterize the distortion-cost region and define the distortion-cost function for encoding, decoding, and the combined encoding decoding problem, respectively.

Finally, a concrete example of a grid world is explored to demonstrate proposed concepts and methods. Other approaches such as information bottleneck have been adapted to semantic communications [16, 17].

4.2.2 Semantic-Native Communication

Semantic communication transfers the meaningful content or significance conveyed by a message, signal, or piece of data. Such information carries meaning or is related to the content and understanding of a particular concept or context. In Shannon's information theory, information is a quantity measure reflecting the receiver's capability of distinguishing a sequence of symbols (messages) from another. In this setup, the amount of information is defined as the number of combinatorial choices that can be used to transmit the message. For example, the information of observing an event x can be defined as $I(x) = -\log_2[p(x)]$, by converting x into its binary representation. The complexity being the number of binaries needed to determine the content of the message. In semantic communication, the transmitter and receiver agree on a "common knowledge" that can be used to convey the meanings. For example, semantic information in linguistics can be referred as the relationship between words and symbols as well as the concepts they represent. The concepts and relations constitute the knowledge, which should not be changed by its symbolic representation, i.e., different languages or syntax. In this context, semantic information has a much better tolerance to errors with a much smaller amount of bits than Shannon information. This makes semantic communication have the potential to deliver better communication efficiency and reliability.

The capacity of a semantic channel varies with the amount of information that can be jointly decoded between the transmitter and receiver. There are two types of semantic information: source semantics and channel semantics. Source semantics relates to the structure of content (i.e., language, image, sound) used by computing devices to accomplish tasks, such as generating more content, extracting features, making decisions, identifying objects, etc. In this context, source semantics reside both in the channel and the device. Therefore, the measure of source semantics includes the capacity of communication and computation. The communication capacity is determined by the physical channel capacity, including the bit transmission rate and bit error rate. The computation capacity is determined by the computing FLOPs, memory size, semantic coding redundancy, and ambiguity. Specifically, the later one relies on the common or emerging knowledge between the transmitter and receiver. Common knowledge is the invariant part of semantics, which should remain stable in a long run for different scenarios and tasks. The emergent knowledge is the equivariant part of semantics, which is supposed to operate on the invariants to produce the results for a specific

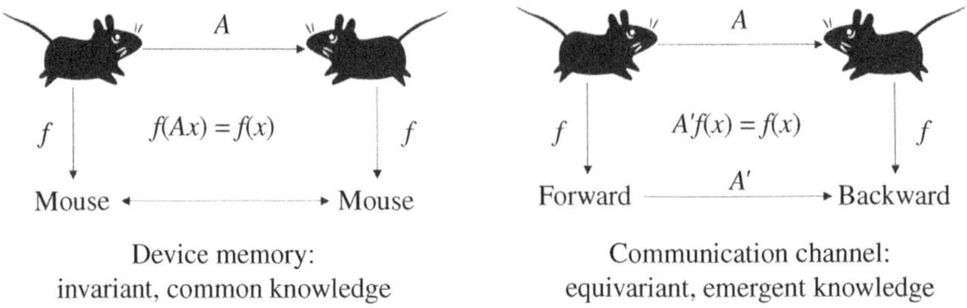

Figure 4.4 Example of common and emergent knowledge for a single entity.

task. An example of a single entity can be illustrated in Figure 4.4. An image of a mouse mapped on the concept "mouse" is the invariant information memorized by devices, and the concept of orientation is the equivariant information exchanged over the channel. In doing so, the communication cost can be reduced compared to sending the raw image each time. The other example for multiple entities is as follows. The concept "mouse, fire, ring" is invariant to different users, whereas the relation "jump, surround" is the equivariant that joins the concepts to produce the raw sentence "A mouse jumps through a ring surrounded by fire."

The channel semantics, on the contrary, describes the environment information when source semantics is sent through the wireless channel. Convention communication system utilizes the channel state information (CSI) to encode the wireless propagation information to optimize the transceivers, such as modulation, coding, equalization, etc. In semantic communication, the semantic encoder and decoder can be independent of the transceivers that see source semantics as raw bits. However, such a mechanism ignores the channel semantics in the transmission scheme and thus is less efficient in the changes of wireless channels. An effective approach is to encode the channel information on the same latent space as the source semantics. In doing so, a semantic encoder can fuse the source and channel semantics and use machine learning to learn the best coding scheme. Encoding channel semantics on a latent space is challenging since it has to characterize rich and irregular channel information. For example, signal and interference power should be embedded with the geographical locations, frequencies, antennas, etc. There are several works with deep learning to encode channel information on a latent embedding space, such as the use of autoencoders, or generative adversarial networks (GANs). However, such methods are trained on the raw data space, which is difficult to generalize in different scenarios. Channel charting has been proposed to learn the geometrical propagation information while extracting the radio features [18]. In the channel charter, the CSI with UE position is embedded into a high-dimensional manifold, such that the logical location of users with regards to channel propagation is learned, that can be used for radio access network (RAN) optimization. On the contrary, joint source and channel semantic coding can be performed on the same manifold.

There are several methods for modeling source and channel semantics. Knowledge graph (KG) is a promising technique that has been used in natural language processing (NLP) to encode language knowledge [19]. In KG, entities (vertices), the relation between objects (edges), or abstracted concepts (subgraphs) are usually described in natural language, making it straightforward to be retrieved or condensed by a semantic generator, i.e., LLM. The abstracted concepts and relations can represent methodologies or mechanisms to execute a specific task, which is efficient for planning. On the one hand, KG allows the language model to effectively inject new knowledge or remove outdated knowledge alongside with the changes in tasks. Furthermore, KG can help the semantic generator to understand and generate more contextually relevant and coherent responses.

Topological space, including group, graph, simplicial complex, and cell complex, can represent the intrinsic structure of data. It can model the higher-order interactions between objects, entities, or concepts. Unlike traditional language or image processing that learns data structure by 1D sequence or 2D regular grid, topological space can exploit more flexible explicit and implicit structures. Encoding such an effective structure can significantly improve the generalization and computation efficiency compared to DNN. A simplicial complex is a candidate topological model representing information with higher-order geometry. As illustrated in Figure 4.5, a simplicial complex can be constructed from a bipartie graph that connects a set of objects U and weights V with edges E. The weights representing raw data are encoded by a cochain function $C(W)$ connecting k objects that constitutes a k-cochain. The k objects constitute a k-simplex carrying the weights of the k-cochain. A simplicial complex comprises a set of simplices with a topology representing the cofaces that can be formulated by a distance or Laplacian matrix. The high-order simplex is a generalized representation of the regular graph that investigates more hidden information inside the raw data. For example, in this simple case, the interaction between several groups of objects

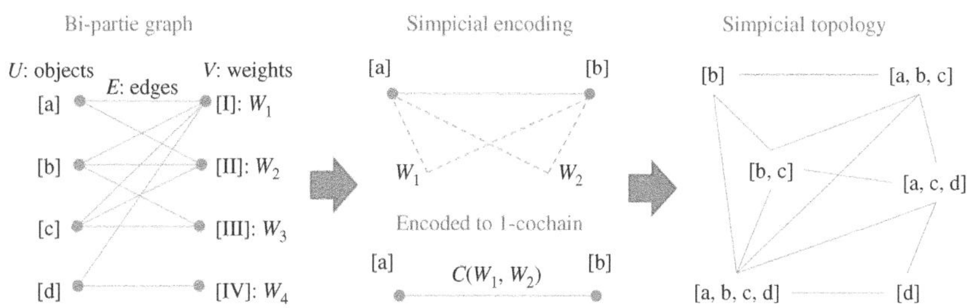

Figure 4.5 Example of semantics on simplicial complex.

can be learned from the topology among simplices. In [20], the simplicial neural network is shown to effectively learn a minimal abstraction from a simplicial complex and produce accurate query responses under high channel distortion that will be detailed in Section 4.3.1.

Semantic-native communication exploits the advantages of communicating semantics of entities enabled by machine learning. The semantic representation conceptualizes and symbolizes the entity of interest that can be decoded as the intended entity at the receiver. System 1 semantic communication recognizes patterns and correlations in raw data space. The semantic encoder and decoder utilize a stochastic model (i.e., deep neural network) to learn the structure of information in specific data domains (i.e., images, natural languages) or environments (i.e., control systems). The semantics of data are evaluated with goal-oriented metrics on the effectiveness of information at the receiver, such as similarity or accuracy. On the other hand, system 2 semantic communication finds the underlying causation of the perceived correlations while reasoning on semantic representations. Contextual reasoning allows the agent to deduce from the logic of the concept where only the effective semantic representation is sent over the channel. In the example case of Figure 4.4, system 1 semantic communication sends all the concepts of $f(x)$, whereas system 2 semantic communication sends A' in which the receiver can deduce $A'f(x)$. New emerging languages can emerge between agents that is specialized for its unique way of semantic coding induced by different tasks and contexts. The communication cost is thus largely reduced by sending the most efficient semantics. System 1 (recognizing patterns and relations) and system 2 (finding causation and reasoning) AI is the foundation of realizing system 1 and system 2 semantic communication, which will be discussed in Section 4.3.2.

4.3 Interplay of AI and Semantic Communication

In this chapter, we discuss the machine learning technologies in realizing semantic communications, from semantic coding to semantic protocols. Example applications such as natural language, image, and sound are given. Thereafter, we introduce a semantic-native network powering multiagent collective intelligence. The rise of generative AI and LLM has the potential to realize Shannon's level 3 effectiveness communication.

4.3.1 AI for Semantic Communication

Machine learning has been deeply studied in data science that is shown to successfully recognize hidden patterns and understand the correlations among different data modalities, to empower effective decision-making, inference, and generation for various applications [21]. In this context, machine learning and in particular deep learning is seen as a promising technology for semantic communications [22]. For example, the convolutional neural network (CNN) can enable the agents to extract the image patterns and embed them in compressed representation for communication [23]. The correlations among different images learned by CNN can help the agents to complete related tasks without transferring all the raw data. Similarly, long short-term memory (LSTM) and transformer neural networks achieved success in extracting semantic information from text messages [22]. It has been demonstrated to provide better bilingual evaluation understudy (BLEU) score than conventional source (i.e., Huffman) channel (i.e., Turbo) coding, especially at a low signal-to-noise ratio (SNR), which benefits from higher tolerance of distortion in semantic channel than physical channel. This has been extensively analyzed in [24].

In the application of semantic communications, a number of research in NLP has been investigated, with wide adoption of transformer neural networks. This includes an adaptive universal transformer proposed in [22, 25], and [26], which designs an autoencoder to encode text sentences on high-dimensional latent representation, with CSI used to adjust the attention weights according to the changing SNR. Similarly, the work of [27] incorporates hybrid automatic repeat request (HARQ) to transmit incremental bits according to reliability requirements. Since the attention layer weights the correlation between words, the embedding with higher weights can be applied with more redundancy in channel coding to guarantee the entire reliability. A further application of a nonlinear transformer on images is proposed in [28]; it maps source image in latent representation and learns the entropy model of latent distribution, such that a variable length source-channel encoder can be applied to reduce the source distortion. Weng et al. [29] studies the problem of speech recognition and synthesis with joint CNN and RNN modules, which recover the text and speech at the receiver. An application of transformer-based SC in massive MIMO is investigated in [30], adding over the dedicated transformers for digital and analog beamformers, showing improved semantic similarity over joint source-channel coding.

A transformer neural network is powerful to learn the latent distance of sequences with its attention mechanism but it also significantly increases the computational complexity. Xie and Qin [31] proposes network sparsification and quantization on model compression for deployment in power-constrained devices. Moreover, the knowledgebase at the transceivers can be distorted, making the receiver unable to decode semantic messages. Thomas and Saad [32] proposed a generative flow network to learn the probabilistic structure generating the observed data, with a measure of semantic distortion reflecting reasoning capability. It shows that the system can transmit at higher reliability with fewer transmitted bits. Some cross-disciplinary studies are carried out including quantum semantic communication [33], which proposes to learn semantic representation in Hilbert space suited for quantum computing and communications. Furthermore, Farshbafan et al. [34] studied a problem in the transmitter sends a hierarchical belief of environment observation to the receiver, which reconstructs the observation and takes an action on the environment, resulting in transition to new states in the next observation. It shows to improve task execution reliability and belief efficiency than traditional reinforcement learning.

Despite successful research of deep-learning-based semantic communication in many applications, very few studies communication with structured data modelling on topological space. For example, graph [35] and hyper-graph [36] neural networks have proven to be and effective tool taking into account the irregular correlations between data. Furthermore, the simplicial complex [37] and cell complex [38] neural networks are also demonstrated to learn higher-order interactions beyond pairwise elements. Such topological models have the potential to extract the hidden patterns in complex datasets. In this context, Zhao et al. [20] proposed a model for semantic communication living in topological spaces of data, which operates on non-Euclidean spaces. It shows significant improvement when encoding semantic information with higher-order simplicial complex to transmit at a low-SNR channel. Furthermore, Zhao et al. [39] studied the systemâs efficiency in transmitting minimal sufficient semantic information on simplicial complexes, with the effectiveness in generating high-accuracy answers to query at the receiver node. It is demonstrated that by memorizing a minimal simplicial complex on the devices, the total communication rounds and the size of transmitted data are largely reduced. This significantly improves the reliability of the semantic channel when the physical channel is poor, with reduced communication resources and latency.

A typical semantic communication system powered by machine learning is illustrated in Figure 4.6. Here a student agent on mobile devices query a teacher agent in the cloud, which holds the world model or world knowledge (i.e., realized by ChatGPT). The student agent first utilizes a language model to convey a text query into semantic concepts. For example, for a query "I want to know the collaboration of Alice and Bob in terms of their co-authored paper," the "co-authors Alice and Bob" can be a semantic concept since it typically needs to be retrieved from an external database. However, it is costly to train or fine-tune a model with the entire semantic scholar database and deploy it on a mobile device. Therefore, the student agent searches its local knowledge base and induce the minimal semantic information needed to ask the teacher agent to complete this query. The teacher agent extracts simplicial complexes from the raw database and

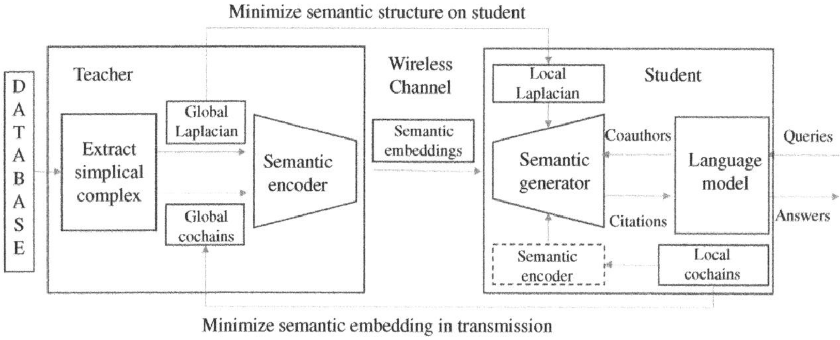

Figure 4.6 Example of machine-learning-powered semantic communication for remote query through simplicial neural network.

encode it to semantic embedding for transmission over the wireless channel. The student agent then decodes it and merges it with local knowledge to generate the response. During the query process, an online optimization is conducted jointly between the two agents. Namely, the teacher agent minimizes the semantics sent to the channel, and the student agent minimizes the semantics stored in the memory. Both minimizations are on condition of the performance of responses. With interactive query, a semantic language can emerge between the agents, which is the minimal simplicial complex to convey the needed information for a query.

4.3.2 Semantic-Native Collective Intelligence

The rise of LLM brings remarkable changes in AI. With large-scale parameters pretrained on massive unlabeled language corpus, LLM has witnessed strong capability of in-context learning, intrusion follow, multistep reasoning, few-shot learning, and generalization. The ChatGPT powered by generative pretrained transformers (GPT) with one trillion parameters (GPT-4) has shown impressive performance in content generation, language understanding, code generation, common sense reasoning, and so on [40]. The race of LLM is growing rapidly following the release of ChatGPT, which can be categorized in three aspects. (i) Larger scales of model size incorporating stronger capabilities with energy-efficient deployment and inference. For example, the open-source Falcon LLM [41] has outperformed GPT-3 with only 40B parameters, 75% training cost, and 20% inference compute budget. (ii) From the conventional natural language setting to the general scenario of multimodal inputs/outputs and corresponding applications. For example, DALL-E2 [42] excels in zero-shot text-to-image generation with a remarkable understanding of the meanings. (iii) From probabilistic model to planning and reasoning with autonomous AI agent. This gives rise to artificial general intelligence (AGI). For example, Auto-GPT [43] is an AI agent that attempts to achieve a goal specified in natural language. It chains together "thoughts" in an infinite loop of reasoning and planning the next action.

LLM is built on the GPT architecture. With the help of multihead attention, GPT could capture the long-range dependencies in parallel by having every word attending to all previous words or every word except the target. Compared to conventional deep-learning approaches, LLMs are first pretrained on vast amount of unlabeled texts for given pretraining tasks such as causal language modeling or denoising autoencoding. Then LLMs are fine-tuned or adapted to downstream tasks through different methods, such as task dataset fine-tuning, instruction tuning, and alignment tuning. The pretrained LLM is proved to be capable of quantifying the semantic importance of data frames, giving the opportunity for semantic communications [44]. Furthermore, LLM has the capability to compress text and code, as well as to recall and manipulate compressed representations of prompts. This shows LLM can be used as a semantic compressor and reconstructor to realize system 1 semantic communications.

A significant revolution of LLM is the capability of planning and reasoning, which is the foundation of system 2 semantic communication. Planning refers to the process of creating actionable tasks over time to achieve a specific goal. This includes breaking down high-level goal into subgoals and actions, as well as self-reflection to evaluate and criticize the past decisions and optimize the policy. On the contrary, reasoning is the process of generating causal and logical relations between different steps of thinking. Even though an LLM is a probabilistic generative model, pretraining on billions of language tokens, many recent works have shown that LLMs do have reasoning abilities via planning multiple steps of thoughts. In a multiagent wireless network, reasoning allows agents to communicate with minimal semantics, while planning enables agents to effectively accomplish the targeted task. In this context, LLM has the potential to achieve semantic and effectiveness level communication via reasoning and planning.

Connecting LLM-powered AI agents through semantic-native communication brings collective intelligence to the wireless network. Semantic communication transforms communication networks from data and model driven toward knowledge and reasoning driven. Specifically, LLM can retrieve from knowledgebase storing semantic representation to plan long-term thoughts, and teach a remote agent to perform fast and accurate reasoning. This helps multiple wireless agents to collaboratively solve tasks or find optimal individual decisions in an interactive environment. Furthermore, since LLM agent continuously collects and generates large amounts of data, semantic communication is particularly important to minimize the computing, communication, and storage costs. With strong semantic understanding and generation capabilities, LLM can largely improve communication efficiency as with a minimal prompt LLM can perform effectively on a wide range of tasks.

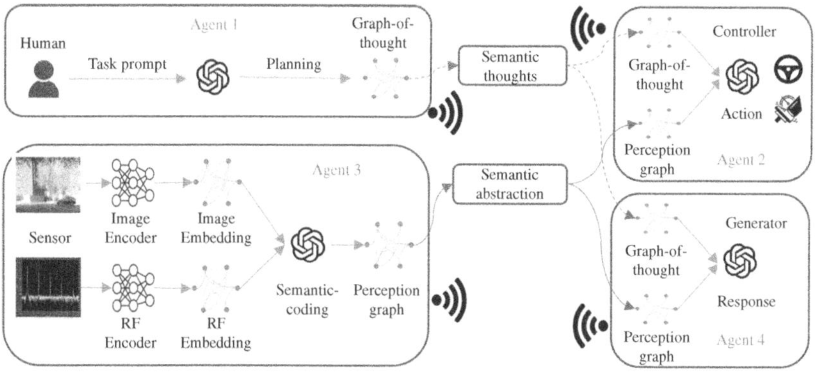

Figure 4.7 Example of semantic-native collective intelligence. Source: Ahmed et al. (2023)/IEEE.

An example paradigm of semantic-native collective intelligent network is shown in Figure 4.7. Consider 4 agents of human device (mobile phone), sensors (camera, radar), controller (vehicle, robot), and content generator (TV, XR) communicate wirelessly to jointly solve tasks. Agent 1 with a human prompt describing the task and goal, LLM plans actionable subtasks and condenses them into a graph-of-thought. In parallel, agent 3 holds a multimodal LLM pretrained on image and RF data, which encodes them onto a common latent space. LLM performs semantic coding to extract the minimal semantics on the graph. Agents 1 and 3 send the graph of thought and graph of perception to targeted agents 2 and 4. The LLM fuses the semantics of perception and planning to perform effective reasoning and decision-making, in controlling a robot-style device or generating content on the user's intent. With semantic communication, on-device LLM with efficient scale and cost can be implemented without the lose of performance through multiagent collaboration. Furthermore, LLM can merge the system 2 type AI and communication, bringing full autonomy to telecom networks and vertical applications.

4.4 Conclusion

In this chapter, we give a comprehensive review of semantic communications powered by machine learning and a forward-looking perspective of semantic-native collective intelligent network. We first review the semantic information theory with a consolidated analysis of the semantic channel capacity. Thereafter, we discuss in detail the source and channel semantics, with examples of semantic representation modeled on topological spaces. Subsequently, we explain the difference between system 1 and system 2 level semantic communications, from recognizing structure and relations to understanding causality and reasoning. In the Section 4.3, we discuss the use of AI in semantic communication, including semantic coding and semantic protocol. We introduce typical deep learning techniques such as autoencoder, GAN with CNN, LSTM, transformer neural network in realizing semantic encoder and decoder. Applications in computer vision, natural language understanding, and robotic control are given. Finally, we introduce our perspective of semantic communication powering wireless collective intelligence by connecting LLM-powered AI agents effectively. We foresee that LLM can realize the system 2 AI and semantic communication with effective planning and reasoning on wireless devices, bringing significant computing, communication, and storage efficiency.

Semantic communication has a number of challenges which gives opportunities for future research. First of all, there is a lack of a consensus on the definition of semantic communications. It is commonly referred to as joint source channel coding, goal-oriented communication and, emergent communication, etc. From this perspective, it is challenging to unify the system design and standardization of semantic communication in 6G. Second, the implementation of semantic communication in wireless systems is unclear. While from some perspective semantic communication can be realized in the application layer, the missing channel semantics can hardly achieve remarkable performance. Moreover, semantic information theory and semantic representation still require significant research effort to build a generalized foundation for the system design. Furthermore, despite the fact that AI and LLM show strong capability in a semantic communications, the solutions are still for specific scenario, data domain, task, or application. There is a lack of a foundation model that can be highly scalable and generalizable. Last but not least, semantic reasoning with minimal computation and communication costs is challenging. LLM achieves "baby AGI" with a large energy cost in training and data collection. Significant advances in AI to power AGI with a world model instead of a probabilistic language model are essential for semantic-native communications.

References

1 Shannon, C.E. (1948). A mathematical theory of communication. *The Bell System Technical Journal* 27 (3): 379–423.

2 Shannon, C.E. and Weaver, W. (1949). *The Mathematical Theory of Communication*. Urbana, IL: University of Illinois Press.

3 Zhang, C., Zou, H., Lasaulce, S. et al. (2022). Goal-oriented communications for the IoT and application to data compression. *IEEE Internet of Things Magazine* 5 (4): 58–63.

4 Zou, H., Zhang, C., Lasaulce, S. et al. (2023). Goal-oriented quantization: analysis, design, and application to resource allocation. *IEEE Journal on Selected Areas in Communications* 41 (1): 42–54.

5 Strinati, E.C. and Barbarossa, S. (2021). 6G networks: beyond Shannon towards semantic and goal-oriented communications. *Computer Networks* 190: 107930.

6 Xu, J., Tung, T.-Y., Ai, B. et al. (2023). Deep joint source-channel coding for semantic communications. *IEEE Communications Magazine* 61 (11): 42–48.

7 Seo, H., Park, J., Bennis, M., and Debbah, M. (2023). Semantics-native communication via contextual reasoning. *IEEE Transactions on Cognitive Communications and Networking* 9 (3): 604–617.

8 Seo, S., Park, J., Ko, S.-W. et al. (2023). Toward semantic communication protocols: a probabilistic logic perspective. *IEEE Journal on Selected Areas in Communications* 41 (8): 2670–2686.

9 Chaccour, C., Saad, W., Debbah, M. et al. (2024). Less data, more knowledge: building next generation semantic communication networks. *IEEE Communications Surveys & Tutorials* https://doi.org/10.1109/COMST.2024.3412852.

10 Bar-Hillel, Y. and Carnap, R. (1952). An Outline of a Theory of Semantic Information. *Technical Report 247*. Cambridge, MA: Research laboratory of electronics. Massachusetts Institute of Technology.

11 Venhuizen, N.J., Crocker, M.W., and Brouwer, H. (2019). Semantic entropy in language comprehension. *Entropy* 21 (12): 1159.

12 Melamed, I.D. (1997). Measuring semantic entropy. In: *Tagging Text with Lexical Semantics: Why, What, and How?*

13 Liu, J., Shao, S., Zhang, W., and Poor, H.V. (2022). An indirect rate-distortion characterization for semantic sources: general model and the case of Gaussian observation. *IEEE Transactions on Communications* 70 (9): 5946–5959.

14 Stavrou, P.A. and Kountouris, M. (2023). The role of fidelity in goal-oriented semantic communication: a rate distortion approach. *IEEE Transactions on Communications* 71 (7): 3918–3931.

15 Shao, Y., Cao, Q., and Gunduz, D. (2022). A theory of semantic communication. *arxiv*. https://arxiv.org/pdf/2212.01485.pdf.

16 Zhu, M., Feng, C., Guo, C. et al. (2022). Semantics-aware remote estimation via information bottleneck-inspired type based multiple access. *arxiv e-prints*. https://arxiv.org/pdf/2212.09337.pdf.

17 Gunduz, D., Qin, Z., Aguerri, I.E. et al. (2022). Beyond transmitting bits: context, semantics, and task-oriented communications. *arxiv*. https://arxiv.org/pdf/2207.09353.pdf.

18 Studer, C., Medjkouh, S., Gonultaş, E. et al. (2018). Channel charting: locating users within the radio environment using channel state information. *IEEE Access* 6: 47682–47698.

19 Xiao, Y., Li, Y., Shi, G., and Poor, H.V. (2022). Reasoning on the air: an implicit semantic communication architecture. *2022 IEEE International Conference on Communications Workshops (ICC Workshops)*, 289–294. IEEE.

20 Zhao, Q., Bennis, M., Debbah, M., and Da Costa, D.B. (2022). Semantic-native communication: a simplicial complex perspective. *2022 IEEE Globecom Workshops (GC Wkshps)*, 1513–1518. IEEE.

21 Belfiore, J.-C. and Bennequin, D. (2021). Topos and stacks of deep neural networks. *arXiv preprint arXiv:2106.14587*.

22 Xie, H., Qin, Z., Li, G.Y., and Juang, B.-H. (2021). Deep learning enabled semantic communication systems. *IEEE Transactions on Signal Processing* 69: 2663–2675.

23 Bourtsoulatze, E., Kurka, D.B., and Gündüz, D. (2019). Deep joint source-channel coding for wireless image transmission. *ICASSP 2019 - 2019 IEEE International Conference on Acoustics, Speech and Signal Processing (ICASSP)*, 4774–4778.

24 Seo, H., Park, J., Bennis, M., and Debbah, M. (2021). Semantics-native communication with contextual reasoning. *arXiv preprint arXiv:2108.05681*.

25 Zhou, Q., Li, R., Zhao, Z. et al. (2021). Semantic communication with adaptive universal transformer. *IEEE Wireless Communications Letters* 11 (3): 453–457.

26 Lu, K., Zhou, Q., Li, R. et al. (2022). Rethinking modern communication from semantic coding to semantic communication. *IEEE Wireless Communications* 30 (1): 158–164.

27 Jiang, P., Wen, C.-K., Jin, S., and Li, G.Y. (2022). Deep source-channel coding for sentence semantic transmission with HARQ. *IEEE Transactions on Communications* 70 (8): 5225–5240.

28 Dai, J., Wang, S., Tan, K. et al. (2022). Nonlinear transform source-channel coding for semantic communications. *IEEE Journal on Selected Areas in Communications* 40 (8): 2300–2316.

29 Weng, Z., Qin, Z., Tao, X. et al. (2023). Deep learning enabled semantic communications with speech recognition and synthesis. *IEEE Transactions on Wireless Communications* 22 (9): 6227–6240.

30 Wang, Y., Gao, Z., Zheng, D. et al. (2022). Transformer-empowered 6G intelligent networks: from massive MIMO processing to semantic communication. *IEEE Wireless Communications* 30 (6): 127–135.

31 Xie, H. and Qin, Z. (2021). A lite distributed semantic communication system for internet of things. *IEEE Journal on Selected Areas in Communications* 39 (1): 142–153.

32 Thomas, C.K. and Saad, W. (2022). Neuro-symbolic artificial intelligence (AI) for intent based semantic communication. *IEEE Global Communications Conference*, Rio de Janeiro, Brazil.

33 Chehimi, M., Chaccour, C., and Saad, W. (2022). Quantum semantic communications: an unexplored avenue for contextual networking. *arXiv preprint arXiv:2205.02422.*

34 Farshbafan, M.K., Saad, W., and Debbah, M. (2023). Curriculum learning for goal-oriented semantic communications with a common language. *IEEE Transactions on Communications* 71 (3): 1430–1446.

35 Bronstein, M.M., Bruna, J., LeCun, Y. et al. (2017). Geometric deep learning: going beyond Euclidean data. *IEEE Signal Processing Magazine* 34: 18–42.

36 Feng, Y., You, H., Zhang, Z. et al. (2019). Hypergraph neural networks. *Proceedings of the AAAI Conference on Artificial Intelligence* 33 (01): 3558–3565.

37 Ebli, S., Defferrard, M., and Spreemann, G. (2020). Simplicial neural networks. *arXiv, vol. abs/2010.03633.*

38 Hajij, M., Istvan, K., and Zamzmi, G. (2020). Cell complex neural networks. *arXiv preprint arXiv:2010.00743.*

39 Zhao, Q., Zou, H., Bennis, M. et al. (2023). Joint semantic-native communication and inference via minimal simplicial structures. *GLOBECOM 2023 - 2023 IEEE Global Communications Conference*, 2233–2238. IEEE.

40 Yenduri, G., Ramalingam, M., Selvi, G.C. et al. (2023). GPT (generative pre-trained transformer)—a comprehensive review on enabling technologies, potential applications, emerging challenges, and future directions. *IEEE Access* 12: 54608–54649.

41 Falcon LLM. https://falconllm.tii.ae (accessed 1 October 2024).

42 Ramesh, A., Dhariwal, P., Nichol, A. et al. (2022). Hierarchical text-conditional image generation with clip latents. *arXiv preprint arXiv:2204.06125.*

43 GitHub, Inc. Auto-GPT: An autonomous GPT-4 experiment. https://github.com/Significant-Gravitas/Auto-GPT (accessed 1 October 2024).

44 Guo, S., Wang, Y., Li, S., and Saeed, N. (2023). Semantic importance-aware communications using pre-trained language models. *IEEE Communications Letters* 27 (9): 2328–2332.

5

Federated Learning for Wireless Communications

Ahmet M. Elbir[1] *and Wei Shi*[2]

[1] *Department of Electrical and Electronics Engineering, Istinye University, Istanbul, Turkey*
[2] *School of Information Technology, Carleton University, Ottawa, Ontario, Canada*

CHAPTER MENU

5.1 Introduction

Compared with cellular communication systems operating in lower frequency bands, millimeter wave (mmWave) (30–300 GHz) and terahertz (THz) (0.1–10 THz) signals face a more intricate propagation environment characterized by increased scattering, heightened penetration losses, and amplified path loss for fixed transmitter and receiver gains [1–3]. To counterbalance these losses, beamforming power gain is provided through a substantial number of antennas at both the transmitter and receiver, employing a multiple-input-multiple-output (MIMO) architecture. However, the implementation of such a large antenna array demands a dedicated radio frequency (RF) chain for each antenna, resulting in a costly system architecture and elevated power consumption. To mitigate these challenges and minimize the number of digital RF components, hybrid analog and baseband beamforming architectures have been introduced. These architectures utilize a small number of phase-only analog beamformers in combination with low-dimensional baseband beamformers, effectively reducing the count of RF chains while preserving sufficient beamforming gain [4, 5].

Artificial Intelligence for Future Networks, First Edition.
Edited by Mohammad A. Matin, Sotirios K. Goudos, and George K. Karagiannidis.

Despite the fewer RF chains, the combination of hybrid beamforming and mmWave transmission incurs high energy consumption and hardware complexity, as highlighted in [6]. To tackle these challenges and optimize wireless network performance sustainably, reconfigurable intelligent surfaces (RISs), also known as intelligent reflecting surfaces, emerge as a promising solution. They offer cost-effective and simpler alternatives, as discussed in [6–10]. An RIS represents a 2D electromagnetic surface comprising numerous passive reconfigurable meta-material elements. These elements reflect incoming signals by applying predetermined phase shifts, which the base station (BS) controls using external signals via a backhaul control link. This real-time manipulation redirects the BS-transmitted signal toward users, thereby enhancing signal energy for remote users and expanding the BS coverage area. Ultimately, the deployment of RIS optimizes signal reception and extends coverage for improved network performance.

In both conventional massive MIMO and RIS-assisted scenarios, the effectiveness of the system architecture heavily hinges on the precision of the instantaneous channel state information (CSI), especially given the highly dynamic nature of the mmWave channel, as noted in [11, 12]. Thus, channel estimation accuracy significantly influences the design of analog and digital beamformers in conventional massive MIMO [13, 14] and the configuration of phase shifts for reflecting beamformers in RIS-assisted scenarios [9, 15].

Moreover, in RIS-assisted massive MIMO setups, signal reception occurs across multiple channels (e.g., BS-RIS, RIS-user, and BS-user), amplifying the complexity of the channel estimation task. Consequently, various channel estimation schemes have been proposed, leveraging compressed sensing [15], angle-domain processing [16], and coordinated pilot-assignment techniques [17]. The efficacy of these methods relies on the quality of the antenna array output to attain reliable channel estimation accuracy.

To fortify these approaches against imperfections or corruptions in an array data, researchers have turned to data-driven techniques like machine learning (ML)-based approaches. These methods aim to unveil nonlinear relationships within data or signals, offering lower computational complexity while achieving improved performance for parameter inference. They also exhibit greater resilience against data imperfections, contributing to enhanced robustness in channel estimation processes. As listed below, ML is more efficient than model-based techniques that largely rely on mathematical models [18]:

1) A learning model establishes a nonlinear relationship between the initial input data and the expected output, providing a robust performance against distortions or shortcomings in wireless channel data.

2) Machine learning (ML) discerns characteristic patterns, swiftly adjusting to new data and environmental shifts. Over time, this leads to decreased computational complexity compared to model-centered optimization.
3) ML-driven solutions boast notably reduced processing times owing to their capacity for parallel processing. Conversely, conventional optimization and signal processing algorithms often encounter challenges in achieving parallel implementations.

In massive MIMO- and RIS-assisted systems, ML has been proven to have higher spectral efficiency and lower computational complexity for problems such as channel estimation [19–21], hybrid beamforming [22–24] and angle-of-arrival (AoA) estimation [25, 26].

In the realm of ML, tackling the channel estimation problem involves training models, such as neural networks (NNs), using local datasets gathered by individual users, as referenced in [19–21]. These trained models establish nonlinear mappings between input data (often received pilot signals) and output data (representing channel information). Previously, centralized learning (CL) schemes dominated, where entire datasets (input–output pairs) were transmitted to the BS (via RIS in RIS-assisted scenarios) for model training, depicted in Figure 5.1a. After training at the BS, model parameters were sent back to users to perform channel estimation by inputting received pilot data into the model. However, this approach incurred substantial communication overhead due to transmitting large datasets from users to the BS. For instance, in LTE, a single frame with 5 MHz bandwidth and 10 ms duration could only accommodate 6000 complex symbols, whereas the dataset size could reach hundreds of thousands of symbols [19, 20, 23, 24, 27]. Consequently, CL-based techniques demanded extensive bandwidth.

To address the high communication overhead of CL schemes, recent advancements introduced federated learning (FL) schemes [28, 29]. FL operates by

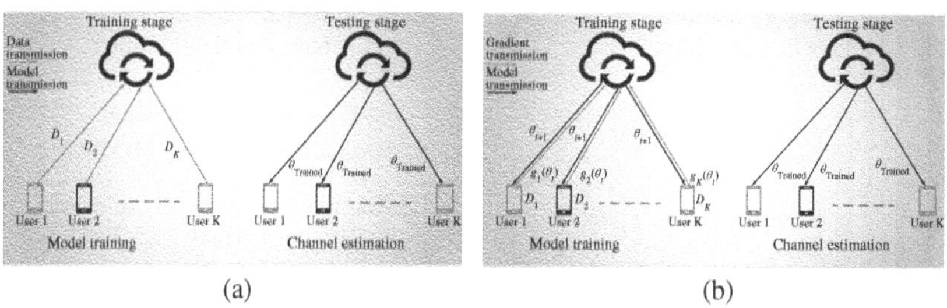

Figure 5.1 The training and testing phases for (a) centralized learning (CL) and (b) federated learning (FL) differ in data transmission. During training, CL involves users sending datasets $\mathcal{D}_{k\in\mathcal{K}}$ to the server, while in FL, users transmit only their model updates $\mathbf{g}_{k\in\mathcal{K}}(\theta t)$. In the testing phase of both CL and FL, the server broadcasts the trained learning models θ_{Trained} to the users.

transmitting only model updates, i.e., gradients of model parameters, instead of the entire dataset, as shown in Figure 5.1b, thereby reducing communication overhead. FL has been applied across various domains, including wireless sensor network scheduling, UAV trajectory planning, task fetching/offloading in vehicular networks, image classification, mmWave and THz channel estimation, and massive MIMO hybrid beamforming design [27, 29–40]. These studies involve multiple edge devices exchanging model updates with a parameter server to train a global model. FL, compared to CL, is more suitable for distributed devices like mobile phones. In addition, employing FL for training significantly reduces communication overhead while maintaining channel estimation performance close to CL.

In this chapter, we explore FL-driven techniques within the physical layer of wireless communication (see Figure 5.2 for the outline). We cover applications like hybrid beamforming and channel estimation across mmWave and THz conventional MIMO systems, along with scenarios involving RIS assistance. Each of these poses specific challenges for FL implementation, such as managing data/model size and handling different input–output data types.

Throughout the chapter, we start by outlining mmWave and THz channel models. Following that, we delve deeply into FL-based channel estimation, discussing data collection and tailored model training for both conventional

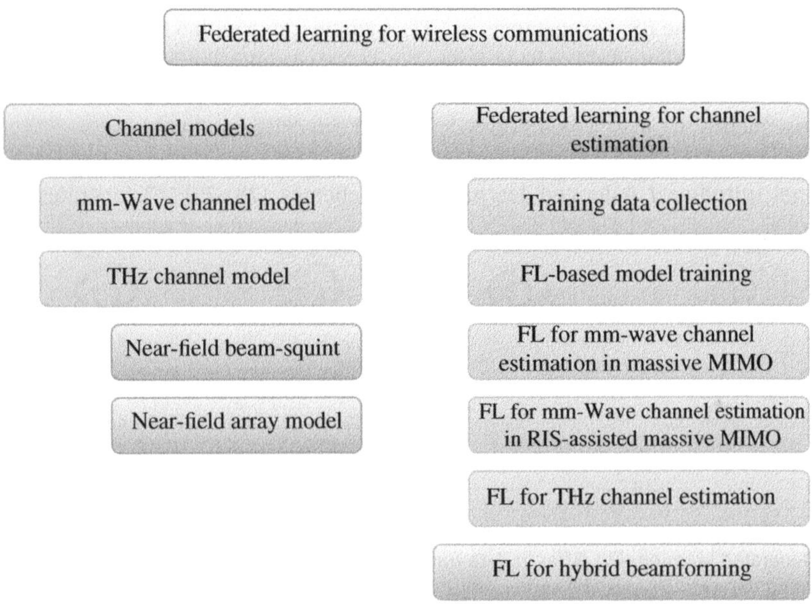

Figure 5.2 The outline of the chapter.

massive MIMO and RIS-assisted systems. In addition, we examine FL-based THz channel estimation, utilizing the spherical wave near-field model.

Subsequently, the chapter introduces FL-based hybrid beamforming techniques and explores their relevance in these contexts. Performance evaluations for each scenario are presented to demonstrate the effectiveness of employing FL methods. Finally, the chapter concludes by summarizing key points and offering insights into potential future research directions.

5.2 Channel Models

In this section, we first introduce the wireless channel model for both mmWave and THz transmission.

5.2.1 mmWave Channel Model

Let us consider a multiuser MIMO–OFDM (orthogonal frequency division multiplexing) system comprising M subcarriers. At the base station (BS), there are N_{BS} antennas for communication with K users, each equipped with N_{MS} antennas. In the downlink, the BS initially applies subcarrier-specific baseband precoders $\mathbf{F}_{BB}[m]$ to precoded symbols $\mathbf{s}[m] = [s_1[m], s_2[m], \dots, s_K[m]]^T \in \mathbb{C}^K$ at each subcarrier ($m \subset \mathcal{M} - 1, \dots, M$). These baseband precoders are $K \times K$ matrices

The precoded signal is then transformed into the time domain through an M-point inverse (IDFT). Subsequently, a cyclic prefix (CP) is added, following which the BS utilizes a subcarrier-independent analog precoder $\mathbf{F}_{RF} \in \mathbb{C}^{N_{BS} \times K}$ to compose the transmitted signal. The analog RF precoder, comprised of analog phase shifters, adheres to constant unit-modulus constraints, meaning that $|[\mathbf{F}_{RF}]i,j|^2 = 1$.

Furthermore, there exists a power constraint given by

$$\sum_{m=1}^{M} \|\mathbf{F}_{RF}\mathbf{F}_{BB}[m]\|_{\mathcal{F}}^2 = MK, \tag{5.1}$$

which is maintained by the normalization of the baseband precoder $\mathbf{F}_{BB}[m]$, $m \in \mathcal{M}$. As a result, the transmitted signal can be represented as

$$\mathbf{x}[m] = \mathbf{F}_{RF} \sum_{k=1}^{K} \mathbf{f}_{BB_k}[m] s_k[m]. \tag{5.2}$$

Before reaching the users, the transmitted signal undergoes transmission through the mmWave channel, which can be depicted using a geometric model emphasizing limited scattering [13]. Let us define $\mathbf{H}_k[m]$ as the $N_{MS} \times N_{BS}$

mmWave channel matrix linking the BS and the kth user. This matrix encapsulates contributions from L paths, each characterized by a time delay $\tau_{k,l}$, a relative angle of arrival (AoA) $\overline{\phi}_{k,l} \in \Theta$ (where $\Theta = [-\frac{\pi}{2}, \frac{\pi}{2}]$), an angle of departure (AoD) $\phi_{k,l} \in \Theta$, and the complex path gain $\alpha_{k,l}$ corresponding to the kth user and lth path. Furthermore, let $p(\tau)$ denote a pulse shaping function designed for T_s-spaced signaling, evaluated at τ seconds. Then, the mmWave delay-d MIMO channel matrix in time domain is given by Elbir et al. [5]

$$\overline{\mathbf{H}}_k[d] = \sqrt{\frac{N_{\text{BS}} N_{\text{MS}}}{L}} \sum_{l=1}^{L} \alpha_{k,l} p(dT_s - \tau_{k,l}) \mathbf{a}_{\text{MS}}(\overline{\phi}_{k,l}) \mathbf{a}_{\text{BS}}^{\text{H}}(\phi_{k,l}), \tag{5.3}$$

where $\mathbf{a}_{\text{MS}}(\overline{\phi}_{k,l})$ and $\mathbf{a}_{\text{BS}}(\phi_{k,l})$ are the $N_{\text{MS}} \times 1$, and $N_{\text{BS}} \times 1$ steering vectors representing the array responses of the antenna arrays at the users and the BS, respectively.

In mmWave systems, where the operating frequency significantly surpasses the bandwidth and subcarrier frequencies are closely spaced ($f_{m_1} \approx f_{m_2}$ for $m_1, m_2 \in \mathcal{M}$), a single effective operating wavelength $\lambda = \lambda_1 = \cdots = \lambda_M = \frac{c_0}{f_c}$ is employed. Here, c_0 denotes the speed of light, and f_c represents the central carrier frequency [13, 14]. This approximation allows the use of a single frequency-independent analog beamformer for each subcarrier. For a uniform linear array (ULA), the antenna array response at the BS, denoted as $\mathbf{a}_{\text{BS}}(\phi)$, can be expressed as follows

$$\mathbf{a}_{\text{BS}}(\phi) = \left[1, e^{j\frac{2\pi}{\lambda} d_{\text{BS}} \sin(\phi)}, \dots, e^{j\frac{2\pi}{\lambda}(N_{\text{BS}}-1) d_{\text{BS}} \sin(\phi)}\right]^{\text{T}}, \tag{5.4}$$

where $d_{\text{BS}} = \lambda/2$ represents the antenna spacing. Similarly, the nth element of $\mathbf{a}_{\text{MS}}(\overline{\phi})$ can be defined as $\left[\mathbf{a}_{\text{MS}}(\overline{\phi})\right]_n = e^{j\pi(n-1)\sin(\overline{\phi})}$, $n = 1, \dots, N_{\text{MS}}$. Upon performing an M-point discrete Fourier transform (DFT) of the delay-d channel model depicted in (5.3), the channel matrix for the kth user at subcarrier m, denoted as $\mathbf{H}_k[m]$, can be formulated as

$$\mathbf{H}_k[m] = \sum_{d=0}^{D-1} \overline{\mathbf{H}}_k[d] e^{-j\frac{2\pi m}{M} d}, \tag{5.5}$$

where $D \leq M$ signifies the cyclic prefix (CP) length. This frequency domain channel representation in (5.5) is utilized in multiple-input multiple-output orthogonal frequency division multiplexing (MIMO–OFDM) systems, ensuring orthogonality among each subcarrier such that $||\mathbf{H}_k^{\text{H}}[m_1]\mathbf{H}_k[m_2]||\mathcal{F}^2 = 0$ for $m_1, m_2 \in \mathcal{M}$, and $m_1 \neq m_2$.

With the aforementioned block-fading channel model [13], the received signal at the kth user before analog processing at subcarrier m is $\tilde{\mathbf{y}}_k[m] = \sqrt{\rho}\mathbf{H}_k[m]\mathbf{x}[m]$, i.e.,

$$\tilde{\mathbf{y}}_k[m] = \sqrt{\rho}\mathbf{H}_k[m]\mathbf{F}_{\text{RF}}\mathbf{F}_{\text{BB}}[m]\mathbf{s}[m] + \mathbf{n}[m]. \tag{5.6}$$

In this setup, an average received power denoted by ρ, where $\mathbf{n}[m] \sim \mathcal{CN}(0, \sigma^2 \mathbf{I}_{N_{\text{MS}}})$ represents an additive white Gaussian noise (AWGN) vector. At the kth user, the received signal is initially handled by the analog combiner $\mathbf{w}_{\text{RF},k} \in \mathbb{C}^{N_{\text{MS}}}$. Following this, the processed signal has its cyclic prefix removed and undergoes M-point discrete Fourier transforms (DFTs) to be converted into the frequency domain. Then, the received baseband signal for mmWave becomes

$$\bar{y}_k[m] = \sqrt{\rho}\mathbf{w}_{\text{RF},k}^{\text{H}}\mathbf{H}_k[m]\mathbf{F}_{\text{RF}}\mathbf{F}_{\text{BB}}[m]\mathbf{s}[m] + \mathbf{w}_{\text{RF},k}^{\text{H}}\mathbf{n}[m], \tag{5.7}$$

where the analog combiner $\mathbf{w}_{\text{RF},k}$ has the constraint $\left[\mathbf{w}_{\text{RF},k}\mathbf{w}_{\text{RF},k}^{\text{H}}\right]_{i,i} = 1$, similar to the RF precoder. Once the received symbols, denoted as $y_k[m]$, are acquired at the kth user, they undergo demodulation based on their respective modulation schemes. This process allows for the recovery of information bits for each subcarrier. To accurately retrieve the data streams $\mathbf{s}[m]$ as in (5.7), it is essential to estimate the channel matrix $\mathbf{H}_k[m]$. Typically, this estimation involves using pilot signals during the preamble stage, as detailed in previous works [19, 41]. In this stage, beamformers such as $\mathbf{F}_{\text{RF}}$, $\mathbf{F}_{\text{BB}}$, and $\mathbf{w}_{\text{RF},k}$ are designed accordingly (Refer to Section 5.3.3).

5.2.2 THz Channel Model

Consider a wideband THz MIMO–OFDM setup employing hybrid analog/digital beamforming across M subcarriers, as discussed in previous studies [36, 37, 42, 43]. In this scenario, we assume the base station (BS) consists of N antennas and N_{RF} RF chains to cater to K users who are equipped with single antennas. The data symbols, denoted as $\mathbf{s}[m] = [s_1[m], \cdots, s_K[m]]^{\text{T}}$, where m ranges within $M = 1, \cdots, M$, undergo processing via a $K \times K$ subcarrier-specific baseband beamformer $\mathbf{F}_{\text{BB}}[m] = [\mathbf{f}_{\text{BB}_1}[m], \cdots, \mathbf{f}_{\text{BB}_K}[m]]$. To direct the generated beams toward users in the downlink, an $N \times N_{\text{RF}}$ subcarrier-independent analog beamformer $\mathbf{F}_{\text{RF}}$ is employed ($N_{\text{RF}} = K < N$). Given that these analog beamformers are constructed using phase shifters, they adhere to a constant-modulus constraint, specifically $|[\mathbf{F}_{\text{RF}}]_{i,j}| = \frac{1}{\sqrt{N}}$. Then, the $N \times 1$ transmitted signal becomes

$$\tilde{\mathbf{x}}[m] = \mathbf{F}_{\text{RF}}\mathbf{F}_{\text{BB}}[m]\mathbf{s}[m]. \tag{5.8}$$

The kth user receives the broadcast signal at the mth subcarrier after it has traveled via the wireless channel as

$$y_k[m] = \mathbf{h}_k^{\text{T}}[m]\tilde{\mathbf{x}}[m] + w_k[m] = \mathbf{h}_k^{\text{T}}[m]\sum_{i=1}^{K}\mathbf{F}_{\text{RF}}\mathbf{f}_{\text{BB}_i}[m]s_i[m] + w_k[m], \tag{5.9}$$

where $w_k[m] \in \mathbb{C}$ represents the complex AWGN with variance of σ_n^2, i.e., $w_k[m] \sim \mathcal{CN}(0, \sigma_n^2)$.

The THz channel is typically constructed by combining a primary line-of-sight (LoS) path with a few supporting non-line-of-sight (NLoS) paths due to the limited presence of reflected path elements and minimal scattering in the THz spectrum [44–46]. Multipath channel models are also frequently used, particularly for indoor applications [44]. We thus examine a general case in which the $N \times 1$ channel matrix for the kth user at the mth subcarrier is made up of a mixture of L pathways, as previously shown in studies [44] as

$$\mathbf{h}_k[m] = \sqrt{\frac{N}{L}} \sum_{l=1}^{L} \alpha_{k,m,l} \mathbf{a}(\phi_{k,l}, r_{k,l}) e^{-j2\pi\tau_{k,l} f_m}, \tag{5.10}$$

where the time delay of the lth path that corresponds to the array origin is represented by $\tau_{k,l}$. The steering vector $\mathbf{a}(\phi_{k,l}, r_{k,l}) \in \mathbb{C}^N$ represents the physical direction $\phi_{k,l}$ and range. The complex path gain is represented by $r_{k,l}$. $\alpha_{k,m,l} \in \mathbb{C}$, and the expected value of its magnitude for the indoor THz multipath model is given by

$$\mathbb{E}\{|\alpha_{k,m,l}|^2\} = \left(\frac{c_0}{4\pi f_m r_{k,l}}\right)^2 e^{-k_{\text{abs}}(f_m) r_{k,l}}, \tag{5.11}$$

where the $k_{\text{abs}}(f_m)$ is the mth subcarrier frequency, c_0 is the speed of light, and $r_{k,l}$ is the distance between the kth user and the array origin. $k_{\text{abs}}(f_m)$ represents the medium absorption coefficient that is dependent on subcarriers [44, 47]. In addition, we get $f_m = f_c + \frac{B}{M}(m - 1 - \frac{M-1}{2})$, where B and f_c represent the bandwidth and carrier frequency, respectively.

5.2.2.1 Near-Field Array Model

Due to its high-frequency operations and utilization of extremely short wavelengths, THz-band transmission may experience near-field effects, especially for users in close proximity. Typically, the far-field model assumes the received signal at users resembles a plane wave. However, in the near-field, when the transmission range is shorter than the Fraunhofer distance $F = \frac{2D^2}{\lambda}$ (where D represents the array aperture and $\lambda = \frac{c_0}{f_c}$ is the wavelength) [48], the wavefront of the plane becomes spherical. Figure 5.3 illustrates the beam squint phenomenon for both far-field and near-field models. For a uniform linear array (ULA), the array aperture is $D = (N - 1)d$, with $d = \frac{\lambda}{2}$ as the element spacing. Consequently, in THz-band transmission scenarios, the proximity of the users results in distances smaller than F, requiring the application of a near-field signal model. For instance, at $f_c = 300$ GHz and $N = 256$, the Fraunhofer distance calculates to $F = 32.76$ m [49, 50].

We define the near-field steering vector $\mathbf{a}(\phi_{k,l}, r_{k,l}) \in \mathbb{C}^N$ corresponding to the physical direction-of-arrivals (DoAs) $\phi_{k,l}$ and range $r_{k,l}$ as

$$\mathbf{a}(\phi_{k,l}, r_{k,l}) = \frac{1}{\sqrt{N}} [e^{-j2\pi \frac{d}{\lambda} r_{k,l}^{(1)}}, \dots, e^{-j2\pi \frac{d}{\lambda} r_{k,l}^{(N)}}]^{\mathsf{T}}, \tag{5.12}$$

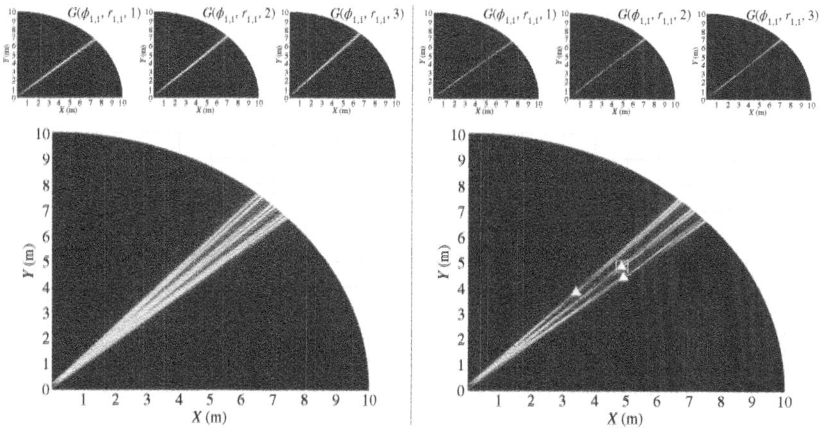

Figure 5.3 For a single user ($K = 1, L = 1$) situated in the far-field ($45°, 800$ m) (left) and the near-field ($45°, 6$ m) (right), respectively, the array gains $G(\phi_{1,1}, r_{1,1}, m)$ in Cartesian coordinates. In this case, $B = 30$ GHz, $f_c = 300$ GHz, and $M = 3$. The gain for each subcarrier is displayed in the top panel and is added together to provide a composite array gain for the near- and far-field scenarios at the bottom, which amply illustrates the beam squint. The triangles correspond to the spatial locations (where the maximum array gain is attained) at various subcarriers, while the square represents the user location. The near-field beam squint (NB) spans both range and angular domains, in contrast to the far-field beam squint, which is solely angular.

where $\phi_{k,l} = \sin \phi_{k,l}$ with $\phi_{k,l} \in [-\frac{\pi}{2}, \frac{\pi}{2}]$, and $r_{k,l}^{(n)}$ is the distance between the kth user and the nth antenna as

$$r_{k,l}^{(n)} = \left(r_{k,l}^2 + 2(n-1)^2 d^2 - 2r_{k,l}(n-1)d\phi_{k,l}\right)^{\frac{1}{2}}. \tag{5.13}$$

As a function of the antenna index n, $r_{k,l}^{(n)}$ in (5.13) contains complex terms that complicate the analysis of the spherical wave model. The Fresnel approximation can be used to simplify (5.13) [48, 51, 52]. In particular [48, 52], the Fresnel approximation allows modeling the received spherical wave as an approximated quadric surface. Thus, (5.13) becomes

$$r_{k,l}^{(n)} \approx r_{k,l} - (n-1)d\phi_{k,l} + (n-1)^2 d^2 \zeta_{k,l}, \tag{5.14}$$

where $\zeta_{k,l} = \frac{1-\phi_{k,l}^2}{2r_{k,l}}$. Rewrite (5.12) as

$$\mathbf{a}(\phi_{k,l}, r_{k,l}) \approx e^{-j2\pi \frac{f_c}{c_0} r_{k,l}} \tilde{\mathbf{a}}(\phi_{k,l}, r_{k,l}), \tag{5.15}$$

where the nth element of $\tilde{\mathbf{a}}(\phi_{k,l}, r_{k,l}) \in \mathbb{C}^N$ is

$$[\tilde{\mathbf{a}}(\phi_{k,l}, r_{k,l})]_n = \frac{1}{\sqrt{N}} e^{j2\pi \frac{f_c}{c_0}\left((n-1)d\phi_{k,l} - (n-1)^2 d^2 \zeta_{k,l}\right)}. \tag{5.16}$$

The steering vector in (5.15) corresponds to the physical location $(\phi_{k,l}, r_{k,l})$, which deviates from the spatial location $(\overline{\phi}_{k,m,l}, \overline{r}_{k,m,l})$ in the beamspace arising from the absent of subcarrier-dependent analog beamformers. Then, the nth entry of the deviated steering vector in (5.16) for the spatial location is

$$[\tilde{\mathbf{a}}(\overline{\phi}_{k,m,l}, \overline{r}_{k,m,l})]_n = e^{\mathrm{j}2\pi\frac{f_m}{c_0}\left((n-1)d\overline{\phi}_{k,m,l}-(n-1)^2d^2\overline{\zeta}_{k,m,l}\right)}. \tag{5.17}$$

5.2.2.2 Near-Field Beam Squint

In the THz-band, the bandwidth is notably broader compared to mmWave frequencies, which renders the assumption of a single wavelength for beamforming inadequate. This discrepancy results in the distortion or deviation of the physical DoA and ranges, denoted as $\phi_{k,l}, r_{k,l}$, within the spatial domain. Consequently, we establish the connection between physical and spatial DoA/ranges in the subsequent theorem:

Proposition 5.1 Let $\mathbf{u} \in \mathbb{C}^N$ and $\mathbf{v}_m \in \mathbb{C}^N$ denote the arbitrary near-field steering vectors associated with the physical locations (i.e., $\phi k, l, r_{k,l}$) and spatial locations (i.e., $\overline{\phi}_{k,m,l}, \overline{r}_{k,m,l}$) as outlined in (5.16) and (5.17), respectively. In the spatial domain at subcarrier frequency f_m, the array gain achieved by $\mathbf{u}^{\mathrm{H}}\mathbf{v}_m$ is maximized, directing the generated beam towards the location $\overline{\phi}_{k,m,l}$, and $\overline{r}_{k,m,l}$. This alignment satisfies the relationship expressed as follows:

$$\overline{\phi}_{k,m,l} = \eta_m\phi_{k,l}, \quad \overline{r}_{k,m,l} = \frac{1-\eta_m^2\phi_{k,l}^2}{\eta_m(1-\phi_{k,l}^2)}r_{k,l}, \tag{5.18}$$

where $\eta_m = \frac{f_c}{f_m}$ signifies the proportional deviation of the DoA/ranges.

Proof: Define the array gain achieved by $\mathbf{v}_m$ on an arbitrary user location $\{\phi_{k,l}, r_{k,l}\}$ with steering vector $\mathbf{u}$ as

$$\begin{aligned}
G(\phi_{k,l}, r_{k,l}, m) &= \frac{|\mathbf{u}^{\mathrm{H}}\mathbf{v}_m|^2}{N^2} \\
&= \frac{1}{N^2}\left|\sum_{n=0}^{N-1} e^{\mathrm{j}\frac{2\pi}{c_0}\left[nd\left(f_m\overline{\phi}_{k,m,l}-f_c\phi_{k,l}\right)-n^2d^2(f_m\overline{\zeta}_{k,m,l}-f_c\zeta_{k,l})\right]}\right|^2 \\
&= \frac{1}{N^2}\left|\sum_{n=0}^{N-1} e^{\mathrm{j}\frac{2\pi n}{c_0}(f_m\kappa_m-f_c\kappa)}\right|^2 \\
&= \frac{1}{N^2}\left|\frac{1-e^{-\mathrm{j}2\pi N(f_m\kappa_m-f_c\kappa)}}{1-e^{-\mathrm{j}2\pi(f_m\kappa_m-f_c\kappa)}}\right|^2 \\
&= \frac{1}{N^2}\left|\frac{\sin(\pi N(f_m\kappa_m-f_c\kappa))}{\sin(\pi(f_m\kappa_m-f_c\kappa))}\right|^2 \\
&= |\Sigma(f_m\kappa_m-f_c\kappa)|^2.
\end{aligned} \tag{5.19}$$

Furthermore, we define

$$\overline{\zeta}_{k,m,l} = \eta_m \zeta_{k,l} = \frac{1 - \overline{\phi}^2_{k,m,l}}{2\overline{r}_{k,m,l}}, \tag{5.20}$$

$$\kappa_m = d(\overline{\phi}_{k,m,l} - nd\overline{\zeta}_{k,m,l}), \tag{5.21}$$

$$\kappa = d(\phi_{k,l} - nd\zeta_{k,l}). \tag{5.22}$$

The expression in (5.19) demonstrates that a substantial amount of power is concentrated within a narrow segment of the beamspace due to the power-focusing property of $\Sigma(a)$. This ability to focus power diminishes noticeably as the difference $|f_m - f_c|$ between subcarriers increases [12]. In addition, $|\Sigma(a)|^2$ reaches its peak when $a = 0$, specifically when $f_m \overline{\phi}_{k,m,l} = f_c \phi_{k,l}$ and $f_m \overline{\zeta}_{k,m,l} = f_c \zeta_{k,l}$. Consequently, this correlation results in:

$$\overline{\phi}_{k,m,l} = \eta_m \phi_{k,l}. \tag{5.23}$$

Then, by using $\overline{\zeta}_{k,m,l} = \eta_m \zeta_{k,l}$, we get

$$\overline{r}_{k,m,l} = \frac{1 - \eta_m^2 \phi^2_{k,l}}{\eta_m (1 - \phi^2_{k,l})} r_{k,l}. \tag{5.24}$$

Finally, by combining (5.14), (5.18), and (5.24), we define the NB in terms of DoAs and the ranges of the users as

$$\Delta(\phi_{k,l}, m) = \overline{\phi}_{k,m,l} - \phi_{k,l} = (\eta_m - 1)\phi_{k,l}, \tag{5.25}$$

$$\Delta(r_{k,l}, m) = \overline{r}_{k,m,l} - r_{k,l} = \left(\frac{1 - \eta_m^2 \phi^2_{k,l}}{\eta_m (1 - \phi^2_{k,l})} - 1 \right) r_{k,l}. \tag{5.26}$$

In Figure 5.3, the computation of array gain is conducted for both far-field and near-field scenarios. It is evident that, due to beam squint, distinct coordinates achieve maximum array gain across various subcarriers. As a result, there are variations in the locations between the physical direction/range $(\phi_{k,l}, r_{k,l})$ and $(\overline{\phi}_{k,m,l}, \overline{r}_{k,m,l})$ at the mth subcarrier. Although there is a linear correlation between the physical and spatial directions, denoted as $\overline{\phi}_{k,m,l} = \eta_m \phi_{k,l}$, the spatial range information does not follow a linear pattern. Instead, it deviates nonlinearly, particularly expressed as

$$\overline{r}_{k,m,l} = \frac{1 - \eta_m^2 \phi^2_{k,l}}{\eta_m (1 - \phi^2_{k,l})} r_{k,l}.$$

5.3 Federated Learning for Channel Estimation

Let $\mathcal{D}_k$ denote the local dataset at the kth user, consisting of input–output pairs $\mathcal{D}_k^{(i)} = (\mathcal{X}_k^{(i)}, \mathcal{Y}_k^{(i)})$, where $i = 1, \ldots, \mathsf{D}_k$ and $\mathsf{D}_k = |\mathcal{D}_k|$ represents the size of the local dataset $\mathcal{D}_k$. Here, $\mathcal{X}_k^{(i)}$ represents the ith input data, corresponding to the received pilot signals, while $\mathcal{Y}_k^{(i)}$ denotes the ith output or label data, specifically the channel matrix. This applies for $k \in \mathcal{K}$, where $\mathcal{K} = 1, \ldots, K$. Therefore, within an input–output pair $(\mathcal{X}, \mathcal{Y})$, a learning model establishes a nonlinear connection between the input and output data as $f(\mathcal{X}|\theta) = \mathcal{Y}$, where $\theta \in \mathbb{R}^P$ denotes the learnable parameters. To elaborate further, we first delve into the methodology for collecting training data relevant to channel estimation. Subsequently, we introduce the FL algorithms and outline the obtained outcomes for different channel models.

5.3.1 Training Data Collection

In Figure 5.4, we depict the communication interval during two consecutive data transmission blocks. At the start of each transmission block, the user collects and processes received pilot signals specifically aimed at channel estimation. This procedure typically involves the use of various analytical channel estimation techniques, including those based on compressed sensing [53], angle-domain processing [16], or coordinated pilot-assignment methods [17]. Notably, the analytical approach primarily serves the purpose of gathering training data, which tends to be relatively smaller compared to the prediction stage, as supported by previous studies [54]. This emphasizes the long-term advantages of employing machine learning/federated learning (ML/FL) techniques in the prediction stage over analytical methods.

Indeed, there is an alternative method for acquiring training data through offline datasets generated from field measurements. In a study by the authors in [55], a channel estimation dataset is presented, created using electromagnetic simulation tools. While this method is viable, offline-collected data may not always perfectly

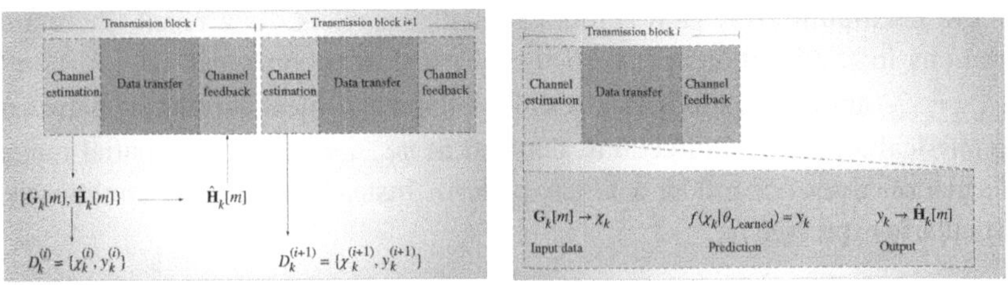

(a) (b)

Figure 5.4 (a) Training data collection and (b) channel estimation with the trained model.

represent the channel characteristics or imperfections present in the mmWave channel. In our research, we assess the performance of the FL-based approach using datasets containing labels representing both actual and estimated channel data. For the estimated channel data, we assume a training data collection process as outlined in Figure 5.4, employing the angle-domain channel estimation (ADCE) technique [16], known for its close approximation to minimum mean-square-error (MMSE) performance.

After completing the channel estimation, the training data can be assembled by storing both the received pilot data $\mathbf{G}_k[m]$ and the estimated channel data $\hat{\mathbf{H}}_k[m]$ within the internal memory of the user. The process of determining $\mathbf{G}_k[m]$ is elaborated on in Section 5.3.3. Subsequently, the user sends back the estimated channel data to the BS using uplink transmission. This accumulation process builds the local dataset $\mathcal{D}_k$ at the kth user across transmission blocks indexed by $i = 1, \dots, \mathsf{D}_k$. This methodology enables the collection of training data across various channel coherence times, which can be notably brief due to the dynamic nature of the mmWave channel, especially in settings like indoor and vehicular communications [11].

The steps described form the initial stage of the FL-based channel estimation framework as presented in prior work [34]. After gathering the training data, the global model undergoes a training phase (illustrated, for instance, in Figure 5.1b). Subsequent to this training, each user can estimate its specific channel using the trained neural network (NN). This is achieved by inputting $\mathbf{G}_k[m]$ into the NN to obtain $\hat{\mathbf{H}}_k[m]$, following the scheme outlined in Figure 5.1b.

5.3.2 FL-Based Model Training

We establish FL-based model training after first introducing the training concept in conventional CL. As shown in Figure 5.1a, the global NN is trained by collecting the local datasets $\mathcal{D}kk \in \mathcal{K}$ from the users in the context of Channel Estimation using Collaborative Learning (CL) models [19–21, 54]. After the base station (BS) has gathered the whole dataset $\mathcal{D}$, training begins with resolving the following issues as

$$\begin{aligned} &\underset{\theta}{\text{minimize}} \quad \mathcal{L}(\theta) \\ &\text{subject to} \quad f(\mathcal{X}^{(i)}|\theta) = \mathcal{Y}^{(i)}, i = 1, \dots, \mathsf{D}, \end{aligned} \tag{5.27}$$

where $\mathsf{D} = |\mathcal{D}|$ is the number of training samples and $\mathcal{L}(\theta)$ denotes the loss function defined as

$$\mathcal{L}(\theta) = \frac{1}{\mathsf{D}} \sum_{i=1}^{\mathsf{D}} \| f(\mathcal{X}^{(i)}|\theta) - \mathcal{Y}^{(i)} \|_{\mathcal{F}}^2, \tag{5.28}$$

which is the MSE between the label data $\mathcal{Y}^{(i)}$ and the prediction of the NN, $f(\mathcal{X}^{(i)}|\theta)$.

Conversely, in FL, the local datasets $\mathcal{D}_{k\in\mathcal{K}}$ remain stored at the users, and they are not sent to the BS. Consequently, the training of FL-based models takes place at the user level as

$$\underset{\boldsymbol{\theta}}{\text{minimize}} \quad \overline{\mathcal{L}}(\boldsymbol{\theta}) = \frac{1}{K}\sum_{k=1}^{K}\mathcal{L}_k(\boldsymbol{\theta})$$
$$\text{subject to} \quad f(\mathcal{X}_k^{(i)}|\boldsymbol{\theta}) = \mathcal{Y}_k^{(i)}, i = 1, \ldots, \mathsf{D}_k, k \in \mathcal{K}, \tag{5.29}$$

where $\mathcal{L}_k(\boldsymbol{\theta}) = \frac{1}{\mathsf{D}_k}\sum_{i=1}^{\mathsf{D}_k} \| f(\mathcal{X}_k^{(i)}|\boldsymbol{\theta}) - \mathcal{Y}_k^{(i)} \|_{\mathcal{F}}^2$. The FL-based model training described in (5.29) is tackled at the user level, while the CL problem articulated in (5.27) is addressed at the BS. To efficiently solve both (5.29) and (5.27), iterative approaches employing gradient descent (GD) are employed. In CL, the gradient is computed across the entire dataset as $\mathbf{g}(\boldsymbol{\theta}_t) = \nabla\mathcal{L}(\boldsymbol{\theta}_t)$, and the parameter update is executed as

$$\boldsymbol{\theta}_{t+1} = \boldsymbol{\theta}_t - \eta\mathbf{g}(\boldsymbol{\theta}_t), \tag{5.30}$$

where η is the learning rate.

In FL, each user computes the gradients individually as $\mathbf{g}_k(\boldsymbol{\theta}_t) = \nabla\mathcal{L}_k(\boldsymbol{\theta}_t)$ to solve (5.29), then send them to the BS, where the model parameters are updated as

$$\boldsymbol{\theta}_{t+1} = \boldsymbol{\theta}_t - \eta\frac{1}{K}\sum_{k=1}^{K}\mathbf{g}_k(\boldsymbol{\theta}_t). \tag{5.31}$$

Sending gradients to the base station (BS) offers higher energy efficiency compared to directly transmitting model parameters, as seen in the `FedAvg` algorithm [56]. Gradients specifically consist of model updates derived from the gradient descent (GD) algorithm, excluding redundant data from previous iterations. Transmitting the entire model includes known data, leading to substantial power wastage across all users [33, 39, 57].

In order to solve (5.29) effectively in the presence of noisy model parameters, we define a regularized loss function $\tilde{\mathcal{L}}_k(\boldsymbol{\theta})$ as

$$\tilde{\mathcal{L}}_k(\boldsymbol{\theta}) = \mathcal{L}_k(\boldsymbol{\theta}) + \sigma_\Delta^2\|\mathbf{g}_k(\boldsymbol{\theta})\|^2, \tag{5.32}$$

which is widely used in stochastic optimization [58]. (5.32) can be obtained via first order Taylor expansion of the expectation-based loss $\mathbb{E}\{\|\mathcal{L}_k(\boldsymbol{\theta} + \Delta)\|^2\}$, which can be approximately written as

$$\begin{aligned}\mathbb{E}\{\|\mathcal{L}_k(\boldsymbol{\theta} + \Delta)\|^2\} &\approx \mathbb{E}\{\|\mathcal{L}_k(\boldsymbol{\theta}) + \Delta\nabla\mathcal{L}_k(\boldsymbol{\theta})\|^2\},\\ &\approx \mathbb{E}\{\|\mathcal{L}_k(\boldsymbol{\theta})\|^2\} + \mathbb{E}\{\|\Delta\|^2\}\mathbb{E}\{\|\nabla\mathcal{L}_k(\boldsymbol{\theta})\|^2\},\\ &\approx \mathbb{E}\{\|\mathcal{L}_k(\boldsymbol{\theta})\|^2\} + \sigma_\Delta^2\|\mathbf{g}(\boldsymbol{\theta})\|^2,\end{aligned} \tag{5.33}$$

where the first term corresponds to the minimization of the loss function with perfect estimation, and the second term is the additional cost due to noise [58, 59].

Using (5.32), the regularized version of the FL-based training problem in (5.29) is given by

$$\underset{\theta}{\text{minimize}} \quad \overline{\mathcal{L}}(\theta) = \frac{1}{K}\sum_{k=1}^{K}\tilde{\mathcal{L}}_k(\theta)$$
$$\text{subject to:} \quad f(\mathcal{X}_k^{(i)}|\theta) = \mathcal{Y}_k^{(i)}, i = 1, \ldots, \mathsf{D}_k, k \in \mathcal{K}, \tag{5.34}$$

which can be effectively solved via GD in the presence of noisy model updates as

$$\tilde{\theta}_{t+1} = \tilde{\theta}_t - \eta\nabla\overline{\mathcal{L}}(\tilde{\theta}), \tag{5.35}$$

where $\nabla\overline{\mathcal{L}}(\tilde{\theta}) = \frac{1}{K}\sum_{k=1}^{K}\overline{\mathbf{g}}_k(\tilde{\theta}_t)$ and

$$\overline{\mathbf{g}}_k(\tilde{\theta}_t) = \nabla\tilde{\mathcal{L}}_k(\tilde{\theta}_t) = \nabla\left[\mathcal{L}_k(\tilde{\theta}_t) + \sigma_\Delta^2\|\mathbf{g}_k(\tilde{\theta}_t)\|^2\right]. \tag{5.36}$$

The impact of noisy gradient transmission causes $\overline{\mathcal{L}}(\theta)$ to converge at a slower pace compared to $\mathcal{L}(\theta)$. In practical settings, the convergence rate is influenced by wireless factors, particularly the signal-to-noise ratio (SNR) of transmitted/received model updates. This slowdown in convergence is attributed to packet errors occurring during training [60]. Moreover, changing channel statistics in each communication round necessitates channel state information (CSI) acquisition for every round. While some recent studies assume a single communication round within a single channel coherence time [27, 33, 39], research in [61] conducts FL-based training within a single long-coherence time, approximating about 40 small-scale fading channel coherence intervals [61].

5.3.3 FL for mmWave Channel Estimation in Massive MIMO

Let us consider the NN model for channel estimation, referred to as ChannelNet. The input to ChannelNet comprises the collected pilot signals during the preamble stage. In the downlink received signal model, as shown in (5.7), suppose the BS activates one RF chain at a time. Let $\bar{\mathbf{f}}_u[m] \in \mathbb{C}^{N_{\mathrm{BS}}}$ represent the resulting beamformer vector, and the pilot signals are denoted as $\bar{s}_u[m]$, where $u = 1, \ldots, M_{\mathrm{BS}}$ and $m \in \mathcal{M}$. On the receiver side, each user activates the RF chain for M_{MS} times and applies the beamformer vector $\overline{\mathbf{w}}_v[m]$, where $v = 1, \ldots, M_{\mathrm{MS}}$, to process the received pilots [19]. Consequently, the total channel utilization during the channel acquisition process equals $M_{\mathrm{BS}}\lceil\frac{M_{\mathrm{MS}}}{N_{\mathrm{RF}}}\rceil$. Thus, the received pilot signal at the kth user is expressed as

$$\overline{\mathbf{Y}}_k[m] = \overline{\mathbf{W}}^{\mathsf{H}}[m]\mathbf{H}_k[m]\overline{\mathbf{F}}[m]\overline{\mathbf{S}}[m] + \tilde{\mathbf{N}}_k[m], \tag{5.37}$$

where

$$\overline{\mathbf{F}}[m] = [\bar{\mathbf{f}}_1[m], \bar{\mathbf{f}}_2[m], \ldots, \bar{\mathbf{f}}_{M_{\mathrm{BS}}}[m]], \tag{5.38}$$

and

$$\overline{\mathbf{W}}[m] = [\overline{\mathbf{w}}_1[m], \overline{\mathbf{w}}_2[m], \dots, \overline{\mathbf{w}}_{M_{\mathrm{MS}}}[m]], \tag{5.39}$$

are $N_{\mathrm{BS}} \times M_{\mathrm{BS}}$ and $N_{\mathrm{MS}} \times M_{\mathrm{MS}}$ beamformer matrices, respectively. $\overline{\mathbf{S}}[m] = \mathrm{diag}\{\overline{s}_1[m], \dots, \overline{s}_{M_{\mathrm{BS}}}[m]\}$ denotes pilot signals and $\tilde{\mathbf{N}}_k[m] = \overline{\mathbf{W}}^{\mathsf{H}}\overline{\mathbf{N}}_k[m]$ is the effective noise matrix, where $\overline{\mathbf{N}}_k[m] \sim \mathcal{N}(0, \sigma^2\mathbf{I}_{M_{\mathrm{MS}}})$. Without loss of generality, we assume that $\overline{\mathbf{F}}[m] = \overline{\mathbf{F}}$ and $\overline{\mathbf{W}}[m] = \overline{\mathbf{W}}$ and $\overline{\mathbf{S}}[m] = \mathbf{I}_{M_{\mathrm{BS}}}$ $\forall m \in \mathcal{M}$. Then, the received signal (5.37) becomes

$$\overline{\mathbf{Y}}_k[m] = \overline{\mathbf{W}}^{\mathsf{H}}\mathbf{H}_k[m]\overline{\mathbf{F}} + \tilde{\mathbf{N}}_k[m]. \tag{5.40}$$

Using $\overline{\mathbf{Y}}_k[m]$, we define the input of ChannelNet $\mathbf{G}_k[m]$ as

$$\mathbf{G}_k[m] = \mathbf{T}_{\mathrm{MS}}\overline{\mathbf{Y}}_k[m]\mathbf{T}_{\mathrm{BS}}, \tag{5.41}$$

where

$$\mathbf{T}_{\mathrm{MS}} = \begin{cases} \overline{\mathbf{W}}, & M_{\mathrm{MS}} < N_{\mathrm{MS}} \\ (\overline{\mathbf{W}}\overline{\mathbf{W}}^{\mathsf{H}})^{-1}\overline{\mathbf{W}}, & M_{\mathrm{MS}} \geq N_{\mathrm{MS}}, \end{cases} \tag{5.42}$$

and

$$\mathbf{T}_{\mathrm{BS}} = \begin{cases} \overline{\mathbf{F}}^{\mathsf{H}}, & M_{\mathrm{BS}} < N_{\mathrm{BS}} \\ \overline{\mathbf{F}}^{\mathsf{H}}(\overline{\mathbf{F}}\overline{\mathbf{F}}^{\mathsf{H}})^{-1}, & M_{\mathrm{BS}} \geq N_{\mathrm{BS}}. \end{cases} \tag{5.43}$$

Here, $\mathbf{T}_{\mathrm{BS}}$ and $\mathbf{T}_{\mathrm{MS}}$ help to remove the effect of unitary matrices $\overline{\mathbf{F}}$ and $\overline{\mathbf{W}}$ in (5.40), respectively. Given that ChannelNet accepts real-valued data, we structure the final form of the input $\mathcal{X}_k$ into three distinct "channel" tensors. Consequently, the first and second "channels" of $\mathcal{X}_k$ represent the real and imaginary components of $\mathbf{G}_k[m]$, denoted as $[\mathcal{X}_k]_1 = \mathrm{Re}\mathbf{G}_k[m]$ and $[\mathcal{X}_k]_2 = \mathrm{Im}\mathbf{G}_k[m]$, respectively. The third "channel" is expressed as $[\mathcal{X}_k]_3 = \angle\mathbf{G}_k[m]$. It is worth noting that employing three "channel" inputs, such as the real, imaginary, and phase angle information of $\mathbf{G}_k[m]$, offers a more effective feature representation [21–23, 25, 41]. Consequently, the input data size becomes $N_{\mathrm{MS}} \times N_{\mathrm{BS}} \times 3$.

The output of ChannelNet is given by a $2N_{\mathrm{BS}}N_{\mathrm{MS}} \times 1$ real-valued vector as

$$\mathcal{Y}_k = \left[\mathrm{vec}\,\{\mathrm{Re}\{\mathbf{H}_k[m]\}\}^{\mathsf{T}}, \mathrm{vec}\,\{\mathrm{Im}\{\mathbf{H}_k[m]\}\}^{\mathsf{T}}\right]^{\mathsf{T}}. \tag{5.44}$$

As a result, ChannelNet maps the received pilot signals $\mathbf{G}_k[m]$ to the channel matrix $\mathbf{H}_k[m]$.

In Figure 5.5, we display the training performance (Figure 5.5a) and the NMSE for channel estimation (Figure 5.5b) using the FL method across various user counts. For this scenario, we maintain a fixed total dataset size D, by setting $G = 20 \cdot \frac{8}{K}$. As the number of users, denoted by K, increases, there is an observed

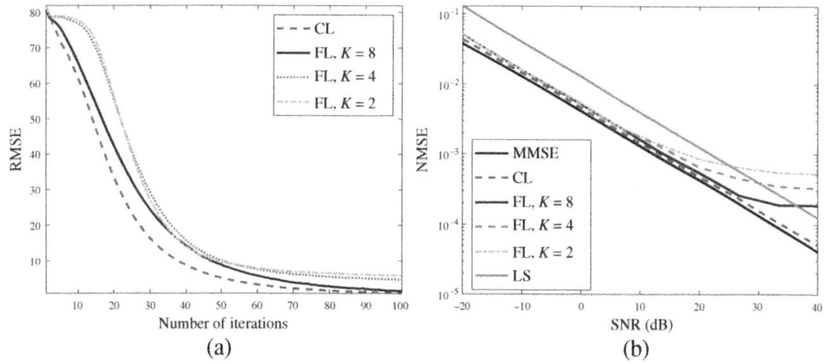

Figure 5.5 Validation RMSE (a) and channel estimation NMSE (b) with respect to K in massive MIMO scenario.

improvement in training performance, bringing it closer to the performance seen in CL. This enhancement occurs because the model updates, when superimposed at the BS, become more resilient against noise with a larger number of users. Conversely, when K decreases, the model aggregation encounters more corruption due to the diversity in the training dataset.

Figure 5.6 displays the training and channel estimation performance under various noise levels added to the transmitted gradient and model data while considering $K = 8$. Specifically, we introduce AWGN to both $\mathbf{g}_k(\theta t)$ and θ_t with respect to SNR_θ, where

$$\text{SNR}_\theta = 20 \log 10 \frac{\|\mathbf{g}_k(\theta_t)\|_2^2}{\sigma_\theta^2}. \tag{5.45}$$

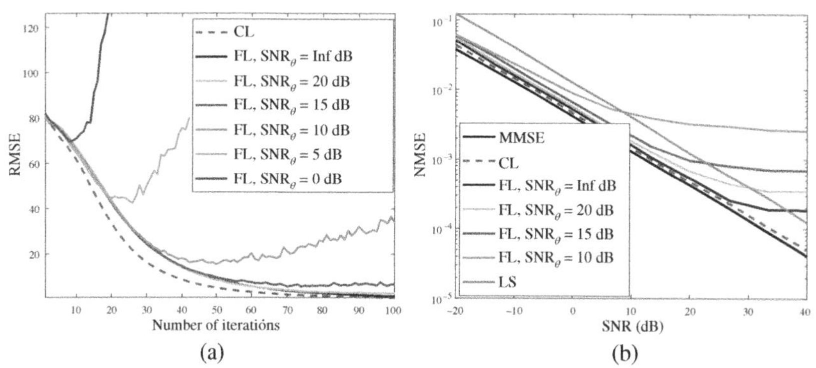

Figure 5.6 Validation RMSE (a) and channel estimation NMSE (b) with respect to SNR_θ in massive MIMO scenarios, respectively.

The results in Figure 5.6a indicate that the training diverges when SNR_θ is low (for instance, $\text{SNR}_\theta \leq 5$ dB), primarily due to corruption in the model parameters. Figure 5.6b portrays the corresponding channel estimation performance, demonstrating that reasonable channel estimation performance, such as NMSE ≤ 0.001, is achieved when the ChannelNet converges, requiring at least $\text{SNR}_\theta \leq 15$ dB.

In Figure 5.7 displays the channel estimation NMSE is displayed for various algorithms under $K = 8$. The comparison between ChannelNet trained with both CL and FL frameworks is depicted, showcasing that CL closely follows MMSE performance. Compared with FL, CL offers superior performance due to access to the entire dataset at once. However, FL demonstrates satisfactory channel estimation despite decentralized training. For SNR $\leq$ 25 dB, FL and CL exhibit similar NMSE, with the performance FL peaking in the high-SNR regime, where the learning model loses precision due to FL training, limiting further improvement. This precision loss is a common challenge in ML-based methods [19, 21]. To improve performance, overtraining might be considered to obtain greater precision, but this can lead to overfitting, where the model memorizes the data and performs poorly on new inputs.

Figure 5.7 also includes a comparison between training with perfect (true channel data) and imperfect (estimated channel via ADCE) labels. Using imperfect labels causes a slight performance drop, still resulting in lower NMSE than SF-CNN [19] and MLP [20]. Other algorithms exhibit similar behavior but perform worse than ChannelNet. This discrepancy is due to SF-CNN relying solely on convolutional layers and MLP exclusively on fully connected layers.

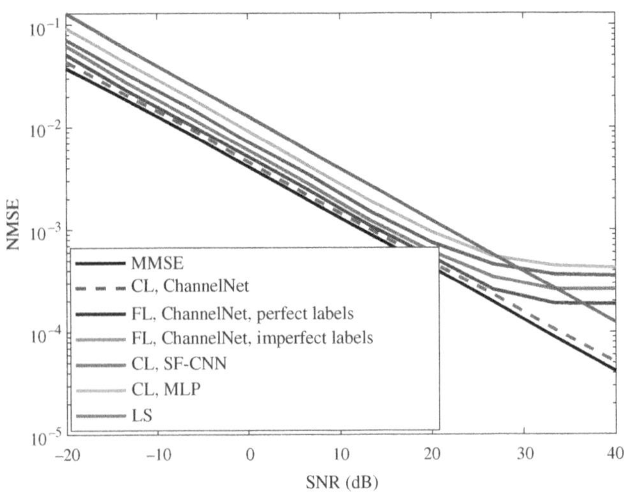

Figure 5.7 Channel estimation NMSE for different algorithms in massive MIMO scenario.

In contrast, ChannelNet encompasses both structures, thereby showcasing improved feature extraction and data mapping capabilities.

5.3.4 FL for mmWave Channel Estimation in RIS-Assisted Massive MIMO

In this section, we delve into the channel estimation challenge within RIS-assisted massive MIMO, illustrated in Figure 5.8. Initially, we introduce the received signal model that encompasses both the direct (BS-user) and cascaded (BS–RIS-user) channels. Subsequently, we elucidate the process of acquiring input–output pairs for ChannelNet in the RIS-assisted scenario as per the findings outlined in [62].

We consider downlink channel estimation, where the BS utilizes N_{BS} antennas to serve K single-antenna users with the aid of a RIS comprising N_{RIS} reflective elements, as depicted in Figure 5.8. As the signal arrives from the BS, it undergoes reflection from the RIS, with each RIS element introducing a phase shift φ_n (for $n = 1, \dots, N_{\mathrm{RIS}}$). These phase shifts are controllable through PIN diodes managed by the RIS controller connected to the BS via the backhaul link. This manipulation enables users to receive the BS-transmitted signal, especially when users are distant from the BS or when obstructions are present.

Let $\overline{\mathbf{S}}_{\mathrm{RIS}} \in \mathbb{C}^{NBS \times M_{\mathrm{BS}}}$ (where $N_{\mathrm{BS}} \leq M_{\mathrm{BS}}$) represent the pilot signals transmitted from the BS. Consequently, the received signal at the kth user can be expressed as

$$\mathbf{y}_k = (\mathbf{h}_{\mathrm{B},k}^{\mathsf{H}} + \psi^{\mathsf{H}}\mathbf{V}_k^{\mathsf{H}})\overline{\mathbf{S}}_{\mathrm{RIS}} + \mathbf{n}_k, \tag{5.46}$$

where $\mathbf{y}_k = [y_{1,k}, \dots, y_{M_{\mathrm{BS}},k}]$ and $\mathbf{n}_k = [n_{1,k}, \dots, n_{M_{\mathrm{BS}},k}]$ represent row vectors of size $1 \times M_{\mathrm{BS}}$, while $\mathbf{h}_{\mathrm{B},k} \in \mathbb{C}^{N_{\mathrm{BS}}}$ stands for the channel between the BS and the kth user. $\psi = [\psi_1, \dots, \psi_{N_{\mathrm{RIS}}}]^{\mathsf{T}} \in \mathbb{C}^{N_{\mathrm{RIS}}}$ denotes the beamforming vector for reflection, where

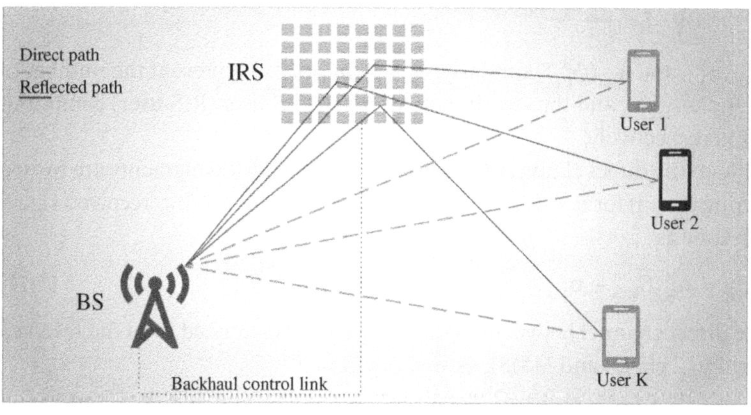

Figure 5.8 RIS-assisted mmWave massive MIMO scenario.

$\psi_n = a_n e^{j\varphi_n}$, with $a_n \in 0, 1$ indicating the on/off status of the nth element of the RIS, and $\varphi_n \in [0, 2\pi]$ is the phase shift introduced by the RIS. In practice, the RIS elements cannot be perfectly turned on/off, hence, they can be modeled as

$$a_n = \begin{cases} 1 - \epsilon_1 & \text{ON} \\ 0 + \epsilon_0 & \text{OFF} \end{cases},$$

for small non-negative values ϵ_1 and ϵ_0, indicating the insertion loss of the reflecting elements [21], (5.46) presents $\mathbf{V}_k \in \mathbb{C}^{N_{\text{BS}} \times N_{\text{RIS}}}$, the combined channel encompassing the BS-RIS-user link. This can be expressed with reference to the channels between the BS–RIS and RIS-user as follows

$$\mathbf{V}_k = \mathbf{H}_{\text{B}} \mathbf{\Lambda}_k, \tag{5.47}$$

where $\mathbf{H}_{\text{B}} \in \mathbb{C}^{N_{\text{BS}} \times N_{\text{RIS}}}$ is the channel between the BS and the RIS, and it can be defined similar to (5.3) as

$$\mathbf{H}_{\text{B}} = \sqrt{\frac{N_{\text{BS}} N_{\text{RIS}}}{L_{\text{RIS}}}} \sum_{l=1}^{L_{\text{RIS}}} \alpha_l^{\text{RIS}} \mathbf{a}_{\text{BS}}(\phi_l^{\text{BS}}) \mathbf{a}_{\text{RIS}}(\phi_l^{\text{RIS}})^{\text{H}}, \tag{5.48}$$

where L_{RIS} and α_l^{RIS} are the number of received paths and the complex gain, respectively. $\mathbf{a}_{\text{BS}}(\phi_l^{\text{BS}}) \in \mathbb{C}^{N_{\text{BS}}}$ and $\mathbf{a}_{\text{RIS}}(\phi_l^{\text{RIS}}) \in \mathbb{C}^{N_{\text{RIS}}}$ are the steering vectors corresponding to the BS and RIS with the AoA and AoD angles of $\phi_l^{\text{BS}}, \phi_l^{\text{RIS}}$, respectively. In (5.47), $\mathbf{\Lambda}_k = \text{diag}\{\mathbf{h}_{\text{S},k}\}$ and $\mathbf{h}_{\text{S},k} \in \mathbb{C}^{N_{\text{RIS}}}$ represents the channel between the RIS and the kth user. $\mathbf{h}_{\text{S},k}$ and $\mathbf{h}_{\text{B},k}$ have similar structures, and they can be defined as follows

$$\mathbf{h}_{\text{B},k} = \sqrt{\frac{N_{\text{BS}}}{L_{\text{B}}}} \sum_{l=1}^{L_{\text{B}}} \alpha_{k,l}^{\text{B}} \mathbf{a}_{\text{BS}}(\phi_{k,l}^{\text{B}}), \tag{5.49}$$

$$\mathbf{h}_{\text{S},k} = \sqrt{\frac{N_{\text{RIS}}}{L_{\text{S}}}} \sum_{l=1}^{L_{\text{S}}} \alpha_{k,l}^{\text{S}} \mathbf{a}_{\text{RIS}}(\phi_{k,l}^{\text{S}}), \tag{5.50}$$

where, L_{B}, $\alpha_{k,l}^{\text{B}}$, and $\mathbf{a}_{\text{BS}}(\phi_{k,l}^{\text{B}})$ (similarly, L_{S}, $\alpha_{k,l}^{\text{S}}$, $\mathbf{a}_{\text{RIS}}(\phi_{k,l}^{\text{B}})$) represent the number of paths, complex gain, and the steering vector for the BS-user (RIS-user) communication link, respectively.

To estimate the direct channel $\mathbf{h}_{\text{B},k}$, we assume that all RIS elements are turned off, meaning $a_n = 0$ for $n = 1, \dots, N_{\text{RIS}}$. This results in the $1 \times M_{\text{BS}}$ received signal at the kth user as

$$\mathbf{y}_{\text{B},k} = \mathbf{h}_{\text{B},k}^{\text{H}} \overline{\mathbf{S}}_{\text{RIS}} + \mathbf{n}_{\text{B},k}. \tag{5.51}$$

Then, the direct channel between BS-user $\mathbf{h}_{\text{B},k}$ can be estimated from the received pilot signal $\mathbf{y}_{\text{B},k}$ via LS and MMSE estimators as

$$\mathbf{h}_{\text{B},k}^{\text{LS}} = \left(\mathbf{y}_{\text{B},k} \overline{\mathbf{S}}_{\text{RIS}}^{\dagger} \right)^{\text{H}}, \tag{5.52}$$

and

$$\mathbf{h}_{\mathrm{B},k}^{\mathrm{MMSE}} = \mathbf{h}_{\mathrm{B},k}^{\mathrm{LS}} \mathbf{R}_{\mathrm{B},k} \left(\mathbf{R}_{\mathrm{B},k} + \frac{1}{\sigma^2} \overline{\mathbf{S}}_{\mathrm{RIS}} \overline{\mathbf{S}}_{\mathrm{RIS}}^{\mathsf{H}} \right)^{-1}, \tag{5.53}$$

where $\mathbf{R}_{\mathrm{B},k} = \mathbb{E}\{\mathbf{h}_{\mathrm{B},k}\mathbf{h}_{\mathrm{B},k}^{\mathsf{H}}\}$ [19].

We now proceed to estimate the cascaded channel. This involves activating each RIS element individually while keeping all other elements turned off. The BS manages this process by communicating with the RIS through a micro-controller device over the backhaul link, enabling one RIS element to be activated at a time. Then, the reflecting beamformer vector at the nth frame becomes $\boldsymbol{\psi}^{(n)} = [0, \dots, 0, \psi_n, 0, \dots, 0]^{\mathsf{T}}$, where $a_n = \{0 : \tilde{n} = 1, \dots N_{\mathrm{RIS}}, \tilde{n} \neq n\}$ and the received signal is given by

$$\mathbf{y}_{\mathrm{C},k}^{(n)} = (\mathbf{h}_{\mathrm{B},k}^{\mathsf{H}} + \mathbf{v}_k^{(n)^{\mathsf{H}}})\overline{\mathbf{S}}_{\mathrm{RIS}} + \mathbf{n}_k, \tag{5.54}$$

where $\mathbf{v}_k^{(n)} \in \mathbb{C}^{N_{\mathrm{BS}}}$ is the nth column of $\mathbf{V}_k$, i.e., $\mathbf{v}_k^{(n)} = \mathbf{V}_k\boldsymbol{\psi}^{(n)}$, where $\psi_n = 1$. Using the estimate of $\mathbf{h}_{\mathrm{B},k}$ from (5.51), (5.54) can be solved for $\mathbf{v}_k^{(n)}$, $n = 1, \dots, N_{\mathrm{RIS}}$, and the cascaded channel $\mathbf{V}_k$ can be estimated. Then, the received data for $n = 1 \dots, N_{\mathrm{RIS}}$ can be given by $\mathbf{Y}_{\mathrm{C},k} \in \mathbb{C}^{N_{\mathrm{RIS}} \times M_{\mathrm{BS}}}$ as

$$\mathbf{Y}_{\mathrm{C},k} = \begin{bmatrix} \mathbf{y}_{\mathrm{C},k}^{(n)} \\ \vdots \\ \mathbf{y}_{\mathrm{C},k}^{(N_{\mathrm{RIS}})} \end{bmatrix}.$$

In the training of ChannelNet for the RIS-assisted massive MIMO scenario, we define the input–output data pairs as $\mathbf{y}_{\mathrm{B},k}, \mathbf{h}_{\mathrm{B},k}$, and $\mathbf{Y}_{\mathrm{C},k}, \mathbf{V}_k$ representing the direct and cascaded channels, respectively. To simultaneously learn both channels, we construct a unified input for training a single neural network: $\mathbf{\Upsilon}_k = \left[\mathbf{y}_{\mathrm{B},k}\, \mathbf{Y}_{\mathrm{C},k}\right] \in \mathbb{C}^{(N_{\mathrm{RIS}}+1)\times M_{\mathrm{BS}}}$. Similarly to the previous scenario, the three "channel" components of the input data are constructed as $[\mathcal{X}_k]_1 = \mathrm{Re}\mathbf{\Upsilon}_k$, $[\mathcal{X}_k]_2 = \mathrm{Im}\mathbf{\Upsilon}_k$, and $[\mathcal{X}_k]_3 = \angle\mathbf{\Upsilon}_k$. We define the output data as $\mathbf{\Sigma}_k = \left[\mathbf{h}_{\mathrm{B},k}, \mathbf{V}_k\right] \in \mathbb{C}^{N_{\mathrm{BS}}\times(N_{\mathrm{RIS}}+1)}$, and consequently, the output label is represented by a $2N_{\mathrm{BS}}(N_{\mathrm{RIS}}+1) \times 1$ real-valued vector.

$$\mathcal{Y}_k = \left[\mathrm{vec}\,\{\mathrm{Re}\{\mathbf{\Sigma}_k\}\}^{\mathsf{T}}, \mathrm{vec}\,\{\mathrm{Im}\{\mathbf{\Sigma}_k\}\}^{\mathsf{T}}\right]^{\mathsf{T}}. \tag{5.55}$$

Consequently, we have the sizes of $\mathcal{X}_k$ and $\mathcal{Y}_k$ are $(N_{\mathrm{RIS}}+1) \times M_{\mathrm{BS}} \times 3$ and $2N_{\mathrm{BS}}(N_{\mathrm{RIS}}+1) \times 1$, respectively. Then, the FL-based training can be employed for RIS channel estimation scenario given this dataset.

In Figure 5.9, we display the validation RMSE and the channel estimation NMSE in the RIS-assisted scenario. The NMSE, calculated for both the direct and cascaded channels jointly, is depicted in a single plot. Consistent with the training phase, the results show divergence when $\mathrm{SNR}_\theta \leq 5$ dB, while the channel estimation NMSE reaches relatively low values at $\mathrm{SNR}_\theta \geq 15$ dB.

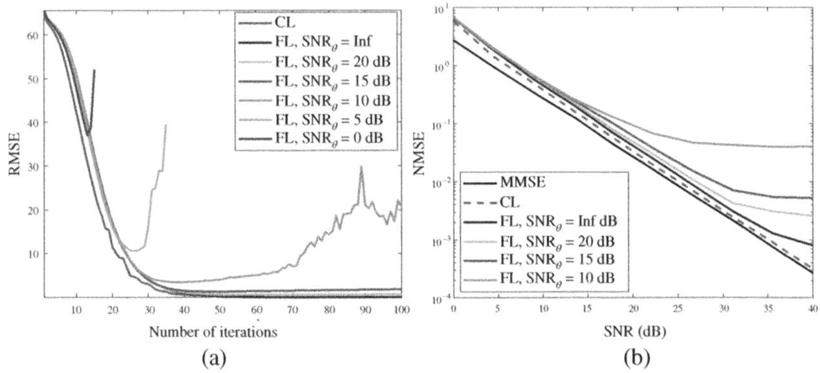

Figure 5.9 Validation RMSE (a) and channel estimation NMSE (b) with respect to SNR$_\theta$ in RIS-assisted massive MIMO scenario, respectively.

5.3.5 FL for THz Channel Estimation

In this section, we initially outline the THz channel estimation utilizing model-based methodologies before introducing FL-based techniques.

To address the challenges posed by NB, we propose an NB-aware (NBA) model-based approach called NBA–OMP [43]. This method incorporates parameter estimation within an NBA dictionary design.

The central concept behind the NBA dictionary design involves leveraging prior knowledge of η_m to acquire beam-squint-adjusted steering vectors. Essentially, given an arbitrary physical DoA and range, we can readily determine the spatial DoA and ranges as $\overline{\phi} = \eta_m \phi$ and $\overline{r} = \frac{1-\eta_m^2\phi^2}{\eta_m(1-\phi^2)} r$. Leveraging this insight, we craft the NBA dictionary C_m that comprises steering vectors $\mathbf{c}(\phi_m, r_m) \in \mathbb{C}^N$, expressed as:

$$C_m = \{\mathbf{c}(\phi_m, r_m) | \phi_m \in [-\eta_m, -\eta_m], r_m \in \mathbb{R}^+\}, \tag{5.56}$$

where the nth element of $\mathbf{c}(\phi_m, r_m)$ is

$$[\mathbf{c}(\phi_m, r_m)]_n = e^{\mathrm{j}2\pi \frac{f_m}{c_0}((n-1)d\phi_m - (n-1)^2 d^2 \zeta_m)}, \tag{5.57}$$

where $\zeta_m = \frac{1-\phi_m^2}{2r_m}$. It is worth noting that the NBA dictionary design requires the approximated steering vector models in order to easily define and compensate the beam squint.

The NBA dictionary C_m replaces the subcarrier-independent steering vectors $\mathbf{a}_(\phi, r)$ with subcarrier-specific virtual steering vectors $\mathbf{c}_(\phi_m, r_m)$, utilized within the OMP algorithm. This alteration allows for the computation of beamspace spectra via OMP, enabling the identification of sparse channel support associated with subcarrier-specific spatial directions of arrival (DoAs) and ranges. Subsequently, these findings enable the direct derivation of physical DoAs and ranges as $\phi = \phi_m/\eta_m$ and $r = \frac{\eta_m(1-\phi^2)}{1-\eta_m^2\phi^2} r_m$, for all $m \in \mathcal{M}$.

The NBA dictionary C_m also holds spatial orthogonality as

$$\lim_{N\to+\infty} |\mathbf{c}^{\mathrm{H}}(\phi_{m,i}, r_{m,i})\mathbf{c}(\phi_{m,j}, r_{m,j})| = 0, \forall i \neq j. \tag{5.58}$$

Now, we outline the procedure for channel estimation utilizing the NBA dictionary. During the downlink phase, the channel estimation process is concurrently executed by all users, denoted as $k \in \mathcal{K} = 1, \cdots, K$. As the BS operates with a hybrid beamforming architecture, it activates a single RF chain in each channel use solely for transmitting pilot signals during channel acquisition, a method described in literature [1]. Denoted by $\tilde{\mathbf{S}}[m] = \mathrm{diag}\{\tilde{s}_1[m], \cdots, \tilde{s}_P[m]\}$, the $P \times P$ orthogonal pilot signal matrix is utilized. Consequently, the BS employs P beamformer vectors, denoted as $\tilde{\mathbf{F}} = [\tilde{\mathbf{f}}_1, \cdots, \tilde{\mathbf{f}}_P] \in \mathbb{C}^{N\times P}$ (where $|\tilde{\mathbf{f}}_p| = 1/\sqrt{N}$), to dispatch P orthogonal pilots, subsequently captured by the kth user as

$$\mathbf{y}_k[m] = \tilde{\mathbf{S}}[m]\overline{\mathbf{F}}[m]\mathbf{h}_k[m] + \mathbf{w}_k[m], \tag{5.59}$$

where $\overline{\mathbf{F}}[m] = \tilde{\mathbf{F}}^{\mathrm{T}}[m] \in \mathbb{C}^{P\times N}$. Assume that $\overline{\mathbf{F}}[m] = \mathbf{F} \in \mathbb{C}^{P\times N}$ and $\tilde{\mathbf{S}}[m] = \mathbf{I}_P, \forall m \in \mathcal{M}$ [63–65], we get

$$\mathbf{y}_k[m] = \mathbf{F}\mathbf{h}_k[m] + \mathbf{w}_k[m]. \tag{5.60}$$

Note that the solution via traditional techniques, e.g., the least squares (LS) and MMSE estimator can be readily given respectively as

$$\mathbf{h}_k^{\mathrm{LS}}[m] = (\mathbf{F}^{\mathrm{H}}\mathbf{F})^{-1}\mathbf{F}^{\mathrm{H}}\mathbf{y}_k[m], \tag{5.61}$$

$$\mathbf{h}_k^{\mathrm{MMSE}}[m] = \left(\mathbf{R}_k^{-1}[m] + \mathbf{F}^{\mathrm{H}}\mathbf{R}_k^{-1}[m]\mathbf{F}\right)^{-1}\mathbf{F}^{\mathrm{H}}\mathbf{y}_k[m], \tag{5.62}$$

where $\mathbf{R}_k[m] = \mathbb{E}\{\mathbf{h}_k[m]\mathbf{h}_k^{\mathrm{H}}[m]\}$ represents the channel covariance matrix. However, these techniques necessitate prior information (such as MMSE) or demonstrate inadequate estimation performance (as seen in LS) [34, 66]. Moreover, these methods require at least $P \geq N$ pilot signals, which could result in substantial overhead during channel training. Conversely, the OMP approach significantly reduces the burden of channel training.

By leveraging the sparsity inherent in the THz channel, the channel in the THz domain can be depicted through a sparse representation using the support vector $\mathbf{x}_k[m] \in \mathbb{C}^Q$. This vector is characterized by being sparsely populated, containing L nonzero elements that correspond to the following set, i.e.,

$$\{x_{k,l}[m] | x_{k,l}[m] \triangleq \sqrt{\frac{N}{L}}\alpha_{k,m,l}e^{-j2\pi\tau_{k,l}f_m}\}. \tag{5.63}$$

Then, the received signal in (5.60) is rewritten in sparse domain as

$$\mathbf{y}_k[m] = \mathbf{F}\mathbf{C}_m\mathbf{x}_k[m] + \mathbf{w}_k[m], \tag{5.64}$$

where $\mathbf{F}$ stands for an $P \times N$ fixed matrix that corresponds to the hybrid beamformer weights employed during data transmission. The sparse channel $\mathbf{h}_k[m]$ is expressed as $\mathbf{h}_k[m] = \mathbf{C}_m\mathbf{x}_k[m]$, where $\mathbf{C}_m$ represents the $N \times Q$ NBA dictionary matrix that spans the spatial domain. This matrix encompasses values for $\phi_{m,q}$ within the range of $[-\eta_m, \eta_m]$ ($\phi \in [-1, 1]$) and $r_{m,q}$ within the range of $[0, F]$ for $q = 1, \cdots, Q$, defined as

$$\mathbf{C}_m = [\mathbf{c}(\phi_{m,1}, r_{m,1}), \cdots, \mathbf{c}(\phi_{m,Q}, r_{m,Q})]. \tag{5.65}$$

By making use of the NBA dictionary matrix $\mathbf{C}_m$, we apply the OMP algorithm to effectively recover the support of the sparse channel. The NBA-OMP technique is outlined in Algorithm 5.1, taking $\mathbf{y}_k[m]$, $\mathbf{C}_m$, and η_m as inputs. Its output includes the estimated channel $\hat{\mathbf{h}}_k[m]$, and the NB, expressed in terms of DoAs denoted as $\hat{\Delta}(\phi_{k,l}, m)$, and ranges $\hat{\Delta}(r_{k,l}, m)$. Specifically, in steps 4–6 of Algorithm 5.1, it verifies the orthogonality between the residual observation vector $\mathbf{r}_l[m]$ and the columns of the dictionary matrix as

$$q^{\star} = \underset{q}{\operatorname{argmax}} \sum_{m=1}^{M} |\mathbf{c}^{\mathrm{H}}(\phi_{m,q}, r_{m,q})\mathbf{F}^{\mathrm{H}}\mathbf{r}_{l-1}[m]|, \tag{5.66}$$

where $q^{\star}$ denotes the support index for the lth path iteration. Then, the DoA and range estimates can be found as $\hat{\phi}_{k,l} = \frac{\phi_{m,q^{\star}}}{\eta_m}$ and $\hat{r}_{k,l} = \frac{\eta_m(1-\phi^2_{m,q^{\star}})}{1-\eta_m^2\phi_{m,q^{\star}}} r_{m,q^{\star}}$, respectively. In a similar way, the corresponding beam squint terms are $\hat{\Delta}(\phi_{k,l}, m) = (\eta_m - 1)\hat{\phi}_{k,l}$ and $\hat{\Delta}(r_{k,l}, m) = (\eta_m - 1)\frac{1-\eta_m^2\hat{\phi}^2_{k,l}}{\eta_m(1-\hat{\phi}^2_{k,l})}\hat{r}_{k,l}$. Once the DoA and ranges are obtained, the THz channel is constructed from the estimated support as $\hat{\mathbf{h}}_k[m] = \mathbf{\Xi}_k\hat{\mathbf{u}}_k[m]$ where $\mathbf{\Xi}_k = [\mathbf{a}(\hat{\phi}_{k,1}, \hat{r}_{k,1}), \cdots, \mathbf{a}(\hat{\phi}_{k,L}, \hat{r}_{k,L})]$ and $\hat{\mathbf{u}}_k[m] = \mathbf{\Psi}_m^{\dagger}(\mathcal{I}_L)\mathbf{y}_k[m]$, wherein $\mathbf{\Psi}_m(\mathcal{I}_L) = \mathbf{F}\mathbf{C}_m(\mathcal{I}_L)$ and $\mathcal{I}_L$ includes the estimated support indices.

Because of the use of small wavelengths and the substantial path loss inherent in THz communications, the utilization of ultra-massive arrays is anticipated. This leads to the processing of an exceedingly large amount of array data. For example, when employing learning-based methods, the dataset size scales with the number of antennas. Consequently, significant communication overhead arises due to transmitting these datasets in methods based on centralized learning (CL) [18, 38, 56]. To mitigate this overhead, we introduce an FL-based scheme for near-field THz channel estimation in this section.

Let $\theta \in \mathbb{R}^Z$ denote the vector of learnable parameters, and $\mathcal{D}_k$ represent the local dataset for the kth user. The relationship $f(\theta)$ signifies the connection between the input of the learning model, represented by θ, and its prediction denoted as

Algorithm 1 NBA–OMP

Input: Dictionary $\mathbf{C}_m$, observation $\mathbf{y}_k[m]$, $\mathbf{F}$ and η_m

Output: $\hat{\mathbf{h}}_k[m]$, $\hat{\Delta}(\phi_{k,l}, m)$ and $\hat{\Delta}(r_{k,l}, m)$.

for $k \in \mathcal{K}$
  $l = 1$, $\mathcal{I}_{l-1} = \emptyset$, $\mathbf{r}_{l-1}[m] = \mathbf{y}_k[m], \forall m \in \mathcal{M}$.
  while $l \leq L$ **do**
    $q^\star = \arg\max_q \sum_{m=1}^{M} |\mathbf{c}^{\mathrm{H}}(\phi_{m,q}, r_{m,q})\mathbf{F}^{\mathrm{H}}\mathbf{r}_{l-1}[m]|$.
    $\mathcal{I}_l = \mathcal{I}_{l-1} \bigcup \{q^\star\}$.
    $\hat{\phi}_{k,l} = \frac{\phi_{m,q^\star}}{\eta_m}$.
    $\hat{r}_{k,l} = \frac{\eta_m(1-\phi^2_{m,q^\star})}{1-\eta_m^2\phi_{m,q^\star}} r_{m,q^\star}$.
    $\hat{\Delta}(\phi_{k,l}, m) = (\eta_m - 1)\hat{\phi}_{k,l}, \forall m \in \mathcal{M}$.
    $\hat{\Delta}(r_{k,l}, m) = (\eta_m - 1)\frac{1-\eta_m^2\hat{\phi}_{k,l}^2}{\eta_m(1-\hat{\phi}^2_{k,l})}\hat{r}_{k,l}, \forall m \in \mathcal{M}$.
    $\mathbf{\Psi}_m(\mathcal{I}_l) = \mathbf{F}\mathbf{C}_m(\mathcal{I}_1)$.
    $\mathbf{r}_l[m] = \left(\mathbf{I}_P - \mathbf{\Psi}_m(\mathcal{I}_l)\mathbf{\Psi}_m^\dagger(\mathcal{I}_l)\right)\mathbf{y}_k[m]$.
    $l = l + 1$.
  end while
  $\mathbf{\Xi}_k = [\mathbf{a}(\hat{\phi}_{k,1}, \hat{r}_{k,1}), \cdots, \mathbf{a}(\hat{\phi}_{k,L}, \hat{r}_{k,L})]$.
  for $m \in \mathcal{M}$
    $\hat{\mathbf{u}}_k[m] = \mathbf{\Psi}_m^\dagger(\mathcal{I}_{l-1})\mathbf{y}_k[m]$.
    $\hat{\mathbf{h}}_k[m] = \mathbf{\Xi}_k\hat{\mathbf{u}}_k[m]$
  end for
end for

$\mathcal{Y}_k^{(i)} = f(\theta)\mathcal{X}_k^{(i)}$. The sample index within the local dataset is $i = 1, \ldots, \mathsf{D}_k$, where $\mathsf{D}_k = |\mathcal{D}_k|$ signifies the number of samples in the kth local dataset. In addition, $\mathcal{X}_k^{(i)}$ and $\mathcal{Y}_k^{(i)}$ stand for the input and output, respectively, constituting the input–output pair of the local dataset as $\mathcal{D}_k^{(i)} = (\mathcal{X}_k^{(i)}, \mathcal{Y}_k^{(i)})$.

The input data is the received pilots, i.e., $\mathbf{y}_k[m]$. For the output (label) data, we consider the channel estimates obtained via the NBA–OMP approach in Algorithm 5.1. Then, by using $\hat{\mathbf{h}}_k[m]$, the output data is constructed as

$$\mathcal{Y}_k = \left[\mathrm{Re}\{\hat{\mathbf{h}}_k[m]\}^{\mathsf{T}}, \mathrm{Im}\{\hat{\mathbf{h}}_k[m]\}^{\mathsf{T}}\right]^{\mathsf{T}} \in \mathbb{R}^{2N}. \tag{5.67}$$

In formulating the input data, the real, imaginary, and angle information of $\mathbf{y}_k[m]$ are combined to enhance feature extraction capabilities of the learning model. This results in each input sample adopting a "three-channel" structure sized $N_{\mathrm{RF}} \times 3$ [18]. Specifically, we organize $[\mathcal{X}_k]_1 = \mathrm{Re}\,\mathbf{y}_k[m]$, $[\mathcal{X}_k]_2 = \mathrm{Im}\,\mathbf{y}_k[m]$, and $[\mathcal{X}_k]_3 = \angle\mathbf{y}_k[m]$ within this dataset structure. Leveraging this dataset configuration allows for conducting FL-based training as outlined in Section 5.3.2.

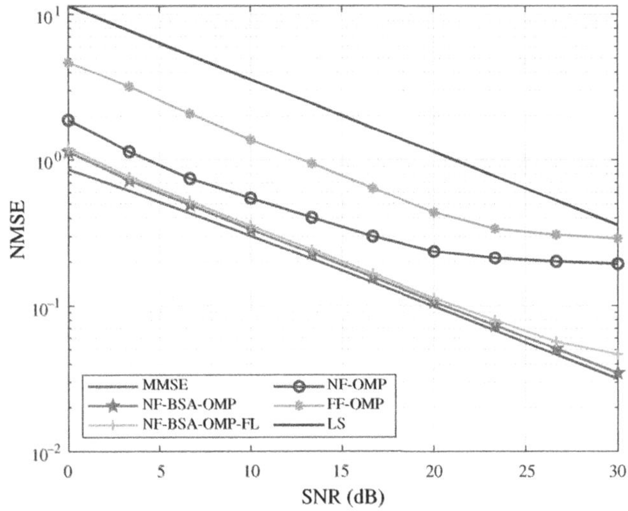

Figure 5.10 THz wideband channel estimation NMSE vs. SNR. $N = 256, f_c = 300$ GHz, $M = 128$ and $B = 30$ GHz.

Figure 5.10 displays the normalized MSE (NMSE) of channel estimation across various SNR levels. Both NBA–OMP and NBA–OMP–FL methods exhibit superior performance compared to other techniques and closely track the MMSE performance. The effectiveness of NBA–OMP stems from its precise compensation for beam squint effects across direction and range parameters using the NBA dictionary, comprised of subcarrier-dependent steering vectors. In contrast, other methods lack this precision, especially in high SNR scenarios. This shortfall arises from their disregard for the NB or their sole consideration of the far-field signal model [13, 65, 67].

5.4 FL For Hybrid Beamforming

In this setup, we are exploring a multiuser mmWave MIMO scenario, where a BS with N_{T} antennas communicates with K single-antenna users. In the downlink, the BS initially encodes K data symbols $\mathbf{s} = [s_1, s_2, \dots, s_K]^{\mathrm{T}} \in \mathbb{C}^K$ using baseband precoders $\mathbf{F}_{\mathrm{BB}} = [\mathbf{f}_{\mathrm{BB},1}, \dots, \mathbf{f}_{\mathrm{BB},K}] \in \mathbb{C}^{K \times K}$. Then, it employs K RF precoders $\mathbf{F}_{\mathrm{RF}} = [\mathbf{f}_{\mathrm{RF},1}, \dots, \mathbf{f}_{\mathrm{RF},K}] \in \mathbb{C}^{N_{\mathrm{T}} \times K}$ to create the transmitted signal. These RF precoders are composed of analog phase shifters, ensuring constant unit-modulus elements ($|[\mathbf{F}_{\mathrm{RF}}]_{i,j}|^2 = 1$), resulting in $|\mathbf{F}_{\mathrm{RF}}\mathbf{F}_{\mathrm{BB}}|_{\mathcal{F}}^2 = K$. As a consequence, the $N_{\mathrm{T}} \times 1$ transmit signal is defined as $\mathbf{x} = \mathbf{F}_{\mathrm{RF}}\mathbf{F}_{\mathrm{BB}}\mathbf{s}$, and the signal received at the kth user is expressed as $y_k = \mathbf{h}_k^{\mathsf{H}} \sum_{n=1}^{K} \mathbf{F}_{\mathrm{RF}}\mathbf{f}_{\mathrm{BB},n} s_n + n_k$, where $\mathbf{h}_k \in \mathbb{C}^{N_{\mathrm{T}}}$ denotes the mmWave channel for the kth user with $|\mathbf{h}_k|_2^2 = N_{\mathrm{T}}$, and $n_k \sim \mathcal{CN}(0, \sigma^2)$ represents the AWGN vector.

We employed a clustered channel model, where the mmWave channel $\mathbf{h}_k$ is characterized by a cluster consisting of L line-of-sight (LoS) received path rays [4, 63], i.e.,

$$\mathbf{h}_k = \beta \sum_l \alpha_{k,l} \mathbf{a}(\varphi^{(k,l)}), \tag{5.68}$$

where $\beta = \sqrt{N_T/L}$, $\alpha_{k,l}$ represents the complex channel gain, $\varphi^{(k,l)}$ falls within the range $[\frac{-\pi}{2}, \frac{\pi}{2}]$ indicating the direction of the transmitted paths, and $\mathbf{a}(\varphi^{(k,l)})$ denotes the $N_T \times 1$ array steering vector. Its mth element is defined as $\left[\mathbf{a}(\varphi^{(k,l)})\right]_m = \exp\{-j\frac{2\pi}{\lambda}d(m-1)\sin(\varphi^{(k,l)})\}$, where λ represents the wavelength and d signifies the array element spacing.

Assuming Gaussian signaling, then the achievable rate for the kth user is

$$R_k = \log_2 \left| 1 + \frac{\frac{1}{K}|\mathbf{h}_k^{\mathsf{H}} \mathbf{F}_{\mathrm{RF}} \mathbf{f}_{\mathrm{BB}_k}|^2}{\frac{1}{K}\sum_{n \neq k} |\mathbf{h}_n^{\mathsf{H}} \mathbf{F}_{\mathrm{RF}} \mathbf{f}_{\mathrm{BB}_n}|^2 + \sigma^2} \right|, \tag{5.69}$$

and the sum-rate is $\bar{R} = \sum_{k \in \mathcal{K}} R_k$ for $\mathcal{K} = \{1, \dots, K\}$ [63].

Let us define the parameter set of the central model at the BS as $\boldsymbol{\theta} \in \mathbb{R}^P$, which contains P real-valued learnable parameters. The learning model establishes a nonlinear connection between the input $\mathcal{X}$ and the output $\mathcal{Y}$, expressed as $f(\boldsymbol{\theta}, \mathcal{X}) = \mathcal{Y}$.

In this study, our focus lies on the training phase of the global NN model. Thus, we assume each user possesses available training data pairs, such as the channel vector $\mathbf{h}_k$ (input) and the corresponding RF precoder $\mathbf{f}_{\mathrm{RF},k}$ (output label). It is noteworthy that channel and beamformer estimation can be conducted using both ML [23, 24, 54] and non-ML [4, 63] algorithms.

The training dataset for the kth user is denoted by

$$\mathcal{D}_k = (\mathcal{X}_k^{(1)}, \mathcal{Y}_k^{(1)}), \dots, (\mathcal{X}_k^{(\mathsf{D}_k)}, \mathcal{Y}_k^{(\mathsf{D}_k)}), \tag{5.70}$$

where $\mathsf{D}_k = |\mathcal{D}_k|$ represents the dataset size, and $\mathcal{X} = \bigcup_{k \in \mathcal{K}} \mathcal{X}_k$, $\mathcal{Y} = \bigcup_{k \in \mathcal{K}} \mathcal{Y}_k$. In works using ML [22–24, 68], the training of the global model occurs at the BS by gathering datasets from all users (as depicted in Figure 5.1). Subsequently, FL can be readily applied to these distributed datasets $\mathcal{D} = \cup_k \mathcal{D}_k$.

In Figure 5.11, the accuracy of corrupted channel data for FL-based hybrid beamforming (FLHB) is depicted. The degree of channel corruption is denoted by $\mathrm{SNR}_{\mathrm{TEST}}$, similar to $\mathrm{SNR}_{\mathrm{TRAIN}}$. It is evident that ML demonstrates superior performance compared to FLHB. In addition, the performance of FLHB diminishes as the value of K increases due to model aggregation issues caused by dataset diversity.

In Figure 5.12, the sum-rate for various techniques is illustrated when $\mathrm{SNR}_{\mathrm{TEST}} = 5\,\mathrm{dB}$. FL surpasses other algorithms, including the ML-based multilayer perceptron (MLP) technique. This superiority is attributed to FLHB leveraging the capabilities of CL, thereby providing enhanced feature representation. While utilizing the same neural network structures, CL based on

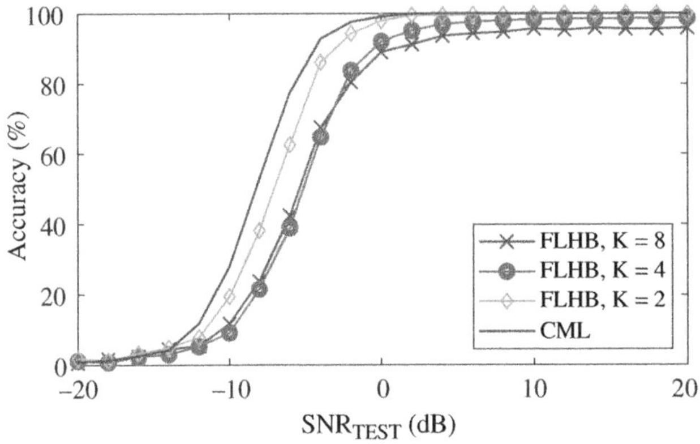

Figure 5.11 FL-based beamforming: Validation accuracy with respect to SNR_{TEST}.

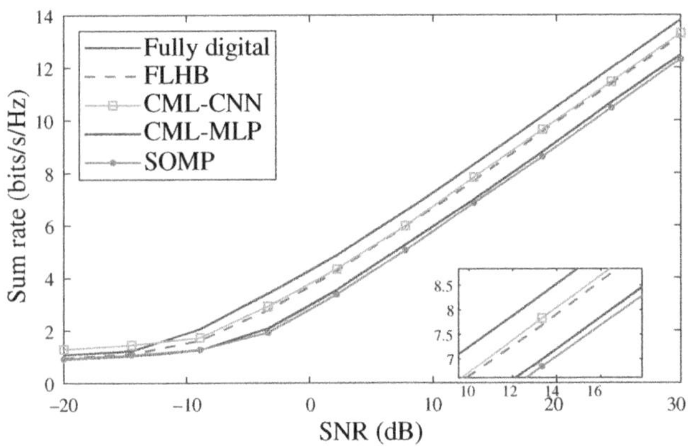

Figure 5.12 Spectral efficiency vs. SNR.

convolutional neural network (CML–CNN) exhibits slightly better performance than FLHB, given the access to the entire dataset at once compared with the decentralized training of FLHB.

5.5 Conclusions

In this chapter, we discussed FL-based implementation of various physical layer design problems, e.g., mmWave and THz channel estimation and hybrid beamforming. The FL framework for channel estimation in conventional and RIS-assisted massive MIMO systems has been presented in detail. While FL-based techniques provide a slight performance degradation in terms of channel estimation NMSE due to being unable to access the whole data at once, they enjoy low communication overhead. Specifically in THz channel estimation, a spherical wave signal model should be employed in order to model the receive channel accurately. It is shown that beam squint phenomenon in near-field can cause significant performance loss. Therefore, additional signal processing techniques are required to first accurately estimate the channel data as labels for FL. In the hybrid beamforming problem, FL can be applied to accurately estimate the user directions, which then leads to the construction of the hybrid beamformers given the CSI. Furthermore, FL-based beamforming can present satisfactory spectral efficiency performance as compared to conventional OMP-based techniques. The investigation of FL for wireless communications invites several future research directions such as

- FL implementation can be integrated with reinforcement learning-based approaches to work effectively when the dataset is not labeled. This can reduce the labor or labeling of the dataset.
- Fast implementation of FL is required, especially in THz applications, wherein the channel coherence time is critical. Furthermore, THz systems are envisioned to employ an ultra-massive number of antennas, which makes FL a more applicable approach than CL due to the large dataset size because of number of antennas.
- Implementation of FL in various emerging research areas introduces interesting application setups, e.g., integrated sensing and communications, vehicular networks as well as cell-free massive MIMO systems.

Acknowledgment

This work was supported in part by the Natural Sciences and Engineering Research Council of Canada (NSERC) and Ericsson Canada.

References

1 Heath, R.W., González-Prelcic, N., Rangan, S. et al. (2016). An overview of signal processing techniques for millimeter wave MIMO systems. *IEEE Journal on Selected Topics in Signal Processing* 10 (3): 436–453.

2 Rusek, F., Persson, D., Lau, B.K. et al. (2013). Scaling up MIMO: opportunities and challenges with very large arrays. *IEEE Signal Processing Magazine* 30 (1): 40–60.

3 Andrews, J.G., Buzzi, S., Choi, W. et al. (2014). What will 5G be? *IEEE Journal on Selected Areas in Communications* 32 (6): 1065–1082.

4 Ayach, O.E., Rajagopal, S., Abu-Surra, S. et al. (2014). Spatially sparse precoding in millimeter wave MIMO systems. *IEEE Transactions on Wireless Communications* 13 (3): 1499–1513.

5 Elbir, A.M., Mishra, K.V., Vorobyov, S.A., and Heath, R.W. (2023 Jun). Twenty-five years of advances in beamforming: from convex and nonconvex optimization to learning techniques. *IEEE Signal Processing Magazine* 40 (4): 118–131.

6 Wu, Q. and Zhang, R. (2020). Towards smart and reconfigurable environment: intelligent reflecting surface aided wireless network. *IEEE Communications Magazine* 58 (1): 106–112.

7 Elbir, AM and Mishra, KV (2020). A survey of deep learning architectures for intelligent reflecting surfaces. *arXiv*. https://arxiv.org/abs/2009.02540v3.

8 Huang, C., Hu, S., Alexandropoulos, G.C. et al. (2020 Jul). Holographic MIMO surfaces for 6G wireless networks: opportunities, challenges, and trends. *IEEE Wireless Communications* 27 (5): 118–125.

9 Huang, C., Zappone, A., Alexandropoulos, G.C. et al. (2019 Jun). Reconfigurable intelligent surfaces for energy efficiency in wireless communication. *IEEE Transactions on Wireless Communications* 18 (8): 4157–4170.

10 Huang, C., Mo, R., and Yuen, C. (2020 Jun). Reconfigurable intelligent surface assisted multiuser MISO systems exploiting deep reinforcement learning. *IEEE Journal on Selected Areas in Communications* 38 (8): 1839–1850.

11 Björnson, E., Van der Perre, L., Buzzi, S., and Larsson, E.G. (2019). Massive MIMO in sub-6 GHz and mmWave: physical, practical, and use-case differences. *IEEE Wireless Communications* 26 (2): 100–108.

12 Elbir, A.M. (2023 Apr). A unified approach for beam-split mitigation in terahertz wideband hybrid beamforming. *IEEE Transactions on Vehicular Technology* 72 (9): 12355–12360.

13 Alkhateeb, A. and Heath, R.W. (2016). Frequency selective hybrid precoding for limited feedback millimeter wave systems. *IEEE Transactions on Communications* 64 (5): 1801–1818.

14 Sohrabi, F. and Yu, W. (2017). Hybrid analog and digital beamforming for mmWave OFDM large-scale antenna arrays. *IEEE Journal on Selected Areas in Communications* 35 (7): 1432–1443.

15 Taha, A., Alrabeiah, M., and Alkhateeb, A. (2019). Enabling large intelligent surfaces with compressive sensing and deep learning. *arXiv preprint arXiv:190410136*.

16 Fan, D., Gao, F., Liu, Y. et al. (2018). Angle domain channel estimation in hybrid millimeter wave massive MIMO systems. *IEEE Transactions on Wireless Communications* 17 (12): 8165–8179.

17 Yin, H., Gesbert, D., Filippou, M., and Liu, Y. (2013). A coordinated approach to channel estimation in large-scale multiple-antenna systems. *IEEE Journal on Selected Areas in Communications* 31 (2): 264–273.

18 Elbir, A.M., Papazafeiropoulos, A.K., and Chatzinotas, S. (2021 Nov). Federated learning for physical layer design. *IEEE Communications Magazine* 59 (11): 81–87.

19 Dong, P., Zhang, H., Li, G.Y. et al. (2019 Sep). Deep CNN-based channel estimation for mmWave massive MIMO systems. *IEEE Journal on Selected Topics in Signal Processing* 13 (5): 989–1000.

20 Huang, H., Yang, J., Huang, H. et al. (2018 Sept). Deep learning for super-resolution channel estimation and DOA estimation based massive MIMO system. *IEEE Transactions on Vehicular Technology* 67 (9): 8549–8560.

21 Elbir, A.M., Papazafeiropoulos, A., Kourtessis, P., and Chatzinotas, S. (2020). Deep channel learning for large intelligent surfaces aided mm-Wave massive MIMO systems. *IEEE Wireless Communications Letters* 9 (9): 1447–1451.

22 Elbir, A.M. (2019). CNN-based precoder and combiner design in mmWave MIMO systems. *IEEE Communications Letters* 23 (7): 1240–1243.

23 Elbir, A.M. and Mishra, K.V. (2020). Joint antenna selection and hybrid beamformer design using unquantized and quantized deep learning networks. *IEEE Transactions on Wireless Communications* 19 (3): 1677–1688.

24 Elbir, A.M. and Papazafeiropoulos, A. (2020). Hybrid precoding for multi-user millimeter wave massive MIMO systems: a deep learning approach. *IEEE Transactions on Vehicular Technology* 69 (1): 552–563.

25 Elbir, A.M., Mishra, K.V., and Eldar, Y.C. (2019). Cognitive radar antenna selection via deep learning. *IET Radar, Sonar & Navigation* 13: 871–880.

26 Elbir, A.M. (2020). DeepMUSIC: Multiple signal classification via deep learning. *IEEE Sensors Letters* 4 (4): 1–4.

27 Amiri, M.M. and Gündüz, D. (2020). Federated learning over wireless fading channels. *IEEE Transactions on Wireless Communications* 19 (5): 3546–3557.

28 Li, T., Sahu, A.K., Talwalkar, A., and Smith, V. (2020). Federated learning: challenges, methods, and future directions. *IEEE Signal Processing Magazine* 37 (3): 50–60.

29 Elbir, A.M., Soner, B., Çöleri, S. et al. (2022). Federated learning in vehicular networks. *2022 IEEE International Mediterranean Conference on Communications and Networking (MeditCom)*, 05–8. IEEE.

30 Wadu, M.M., Samarakoon, S., and Bennis, M. (2020). Federated learning under channel uncertainty: joint client scheduling and resource allocation. *arXiv preprint arXiv:200200802*.

31 Zeng, T., Semiari, O., Mozaffari, M. et al. (2020). Federated learning in the sky: joint power allocation and scheduling with UAV swarms. *arXiv preprint arXiv:200208196*.

32 Batewela, S., Liu, C., Bennis, M. et al. (2020). Risk-sensitive task fetching and offloading for vehicular edge computing. *IEEE Communications Letters* 24 (3): 617–621.

33 Amiri, M.M. and Gündüz, D. (2020). Machine learning at the wireless edge: distributed stochastic gradient descent over-the-air. *IEEE Transactions on Signal Processing* 68: 2155–2169.

34 Elbir, A.M. and Coleri, S. (2021 Nov). Federated learning for channel estimation in conventional and RIS-assisted massive MIMO. *IEEE Transactions on Wireless Communications* 21 (6): 4255–4268.

35 Hou, W., Sun, J., Gui, G. et al. (2021 May). Federated learning for DL-CSI prediction in FDD massive MIMO systems. *IEEE Wireless Communications Letters* 10 (8): 1810–1814.

36 Elbir, A.M., Shi, W., Papazafeiropoulos, A.K. et al. (2023 Apr). Near-field terahertz communications: model-based and model-free channel estimation. *IEEE Access* 11: 36409–36420.

37 Elbir, A.M., Shi, W., Papazafeiropoulos, A.K. et al. (2023 Mar). Terahertz-band channel and beam split estimation via array perturbation model. *IEEE Open Journal of the Communications Society* 4: 892–907.

38 Elbir, A.M., Shi, W., Mishra, K.V., and Chatzinotas, S. (2023). Federated multi-task learning for THZ wideband channel and DoA estimation. *2023 IEEE International Conference on Acoustics, Speech, and Signal Processing Workshops (ICASSPW)*, 1–5. IEEE.

39 Elbir, A.M. and Coleri, S. (2020). Federated learning for hybrid beamforming in mm-Wave massive MIMO. *IEEE Communications Letters* 24 (12): 2795–2799.

40 Elbir, A.M., Coleri, S., and Mishra, K.V. (2021). Federated dropout learning for hybrid beamforming with spatial path index modulation in multi-user mmWave-MIMO systems. *ICASSP 2021 - 2021 IEEE International Conference on Acoustics, Speech and Signal Processing (ICASSP)*, 06–11. IEEE.

41 Elbir, A.M. (2020). A deep learning framework for hybrid beamforming without instantaneous CSI feedback. *IEEE Transactions on Vehicular Technology* 69 (10): 11743–11755.

42 Elbir, A.M. and Chatzinotas, S. (2023 Feb). BSA-OMP: Beam-split-aware orthogonal matching pursuit for THz channel estimation. *IEEE Wireless Communications Letters* 12 (4): 738–742.

43 Elbir, A.M., Mishra, K.V., and Chatzinotas, S. (2023). NBA-OMP: Near-field beam-split-aware orthogonal matching pursuit for wideband THz channel estimation. *ICASSP 2023-2023 IEEE International Conference on Acoustics, Speech and Signal Processing (ICASSP)*, 1–5. IEEE.

44 Sarieddeen, H., Alouini, M.S., and Al-Naffouri, T.Y. (2021). An overview of signal processing techniques for terahertz communications. *Proceedings of the IEEE* 109 (10): 1628–1665.

45 Elbir, A.M., Mishra, K.V., and Chatzinotas, S. (2021 Oct). Terahertz-band joint ultra-massive MIMO radar-communications: model-based and model-free hybrid beamforming. *IEEE Journal of Selected Topics in Signal Processing* 15 (6): 1468–1483.

46 Tan, J. and Dai, L. (2021 Apr). Wideband beam tracking in THz massive MIMO systems. *IEEE Journal on Selected Areas in Communications* 39 (6): 1693–1710.

47 Yuan, H., Yang, N., Yang, K. et al. (2020 Jul). Hybrid beamforming for terahertz multi-carrier systems over frequency selective fading. *IEEE Transactions on Communications* 68 (10): 6186–6199.

48 Björnson, E., Demir, Ö.T., and Sanguinetti, L. (2021). A primer on near-field beamforming for arrays and reconfigurable intelligent surfaces. *arXiv preprint arXiv:211006661*.

49 Elbir, A.M., Celik, A., and Eltawil, A.M. (2023). Near-field hybrid beamforming for terahertz-band integrated sensing and communications. *arXiv:2309.13984*.

50 Elbir, A.M., Mishra, K.V., and Chatzinotas, S. (2023). Spherical wavefront near-field DoA estimation in THz automotive radar. *arXiv:2310.16724*.

51 Cui, M., Dai, L., Wang, Z. et al. (2022). Near-field rainbow: wideband beam training for XL-MIMO. *IEEE Transactions on Wireless Communications* 22 (6): 3899–3912.

52 Elbir, A.M. and Tuncer, T.E. (2014 Sep). Far-field DOA estimation and near-field localization for multipath signals. *Radio Science* 49 (9): 765–776.

53 Bajwa, W.U., Haupt, J., Raz, G., and Nowak, R. (2008). Compressed channel sensing. *Annual Conference on Information Sciences and Systems*, 5–10.

54 Elbir, A.M., Mishra, K.V., Shankar, M.R.B., and Ottersten, B. (2021 Dec). A family of deep learning architectures for channel estimation and hybrid beamforming in multi-carrier mm-Wave massive MIMO. *IEEE Transactions on Cognitive Communications and Networking* 8 (2): 642–656.

55 Klautau, A., Batista, P., González-Prelcic, N. et al. (2018). 5G MIMO data for machine learning: application to beam-selection using deep learning. *2018 Information Theory and Applications Workshop (ITA)*, 1–9.

56 McMahan, H.B., Moore, E., Ramage, D., and Hampson, S. (2016). Arcas BAy. Communication-efficient learning of deep networks from decentralized data. *arXiv:1602.05629*. https://arxiv.org/abs/1602.05629v3.

57 Alistarh, D., Grubic, D., Li, J. et al. (2017). QSGD: Communication-efficient SGD via gradient quantization and encoding. *Advances in Neural Information Processing Systems*, 1709–1720.

58 Bishop, C.M. (1995 Jan). Training with noise is equivalent to Tikhonov regularization. *Neural Computation* 7 (1): 108–116.

59 Ang, F., Chen, L., Zhao, N. et al. (2020 Mar). Robust federated learning with noisy communication. *IEEE Transactions on Communications* 68 (6): 3452–3464.

60 Chen, M., Yang, Z., Saad, W. et al. (2020). A joint learning and communications framework for federated learning over wireless networks. *IEEE Transactions on Wireless Communications* 20 (1): 269–283.

61 Vu, T.T., Ngo, D.T., Tran, N.H. et al. (2020 Jun). Cell-free massive MIMO for wireless federated learning. *IEEE Transactions on Wireless Communications* 19 (10): 6377–6392.

62 Wei, L., Huang, C., Alexandropoulos, G.C. et al. (2023 Sep). Wireless communications empowered by reconfigurable intelligent surfaces: model-based vs model-free channel estimation. *Journal of Information and Intelligence* 1 (3): 253–266.

63 Alkhateeb, A., Leus, G., and Heath, R.W. (2015 Jul). Limited feedback hybrid precoding for multi-user millimeter wave systems. *IEEE Transactions on Wireless Communications* 14 (11): 6481–6494.

64 Dovelos, K., Matthaiou, M., Ngo, H.Q., and Bellalta, B. (2021 Apr). Channel estimation and hybrid combining for wideband terahertz massive MIMO systems. *IEEE Journal on Selected Areas in Communications* 39 (6): 1604–1620.

65 Tan, J. and Dai, L. (2021 May). Wideband channel estimation for THz massive MIMO. *China Communications* 18 (5): 66–80.

66 Wang, J., Lan, Z., Pyo, C.-w. et al. (2009). Beam codebook based beamforming protocol for multi-Gbps millimeter-wave WPAN systems. *IEEE Journal on Selected Areas in Communications* 27 (8): 1390–1399.

67 Wei, X. and Dai, L. (2021). Channel estimation for extremely large-scale massive MIMO: far-field, near-field, or hybrid-field? *IEEE Communications Letters* 26 (1): 177–181.

68 Huang, H., Peng, Y., Yang, J. et al. (2020). Fast beamforming design via deep learning. *IEEE Transactions on Vehicular Technology* 69 (1): 1065–1069.

6

Federated Learning in Mesh Networks

Xu Wang[1], Yuanzhu Chen[1], and Octavia A. Dobre[2]

[1] *School of Computing, Queen's University, Kingston, Ontario, Canada*
[2] *Faculty of Engineering and Applied Science, Memorial University, St. John's, Newfoundland and Labrador, Canada*

CHAPTER MENU

6.1 Introduction

6.1.1 Federated Learning

Artificial intelligence, especially machine learning (ML), is poised to greatly benefit from the substantial amount of data anticipated in the upcoming years. However, data from numerous Internet of Things (IoT) devices is often stored in a distributed manner for various applications like smart grids [1, 2], remote health monitoring, and the Internet of Vehicles (IoVs) [3]. In certain scenarios, collecting data in centralized entities, as traditionally done in ML, becomes impractical due to limited communication resources, data privacy concerns, or country regulations. Federated learning (FL) was introduced to address this issue by enabling participants to collaboratively train a global model while keeping their training data private. In FL, a central orchestration server takes charge of

Artificial Intelligence for Future Networks, First Edition.
Edited by Mohammad A. Matin, Sotirios K. Goudos, and George K. Karagiannidis.

generating and disseminating a global model to all participants [4]. Specifically, clients first train their models using locally available data and then send the model parameters to the central server for modal aggregation. The central server aggregates local modals and generates the global modal.

However, FL is subject to the single point of failure, trust dependencies, and bottlenecks at the server node [5]. Decentralized federated learning (DFL), also known as serverless or distributed FL, was proposed to address the limitations above. DFL revolves around sending locally computed updates from each node, such as model parameters or gradients, to other nodes within the FL network. DFL offers several advantages over traditional FL. In the first place, DFL dramatically reduces communication overhead compared to traditional FL [6]. DFL involves transmitting information between nodes rather than sending information to the central server, leading to more efficient data transmission and reduced network congestion. Besides, if a node fails or goes offline, the overall system remains functional as the local training and updating are performed independently. In contrast, a failure in the central server of traditional FL can disrupt the entire learning process [7]. Furthermore, DFL offers improved scalability [8]. This is because the communication and computational burden in traditional FL increases rapidly as the number of participants joining the computation increases. In contrast, DFL permits every node to conduct its training independently and locally. Every node individually trains its model using its locally available data, and the exchange of information among the nodes is limited to model updates, for example, model parameters or gradients. As such, this localized training approach leads to a reduction in the amount of data transmitted and lowers the overall communication overhead, resulting in improved scalability as the number of participants increases. In addition, DFL enables enhanced data privacy. Each node trains a local model over its own data and shares information with its neighbors rather than transmitting information to the central server. Thus, DFL ensures that sensitive information stays within local nodes, resulting in superior privacy protection in comparison to traditional FL [9]. In summary, DFL presents higher levels of privacy, efficiency, scalability, and robustness, enabling it an appealing substitute for conventional FL in various situations.

6.1.2 Mesh Networks

Mesh networks are decentralized systems where nodes connect directly and dynamically, allowing data to seamlessly hop from one node to its adjacent one until it reaches its intended destination [10]. In contrast to traditional network configurations that depend on a central router, each device in a mesh network is interlinked, creating an intricate web of connections. As we enter the era of the IoT, edge computing, and decentralized architectures, the importance of mesh networks becomes increasingly pronounced, offering resilient and adaptive connectivity solutions for complex and distributed digital ecosystems [11]. Their potential to provide efficient and robust communication pathways makes them indispensable in our rapidly evolving technological landscape.

6.1.3 The Convergence: Federated Learning on Mesh Networks

Mesh networks present an intriguing and fitting infrastructure for the deployment of FL models, owing to several key attributes. One of the primary reasons for this suitability is the distributed nature of both paradigms; FL thrives when executed in decentralized settings [12], and mesh networks intrinsically support this decentralized framework. In addition, the dynamic topologies of mesh networks are advantageous for FL environments [13]; as participants dynamically join or leave the network, mesh nodes can self-organize to optimize data flows, thereby facilitating efficient model training and updates. Besides, the peer-to-peer communication inherent in mesh networks can amplify the efficiency of updates in federated learning, facilitating rapid synchronization between devices [14]. Moreover, one of the foundational principles of FL is local computation, where learning occurs close to where data resides [15]. Mesh networks, by design, empower edge devices to compute closer to their data sources, thereby perfectly aligning with and complementing FL's emphasis on localized computations. This synergy between FL and mesh networks could revolutionize various domains, from healthcare to smart cities, by creating robust, scalable, and efficient learning systems [16]. In essence, the convergence of FL with mesh networks paves the way for more scalable, efficient, and decentralized ML applications.

The integration of FL with mesh networks offers several noteworthy advantages. One of the most notable benefits lies in resilience against failures; mesh networks are renowned for their self-healing capabilities, wherein the network can autonomously reroute data if certain nodes fail [17]. This ensures that FL processes remain robust, as multiple pathways exist to continue the learning even if particular devices or nodes are compromised. Another significant advantage is enhanced data privacy. By emphasizing local data processing, there is a reduced need to transmit raw, sensitive data across the network, adhering to FL's principle of local computation [18]. Furthermore, the decentralized nature of mesh networks means data is not stored in centralized servers, dramatically reducing vulnerability to large-scale data breaches. This convergence leads to efficient resource utilization. The inherent load balancing in mesh networks distributes both computation and communication loads efficiently, while FL intelligently optimizes device participation, taking into account various factors like battery life and computational capabilities [19, 20].

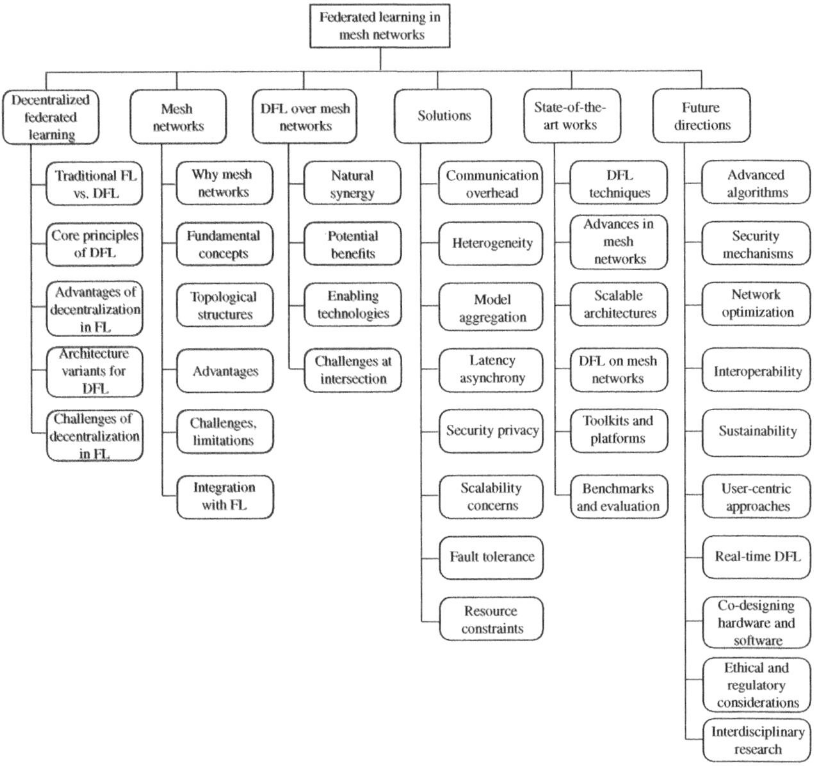

Figure 6.1 Outline of this chapter.

Navigating the convergence of FL and mesh networks presents a myriad of challenges that need to be meticulously addressed. For instance, the communication overhead and latency in such complex networks can undermine real-time data processing requirements. Moreover, scalability, data heterogeneity, and resource constraints further compound the complexity, potentially impeding efficient data exchange and learning. Furthermore, concerns surrounding privacy, security, and the dynamics of network topology, combined with synchronization issues, energy consumption, and potential interference and congestion, add layers of intricacy to the integration. Figure 6.1 illustrates the outline of this chapter.

6.2 Decentralized Federated Learning

FL emerges as a groundbreaking paradigm in ML where models are trained across multiple devices or servers while keeping the data localized. This approach was primarily motivated by the increasing need for enhanced data privacy and the challenges associated with centralized data storage and transmission [21]. As data generation continues to grow exponentially, FL offers a solution that prioritizes both efficiency and privacy, laying the foundation for a new age in decentralized data processing.

6.2.1 Traditional Federated Learning versus Decentralized Federated Learning

Traditional FL operates under a centralized coordinator, wherein numerous devices send their model updates to a primary server which then aggregates and disseminates the information as shown in Figure 6.2a. During the training process, the central roles are played by clients and a cloud server. Let us represent the ensemble of N clients as $\mathcal{N} = 1, \dots, N$, where each client possesses a dataset $D_{i \in \mathcal{N}}$, acquired locally. The dataset $D_i = \left\{\mathbf{x}_i^{(1)}, \mathbf{x}_i^{(2)}, \dots, \mathbf{x}_i^{(n)}\right\}$ for a given client includes n instances of data, with each instance being a d-dimensional vector $\mathbf{x}^{(j)} = \left[x_1^{(j)}, x_2^{(j)}, \dots, x_d^{(j)}\right]$. Client i trains a global model $\mathbf{w}i$ using its dataset D_i and then submits this refined model to the cloud server. This server consolidates the individual models $\mathbf{w} = \cup_{i \in \mathcal{N}} \mathbf{w}_i$ to formulate a unified global model $\mathbf{w}_G$. It is presumed that clients act truthfully, utilizing their authentic locally sourced data for training and sending genuine model updates to the server.

The basic steps for the model training process are as follows:

- *Initialization*: Eligible clients register with the cloud server, which then creates an initial random global model $\mathbf{w}_G^0$ and sets the learning rate η for the training.
- *Model distribution*: The cloud server distributes the global model to randomly chosen energy storage systems to enable distributed learning.
- *Distributed learning*: In each iteration t, where $t = 1, \dots, T$, utilizing the current global model $\mathbf{w}_G^t$, every selected client assesses the training loss.
  The client computes the gradient, g_i, against its dataset with the present model $\mathbf{w}_i^t$ through multiple iterations of stochastic gradient descent. The model is then updated as follows:

$$\mathbf{w}_i^t = \mathbf{w}_i^t - \eta g_i. \tag{6.1}$$

- *Model feedback*: The clients relay their updated models back to the cloud server.
- *Model aggregation*: The server performs a weighted aggregation of the updates, where

$$\mathbf{w}_G^{t+1} = \frac{1}{\sum_{i \in \mathcal{N}} D_i} \sum_{i=1}^{N} D_i \mathbf{w}_i^t. \tag{6.2}$$

In contrast, DFL shown in Figure 6.2b diverges by eliminating the central coordinator, allowing devices or nodes to interact and learn directly from each

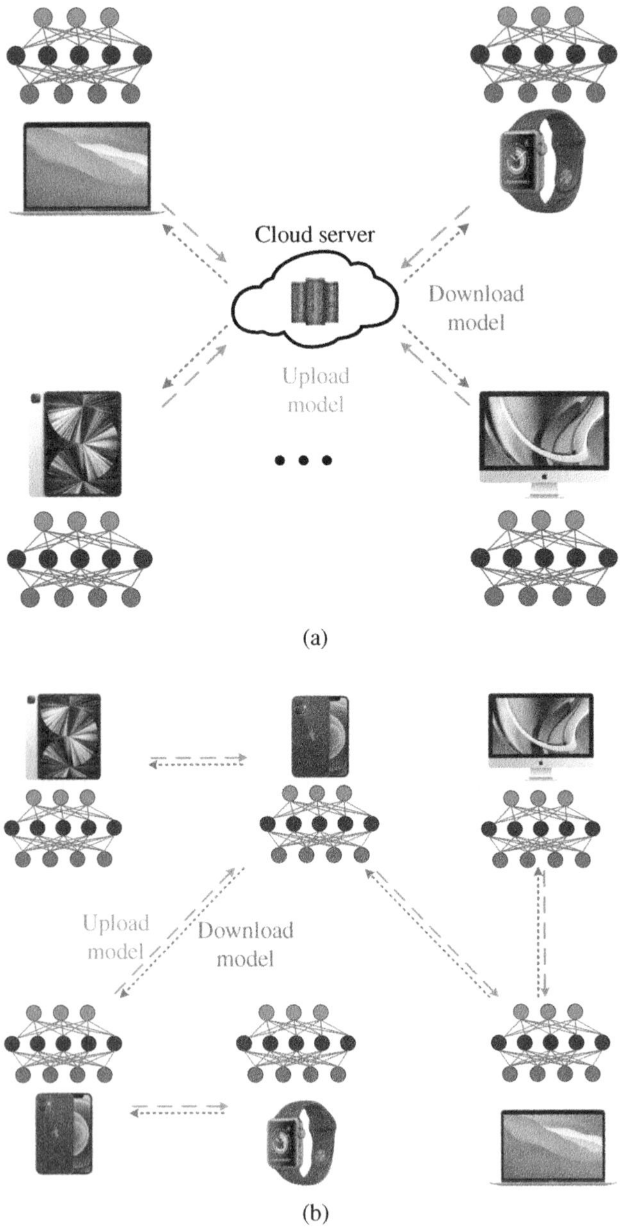

Figure 6.2 Traditional federated learning and decentralized federated learning. (a) Traditional federated learning. (b) Decentralized federated learning.

other [22]. The primary distinction between the two lies in the presence or absence of this centralized entity, which plays a pivotal role in model aggregation in the traditional setup. One of the major shortcomings of centralized FL is that it poses a single point of failure and may suffer from scalability issues, bottlenecks, or even centralized data breaches [23]. Decentralized approaches aim to overcome these limitations by distributing the computational load and responsibilities across all participating nodes, thereby enhancing resilience, security, and scalability [24].

6.2.2 Core Principles of Decentralized Federated Learning

DFL is built on several core principles that enable its functionality and effectiveness. One of these principles is local computation, which means that all computations and training take place on local devices [25]. This is crucial for enhancing data privacy and also reduces the overhead associated with transmitting data over a network. Another fundamental principle is asynchronous updates [26]. Nodes or devices within this setup conduct training and update their models based on their own schedules. They then share these updates when convenient, allowing for adaptability and efficiency in diverse network conditions. Furthermore, peer-to-peer communication is vital to DFL. Devices, instead of connecting through a central entity for model aggregation and dissemination, communicate directly with one another [27]. This mode of communication fosters robustness and resilience in the learning process by preventing potential bottlenecks and eliminating a single point of failure. These principles collectively drive DFL to offer a collaborative and efficient learning environment across devices.

6.2.3 Advantages of Decentralization in Federated Learning

Decentralization in FL offers multiple advantages that improve the efficiency and effectiveness of the learning process. One of its primary advantages is that the decentralized framework inherently offers enhanced robustness [28]. Without a single point of failure, the system becomes more resilient to potential disruptions, ensuring that learning processes remain consistent and reliable across the network. Besides, reduced communication overhead is indeed a critical advantage for DFL; by allowing devices to communicate directly, the system avoids frequent exchanges with a central server [29]. In addition, privacy is inherently increased as local computation allows sensitive data to stay on the originating devices, mitigating the risks associated with data transmission and centralized storage.

6.2.4 Architecture Variants for Decentralized Federated Learning

In DFL, the choice of network topology plays a pivotal role in determining communication patterns and efficiency. Within partially connected networks, nodes are connected to only a subset of other nodes. This approach reduces communication overhead, provides better scalability, and ensures flexibility in adapting to various challenges [30]. Alternatively, node clustering focuses on grouping nodes based on specific criteria, such as data similarities or computational capabilities [9, 31]. By doing so, it allows for faster local aggregations within clusters and minimizes global communication, sharing only the essential updates from representative nodes within clusters. It also aids in optimizing resource allocation and adaptability, enabling the reconfiguration of clusters based on changing requirements [32]. The strategic selection of these topologies has a direct impact on communication efficiency, convergence speed, and overall performance in FL scenarios.

6.2.5 Challenges of Decentralization in Federated Learning

While DFL promises a range of benefits, it also brings forth several challenges. Data heterogeneity is a challenge pervasive in both centralized and DFL. In these setups, distinct data distributions across nodes, stemming from varied user behaviors or device functionalities, can compromise model generalization. Addressing this inconsistency is paramount to ensure the effective training and convergence of federated models [33]. Moreover, model aggregation becomes a complex task, as there is a need to harmonize updates from disparate nodes, which may be operating asynchronously and at different speeds, further complicated by the presence of stragglers [34]. In addition, DFL presents a significant advantage by distributing the data processing workload. Instead of directing traffic to a single central machine, computations are spread across multiple nodes, leading to reduced network stress and potential bottlenecks [35]. This decentralized approach not only enhances efficiency but also promotes resilience and adaptability in the network. These challenges are further compounded by the need for scalable solutions that ensure security and privacy, manage resource constraints effectively, and establish compelling incentive mechanisms to foster active participation.

6.3 Mesh Networks

In a mesh network, devices or nodes are directly and dynamically interconnected, allowing data to be relayed through multiple paths without the need for a centralized hub or server [36]. This structure offers a distinct advantage in terms of resilience and adaptability compared to the traditional network architectures.

6.3.1 Why Mesh Networks

Mesh networks have become increasingly relevant in today's computing and communication environments due to their inherent flexibility and resilience [37]. Their decentralized architecture allows for seamless connectivity in diverse settings, from urban landscapes with dense device deployments to remote areas where traditional infrastructure might be lacking. Moreover, as the world gravitates towards IoT and decentralized systems, mesh networks stand out as an optimal solution, facilitating efficient data exchanges among a myriad of devices [38].

6.3.2 Fundamental Concepts and Terminologies

In mesh networks, nodes denote individual devices or entities, each playing a pivotal role in data transmission, reception, and relay [39]. These nodes connect with one another using links, establishing the communication pathways for data to navigate. The collective arrangement of these interconnected nodes gives rise to what we term the topology of the network. As data travels between nodes, it may traverse multiple intermediary nodes; each such passage is referred to as a hop. The mechanism that charts out the path for data packets across these nodes is called routing [40]. Latency captures the duration it takes for a data packet to transition from its origin to its intended destination. Besides, throughput quantifies the volume of data that the network can handle in a given timeframe, reflecting its overall capacity and efficiency. Both metrics are essential in evaluating and optimizing the performance of a mesh network [20]. Figure 6.3 shows a diagrammatic representation of a mesh network.

6.3.3 Topological Structures

Mesh networks exhibit a unique topology structure that stands out from traditional networking models. Unlike hierarchical or star-based systems, where devices connect to a central node or through a linear pathway, mesh networks are inherently decentralized. In this topology, each node or device is connected to multiple other nodes, allowing for direct peer-to-peer communication. This interconnectedness ensures data can take multiple potential paths from source to destination, enhancing the network's reliability and resilience [41]. If one pathway or node fails, the network can automatically reroute data through alternative routes, ensuring uninterrupted communication. Moreover, mesh networks can dynamically adapt and self-organize, meaning they can reconfigure themselves as nodes are added or removed [42]. This scalability is particularly advantageous for large deployments. Another hallmark of mesh topology

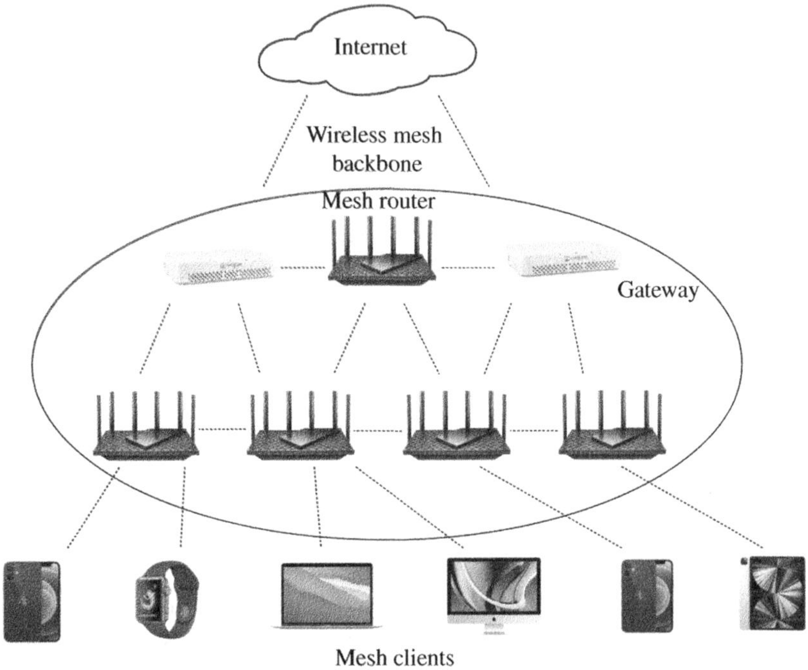

Figure 6.3 A diagrammatic representation of a mesh network.

is its inherent redundancy, with multiple routes available between any two nodes [43]. This redundancy not only fortifies the network against potential points of failure but also optimizes data transmission by reducing bottlenecks and congestion.

6.3.4 Advantages of Mesh Networks

Mesh networks inherently excel in robustness and fault tolerance thanks to their interconnected design [44]. In a mesh network, data traverses through a series of interconnected nodes, allowing it to find multiple potential paths from source to destination. This interconnectedness ensures a resilient and adaptable communication flow, even if certain pathways become unavailable. A defining feature of mesh networks is self-healing, which refers to the network's ability to automatically detect and reroute data when a particular pathway or node becomes unavailable. Instead of relying on a single path, the network dynamically adjusts its routes to ensure continuous data flow [45]. Furthermore, mesh networks present a solution to scalability challenges faced by traditional networks [46]. As more nodes are added, the network can naturally expand and adjust without the need for central infrastructure changes. This makes mesh networks particularly suited for large-scale deployments where the system needs to grow and adapt organically.

6.3.5 Challenges and Limitations

Mesh networks, while innovative and versatile, face their share of challenges. Scalability, for instance, can be a double-edged sword; as the network grows, the complexity of managing connections and routing data can escalate, potentially impacting performance [47]. In wireless mesh networks, interference emerges prominently due to overlapping channels and concurrent transmissions by adjacent nodes. This interference can lead to reduced data throughput, increased packet loss, and unpredictable network behavior [48]. Maintenance complexities arise due to the decentralized nature of these networks; diagnosing and rectifying a fault may require more effort compared to centralized systems [49]. Furthermore, inherent vulnerabilities pose threats to the integrity and security of mesh networks [50]. Eavesdropping, where unauthorized entities might intercept transmissions, is a concern. More sophisticated threats include man-in-the-middle attacks, where an attacker intercepts and potentially alters communications between two nodes, and replay attacks, in which an attacker resends previously transmitted data to cause confusion or unauthorized actions. Sybil attacks involve a malicious node impersonating multiple nodes, while in selective forwarding, a malicious node might deliberately drop packets. The decentralized nature of mesh networks can make them vulnerable to attacks targeting their topology and routing mechanisms.

Despite these challenges, there are methods and protocols designed to enhance the security of mesh networks. Encrypted transmissions can be employed to ensure data privacy and deter eavesdropping [51]. In addition, secure routing protocols can be implemented to prevent routing-based attacks, ensuring that data packets take safe and efficient paths through the network [52]. As mesh networks continue to evolve, the development and application of these security measures will be crucial in maintaining their reliability and trustworthiness.

6.3.6 Integration with Federated Learning

The confluence of mesh networks with DFL emerges as a promising development in modern computing paradigms. Mesh networks, with their inherently decentralized nature, offer an apt infrastructure for FL, where data remain localized and computations are distributed across the network. By pairing these two, there is potential to realize a system where devices collaboratively learn from shared models without central coordination while ensuring that raw data remain s confined to its original device [53]. This fusion not only amplifies the benefits of each approach but also paves the way for more resilient, scalable, and privacy-preserving ML models in real-world applications [54].

6.4 The Intersection: Decentralized Federated Learning over Mesh Networks

6.4.1 Natural Synergy Between Federated Learning and Mesh Networks

Both DFL and mesh networks emerge from a foundational principle of decentralization, operating without a singular, centralized entity controlling the entirety of their processes. In DFL, the decentralized characteristic stands out as computations spread across numerous devices, each processing its own data independently, eliminating the central sharing of raw data [55]. Similarly, mesh networks are designed as systems where every node cooperates, distributing data in a nonhierarchical manner, avoiding the traditional centralized hub-and-spoke model seen in other network architectures [39]. This shared foundation promotes heightened privacy; in the DFL context, raw data remains on the originating device, while mesh networks reduce the need for centralized data storage or transmission hubs that might be more susceptible to breaches [56, 57]. The principle of local computation in FL aligns seamlessly with the peer-to-peer communications intrinsic to mesh networks. Together, they foster an environment where data is processed and communicated efficiently without compromising user privacy or system integrity. Figure 6.4 shows a diagrammatic representation of DFL over mesh networks.

6.4.2 Potential Benefits of the Convergence

The convergence of DFL and mesh networks holds the promise of several significant benefits. One of the foremost advantages is enhanced privacy. DFL inherently promotes data processing at the source without centralized aggregation, ensuring that raw data remain localized [58]. When combined with mesh networks, which operate without a single point of control or data concentration, the distributed nature of both systems synergistically boosts data privacy, making the merged paradigm particularly compelling for applications where data confidentiality is paramount [59]. In addition, the nonhierarchical structure of mesh networks contributes to heightened resilience and redundancy in the learning process [60].

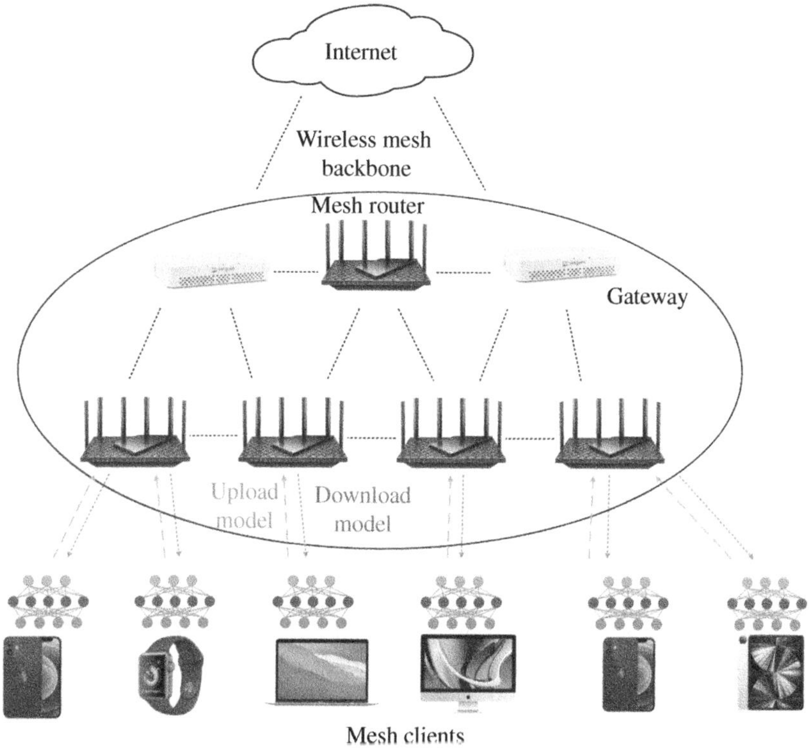

Figure 6.4 A diagrammatic representation of DFL over mesh networks.

In this architecture, if one node fails or becomes compromised, the network can continue to function undisturbed, leveraging alternative pathways. This inherent resilience of mesh networks ensures that the DFL process remains robust, with multiple pathways ensuring continuous learning, even if certain nodes or devices encounter issues. Scalability is another significant advantage of this synergy. Mesh networks allow for the dynamic addition of nodes without the need for central coordination, thus facilitating the expansion of the network in response to varying needs [61]. This characteristic is especially beneficial in DFL scenarios, where the network might need to adapt and scale according to the availability of data sources and computational resources. Localized learning in the context of DFL offers significant benefits when combined with mesh networks. DFL eliminates the need for a central aggregation center, instead allowing nodes to process and learn from local data directly [62]. When paired with the decentralized nature of mesh networks, this approach further minimizes communication overheads, capitalizing on the strengths of each individual node and fostering a more efficient, adaptive, and distributed learning environment.

6.4.3 Enabling Technologies

The seamless intersection of DFL and mesh networks has been made possible through a series of technological advancements. In recent years, we have witnessed a marked increase in the adoption of decentralized systems, transforming the way we view data processing and communications [63, 64]. Alongside this growth, lightweight cryptographic methods have emerged, playing a pivotal role in ensuring data security within these decentralized paradigms [65]. Another breakthrough is the strides made in edge computing, which champions efficient localized data processing, allowing systems to handle data right where they are generated [66]. Complementing this is the evolution of data compression techniques tailored for mesh communications, enhancing the speed and efficiency of data transfer within the network [67, 68]. However, mesh networks, with their distinct communication dynamics, pose certain challenges to FL algorithms. These challenges necessitate modifications to the algorithms, ensuring they are optimized for the mesh environment. Solutions were developed to reduce latency and address potential inconsistencies or data loss, with numerous case studies highlighting scenarios where these enhanced algorithms have significantly outperformed their traditional counterparts [69, 70]. Last, a notable feature that amplifies the efficacy of DFL over mesh is the allowance for dynamic node participation. Traditional systems with a static set of nodes often falter in the ever-evolving mesh environment. However, by permitting nodes to dynamically join or leave the learning process, DFL achieves a level of flexibility and robustness unmatched by static counterparts. This dynamic participation not only bolsters system resilience but also enriches data diversity, leading to more accurate and encompassing learning models [71, 72].

6.4.4 Challenges at the Intersection

As the convergence of DFL and mesh networks progresses, it brings forth a set of intricate challenges that need addressing. Communication overhead, data heterogeneity, and the nuances of model aggregation in a decentralized setup become significant hurdles. In addition, concerns surrounding network latency, security, scalability, robustness, and resource constraints surface prominently. Navigating these challenges is crucial for harnessing the full potential of this intersection.

6.4.4.1 Communication Overhead

Mesh networks can significantly amplify communication overhead, especially in the context of DFL. The frequent exchange of model updates across numerous nodes demands high bandwidth. Furthermore, with more participants in the network, there is an increased likelihood of redundant communications or data transfers. The quality and timeliness of these communications can directly influence the efficacy of DFL [73]. Thus, optimizing communication strategies becomes essential to reduce overhead while ensuring the seamless operation of DFL on mesh networks.

6.4.4.2 Data Heterogeneity and Non-IID Data

Across both traditional and DFL setups, data heterogeneity remains a pervasive challenge [74]. Nodes from different backgrounds, locations, or user groups contribute data with varied distributions, leading to non-IID datasets. Such diversity can introduce biases or lead to models that do not generalize well. In mesh networks, this heterogeneity might be even more pronounced due to the varied nature of the nodes and their data sources. Regardless of the FL type, techniques to address this heterogeneity are essential for achieving reliable and consistent learning outcomes.

6.4.4.3 Model Aggregation in Decentralized Networks

Aggregating models in a decentralized setup, where no single authority oversees the process, is intricate [75]. Nodes in a mesh network might be updating their local models at different rates, leading to discrepancies during aggregation [76]. Furthermore, considering the varying data volumes and qualities across nodes, simple aggregation methods might not suffice. A weighted aggregation based on node credibility or data volume might be needed. Ensuring a holistic and representative global model in such a decentralized scenario requires specialized aggregation strategies.

6.4.4.4 Network Latency and Asynchrony

In vast mesh networks, transmission delays are almost inevitable. Nodes might send model updates at different times, resulting in asynchronous communications [77]. Handling such asynchrony becomes challenging, especially when trying to maintain the consistency of DFL processes. Out-of-order or delayed model updates might affect the learning progress or even introduce errors. Therefore, robust synchronization mechanisms and latency-aware algorithms are essential in such setups.

6.4.4.5 Security and Privacy Concerns

Mesh networks, being decentralized, present unique security challenges. Nodes could be vulnerable to eavesdropping, tampering, or even sophisticated man-in-the-middle attacks [78]. Furthermore, ensuring data privacy while facilitating collaborative learning is a challenge [79]. Incorporating end-to-end encryption and techniques like differential privacy becomes paramount to safeguard user data and model integrity [80]. Balancing security without compromising the efficiency of DFL is a delicate task in these environments.

6.4.4.6 Scalability Concerns

Scaling a DFL setup over an extensive mesh network is not straightforward. As the network grows, so do the complexities related to communication, model aggregation, and data heterogeneity [81]. The increased number of nodes might lead to communication bottlenecks or challenges in maintaining a consistent learning pace across the network. Efficient strategies to manage the growth, such as node clustering or hierarchical setups, become essential. Despite the challenges, ensuring the system's scalability is crucial for real-world applications and broader adoption.

6.4.4.7 Fault Tolerance and Robustness

In the realm of integrating DFL with mesh networks, ensuring fault tolerance and robustness emerges as a pivotal challenge [82]. Due to the inherently distributed nature of mesh networks, they are susceptible to node failures, which can disrupt the DFL process [83]. The decentralized approach, while offering flexibility, lacks a central authority that can swiftly counteract such failures. Consequently, the system must be robust enough to autonomously handle interruptions and recover without significant data loss or latency. Addressing this issue requires sophisticated strategies and protocols to ensure the seamless continuity of learning, even in the face of unpredictable network challenges.

6.4.4.8 Resource Constraints

At the crossroads of DFL and mesh networks, managing resource constraints stands out as a prominent challenge. Given that mesh networks often comprise nodes with varying capabilities, not all nodes are equipped with the same computational power, memory, or battery life [84]. Such disparities can hinder the uniform participation of all nodes in the DFL process. Moreover, some nodes might get overloaded, leading to inefficient model training and updates. Addressing these resource constraints is essential to ensure that the DFL process over mesh networks remains efficient and equitable for all participating nodes.

6.5 Solutions

6.5.1 Communication Overhead

Addressing challenges is crucial for ensuring the effective implementation and broad acceptance of any technological advancement. Within the context of large-scale mesh networks, the communication overhead emerges as a pressing concern [85]. A predominant issue is the escalated bandwidth consumption, often resulting from numerous devices transmitting vast amounts of data. This heavy data traffic can lead to discernible delays as information attempts to traverse through the complex web of interconnected nodes. Another byproduct of this dense communication is congestion, especially when multiple nodes in the network engage in simultaneous data transfers [86]. To mitigate these challenges, data compression techniques are gaining traction, helping reduce the sheer volume of data transmissions, thus easing bandwidth demands [87]. Furthermore, the integration of efficient routing algorithms ensures that data finds the most optimal path, reducing delays and preventing potential congestion, enabling a smoother and more streamlined communication process within the network [88].

6.5.2 Data Heterogeneity and Non-IID Data

In decentralized systems like mesh networks, data heterogeneity emerges as a notable challenge [89, 90]. Different nodes within the network can possess data that vary not only in quality but also in quantity, and its distribution might not always be consistent or representative of the entire network. This divergence, especially when the data are not independently and identically distributed (non-IID), can pose challenges for DFL where model training expects certain uniformity. To combat these inconsistencies, various techniques have been proposed. One promising approach to address this is reinforcement learning. A deep Q-learning-based mechanism is proposed to select specific devices during each communication round, aiming to enhance validation accuracy while minimizing the number of communication rounds used [91]. In addition, leveraging robust aggregation methods, like the geometric median [92] or trimmed mean [93], can provide a more stable and representative global model. These methods are less sensitive to outliers or extreme updates from any node, ensuring a balanced integration of knowledge from all participating nodes.

6.5.3 Model Aggregation in Decentralized Networks

In the landscape of DFL, one of the standout challenges is model aggregation in a decentralized setting. The task of consolidating model updates from a vast array of nodes without the guidance of a central coordinator presents issues like potential inconsistencies in updates and unintended biases. In addition, ensuring the reliability of these aggregated updates in such a dispersed system can be complex. However, innovative solutions have been developed to tackle these challenges. Decentralized consensus algorithms have been introduced to help nodes reach a collective agreement on model updates, ensuring both consistency and integrity of the learning process [94]. Alongside decentralized averaging techniques have proven invaluable [22]. They empower nodes to collaboratively compute a global average of model parameters, integrating the diverse contributions from all nodes and producing a harmonized, aggregated model that is reflective of the entire network's insights.

6.5.4 Latency and Asynchrony

In large-scale mesh networks designed for FL, latency and asynchrony become prominent challenges. As data flows between numerous nodes, inevitable delays can occur, leading to updates that are out-of-sync across the entire network. This asynchronicity, if unaddressed, can compromise the effectiveness and efficiency of the global model's training. To counteract these challenges, asynchronous training methods have been introduced, allowing nodes to process and send updates without having to wait for all other nodes [95]. This method can significantly improve the pace of learning and model aggregation. Alongside this, leveraging adaptive learning rates can be transformative. By dynamically adjusting the learning rates based on the timeliness and quality of incoming updates, the system can effectively accommodate and make the most of asynchronous contributions, ensuring a robust and timely model evolution [96].

6.5.5 Security and Privacy Concerns

In the DFL over mesh networks, security and privacy emerge as paramount concerns. The distributed nature of mesh networks, while offering numerous advantages, also presents potential vulnerabilities. Eavesdropping, where unauthorized entities might intercept communications, poses a significant risk. Similarly, man-in-the-middle attacks, wherein malicious actors secretly relay or even alter the communication between two parties, can jeopardize both data integrity and confidentiality [97]. To counter these challenges, several strategies have been proposed. Encryption protocols can be employed to ensure that data in transit remain s confidential and tamper-proof [98]. Differential privacy provides a means to add noise to the data, ensuring that individual data points cannot be reverse-engineered while still retaining the dataset's overall utility. In addition, secure multiparty computation techniques offer a promising solution, allowing multiple parties to collaboratively compute a function over their inputs while keeping those inputs private, perfectly aligning with the essence of DFL in mesh environments [99].

6.5.6 Scalability Concerns

As DFL on mesh networks continues to expand, scalability remains a pivotal challenge. With the exponential growth in the number of participating nodes, complexities arise in areas such as communication, model aggregation, and resource management, hindering the overall system's performance. However, solutions are being developed to address these challenges. Adopting hierarchical topologies is one such approach, where nodes are organized in structured layers or tiers, facilitating efficient data flow and aggregation [100]. Node clustering is another effective strategy; by grouping similar or proximal nodes together, localized learning can occur within these clusters before contributing to the global model [101]. This not only reduces communication overhead but also streamlines the learning process, enhancing the system's scalability.

6.5.7 Fault Tolerance and Robustness

Within the realm of DFL over mesh networks, ensuring continuous and efficient operation is crucial. However, the distributed nature of these systems brings about inherent vulnerabilities. Mesh networks, for all their advantages, can be susceptible to node failures, either due to technical glitches or external disruptions, leading to potential disruptions in the learning process. In response to such challenges, several solutions have been proposed. Self-healing techniques are instrumental in these scenarios; they allow networks to automatically detect issues and reroute data and communications to ensure uninterrupted operation [102]. Moreover, the integration of proactive node monitoring tools helps in the early detection of potential failures, allowing the system to take preventive measures and guaranteeing the robustness of the FL process across the network [103–105].

6.5.8 Resource Constraints

Within the vast landscape of DFL on mesh networks, each node brings its unique resource profile, leading to diverse challenges. Particularly, some nodes, especially those in edge or IoT settings, face limitations such as restricted battery life. This is compounded by other constraints, like limited computational power or storage space, which can affect their effective participation in the FL process. To circumvent these challenges, innovative solutions have been devised. Energy-aware algorithms have been designed to streamline operations in a manner that conservatively utilizes power, thus extending the effective operational time of nodes with limited battery resources [106]. Another avenue explored is selective participation based on node capabilities; this strategy ensures that nodes partake in the learning process when they meet certain resource criteria, guaranteeing efficiency [107]. In doing so, the mesh network can ensure consistent and high-quality learning while accommodating the diverse capabilities of its nodes.

6.6 State-of-the-Art and Noteworthy Implementations

The rapid evolution of technology consistently brings forth new challenges and breakthroughs. This section aims to shed light on the latest advancements in the realm of DFL on mesh networks. By exploring the state-of-the-art methods and implementations, we will gain a comprehensive understanding of where the field stands currently and the pioneering solutions that are shaping its future.

6.6.1 Decentralized Federated Learning Techniques

DFL, an emerging paradigm in ML, is continuously evolving, driven by innovations in various technical dimensions. As this paradigm grows, understanding the underlying techniques and mechanisms becomes essential for its broader adoption and refinement.

6.6.1.1 Network Topology

Network topology, which refers to the arrangement of different elements in a communication network, is crucial in DFL. The choice of topology can significantly influence data flow, node interaction, and overall system efficiency [108]. The ring topology connects devices in a circular manner, ensuring an ordered data flow, but it might be slower and susceptible to bottlenecks [109]. The Star topology centralizes all nodes around a hub, streamlining data management and communication speed, but there is the risk of a system halt if the hub malfunctions [110]. The hierarchical Tree topology merges groups of star-configured networks linked linearly, enabling DFL tasks to be segmented into manageable sub-tasks [111]. Wrapping up, the Hybrid topology integrates characteristics from various structures, offering flexibility and tailored solutions. The choice of topology in DFL directly influences factors like latency, scalability, and resilience, and decisions often depend on data type, node capabilities, and spatial distribution.

6.6.1.2 Communication Protocols

Optimized communication protocols are fundamental in DFL as they dictate how swiftly and efficiently data is exchanged and processed across nodes. With the surge in DFL applications, there has been a push to minimize communication overhead, which often emerges as a primary bottleneck. To tackle this, gradient compression techniques have gained traction. By sending only the most significant gradients, they dramatically cut down the data exchanged between nodes [112]. In addition, sparsification and quantization methods are employed, where only a subset of updates is transmitted, or updates are mapped to fewer bits, respectively [113, 114].

6.6.1.3 Privacy Enhancements

In the realm of DFL, ensuring privacy without compromising learning quality has garnered significant attention. Differential Privacy stands out as a robust mathematical framework that introduces calibrated noise to model updates, protecting individual data points while maintaining the overall learning trajectory [115]. Another advanced cryptographic tool, homomorphic encryption, enables calculations directly on encrypted data, meaning model updates can be combined in their encrypted state, preventing potential information leaks [116]. Secure multiparty computation enables multiple parties to collaboratively compute a function over their inputs while keeping those inputs private from each other [117]. With this method, model updates from different nodes can be consolidated without any node witnessing the updates from its peers. In addition, strategies such as FL with secure aggregation have emerged, using protocols that encrypt model updates in a manner where only the aggregated update can be decrypted [118]. Another innovative approach is Model Splitting [119]. It divides a ML model into several segments. These segments are distributed and processed across different nodes or devices. This ensures that no single entity has access to the entire model, thereby enhancing privacy and security.

6.6.2 Advances in Mesh Networking Technologies

In the realm of DFL, mesh networking serves as a backbone, connecting a multitude of nodes to foster collaborative learning. Over the past few years, advances in mesh networking technologies have significantly influenced the efficacy of DFL by ensuring faster and more efficient communication between participating nodes.

6.6.2.1 Low-Latency Protocols

One of the crucial requirements for effective DFL is rapid communication between nodes. Newly developed low-latency protocols have been a game-changer in this aspect. By minimizing the time delay in data transmission, these protocols ensure timely model updates and faster convergence. They are particularly beneficial in scenarios where real-time processing is imperative, such as autonomous vehicles or emergency response systems [120]. With reduced latency, the entire learning process becomes more agile, allowing for quicker decision-making based on the updated models [121, 122]. Furthermore, as mesh networks continue to grow in size and complexity, there is a continuous push for the development of more sophisticated protocols. These not only target latency reduction but also focus on optimizing bandwidth usage, enhancing fault tolerance, and ensuring seamless node communication even in highly dense networks. The combination of these factors helps in realizing the full potential of DFL in mesh environments.

6.6.2.2 Scalable Architectures

As data volumes explode and the number of devices within networks multiply, the need for scalable mesh network architectures has become paramount. These architectures are designed to adaptively accommodate growth in network size without a corresponding decline in performance. Advanced routing algorithms in these structures allow for efficient data transfers, ensuring that the increased number of nodes does not lead to bottlenecks or significant latency spikes [123–125]. In addition, the modular nature of scalable architectures means that nodes can be easily added or removed without disrupting the network's overall functionality. This flexibility ensures that as DFL tasks become more complex or demand more computational resources, the underlying mesh network can be adjusted to meet these needs without requiring a complete overhaul. Furthermore, these scalable designs often incorporate redundancy mechanisms, ensuring that the network remains robust and resilient even as its size and complexity grow [126].

6.6.2.3 Security Enhancements

Given the decentralized nature of mesh networks and the sensitive data typically involved in DFL, ensuring robust security is paramount. Mesh networks utilize a hybrid encryption approach, combining both symmetric and asymmetric key encryptions, enhancing the network's performance [127]. This combined encryption technique optimizes the network by decreasing data transmission size, lessening computational burdens, and quickening encryption–decryption processes. Furthermore, there is a growing interest in applying blockchain technology to mesh networking [128, 129]. By leveraging the decentralized ledger and cryptographic features of blockchain, it provides a tamper-proof history of data transactions across the network, ensuring both integrity and authenticity. Another avenue of improvement is in the realm of secure key distribution and management, ensuring that only authorized nodes have access to the shared data in the network [130, 131]. Last, continuous monitoring and intrusion detection systems have been incorporated to promptly detect and respond to any potential threats, thereby fortifying the security perimeter around the mesh network supporting DFL [132]. These enhancements not only instill confidence in using mesh networks for DFL but also lay a robust foundation for future innovations.

6.6.3 Decentralized Federated Learning on Mesh Networks: Integrated Approaches

DFL on mesh networks has been a focal point of numerous research studies, given the synergistic potential they hold. In this section, we spotlight several pivotal research papers, projects, and technologies that have notably advanced the integration of DFL with mesh networks.

DFL operates without a central server, allowing participants to exchange model parameter updates with neighbors in a peer-to-peer manner. When integrated with the Slotted ALOHA wireless mesh networking protocol, this framework enhances direct communication between participants, facilitating efficient parameter sharing and model updates [69]. This combination ensures efficient communication, reduced latency, and enhanced data privacy in a mesh network topology, where devices communicate directly with one-hop neighbors. Besides, Savazzi et al. [133] introduce a real-time simulation platform tailored to evaluate the efficacy of decentralized FL across wireless mesh networks based on the industry-recognized Time Slotted Channel Hopping MAC layer.

Pinyoanuntapong et al. [134] present a DFL framework capable of functioning in both synchronous and asynchronous modes to mitigate the intense communication traffic surrounding the central server. The framework is experimented in a physical testbed. The physical IoT network testbed described in the paper is based on a multihop wireless mesh network. It is composed of 10 wireless edge computing nodes, also referred to as IoT workers. Each node is integrated with a wireless embedded router to facilitate communication. These routers are equipped with three wireless interface cards, enabling the creation of a multiradio wireless node environment. Every mesh router operates in a mesh point mode.

Real-world mesh networks often exhibit local rather than global connectivity, with groups of nodes sometimes isolated or sparsely connected due to their limited communication range. Zecchin et al. [135] investigate the role of Unmanned Aerial Vehicles in supporting decentralized learning over ground mesh networks.

6.6.4 Toolkits and Platforms

In the dynamic landscape of DFL on mesh networks, a slew of toolkits and platforms are taking center stage, each contributing pioneering techniques and functionalities. Among the leaders is Google's TensorFlow Federated, which rides on the success of TensorFlow, but fine-tunes its capabilities for the unique demands of FL, emphasizing data security and privacy [136]. Not to be overlooked, OpenMined's PySyft is becoming indispensable for its innovative take on encrypted, privacy-centric computations, augmenting the safety protocols inherent in DFL [137]. One platform showing significant promise in the domain, especially for mobile deployments, is FedML [138]. With its optimization strategies tailored for mobile devices, it recognizes and harnesses the power and ubiquity of mobile computing in today's digital framework. Furthermore, PaddleFL stands out for its holistic approach to FL, boasting an adaptable architecture well-suited for diverse DFL applications. IBM's FL platform, on the other hand, merges the trustworthiness and innovation associated with IBM, offering scalable solutions without compromising data integrity [139]. Last, platforms like Microsoft's Azure IoT Edge [140] and Federated AI technology enablers are making waves in their respective niches. Azure IoT Edge, renowned for its IoT functionalities, is progressively integrating capabilities tailored for real-time DFL scenarios, bridging the gap between cloud expertise and edge devices. Together, these platforms and toolkits are shaping the future of DFL implementations on mesh networks, providing researchers and developers the arsenal they need to push the boundaries further.

6.6.5 Benchmarks and Evaluation

Benchmarks serve as a pivotal aspect in the realm of DFL on mesh networks, allowing practitioners to evaluate and compare the efficiency, security, and overall performance of various implementations. As the field matures, several benchmarks have been established that cover a wide array of domains, each tailored to reflect real-world conditions and challenges pertinent to DFL on mesh networks.

FedML [138] serves as a pivotal benchmarking tool in the FL domain. It is a versatile platform offering baseline implementations for various ML models and FL algorithms like FedAvg, FedNAS, Vertical FL, and split learning. Notably, FedML supports multiple computational setups, from distributed training to mobile on-device training and independent simulations. While some experimental outcomes are still being refined, its multifaceted capabilities have positioned FedML as a key benchmark framework for FL enthusiasts and researchers.

FedEval [141] serves as a unique evaluation platform tailored specifically for FL. It meticulously assesses FL approaches based on multiple pivotal metrics, including accuracy, communication efficiency, time consumption, privacy, and robustness. Leveraging the capabilities of Docker containers, FedEval ensures a consistent and isolated evaluation environment, which is especially crucial given potential hardware limitations. Currently, the framework is compatible with algorithms like FedAvg and has been tested using models such as MLP and LeNet, marking its significance in the FL.

The Open Application Repository for Federated Learning (OARF) [142] is a benchmark suite designed for FL. Unlike previous benchmarks that primarily relied on synthetic datasets, OARF uses publicly available datasets to simulate real-world application scenarios across various domains like image, text, and structured data. The suite is characterized by its diversity in data size, distribution, feature distribution, and learning task complexity. OARF provides reference implementations and evaluates key aspects of FL, such as model accuracy, communication cost, throughput, and convergence time.

Edge AIBench [143] is a comprehensive end-to-end AI edge computing benchmarking suite. It investigates typical Edge AI scenarios and uses representative real-world datasets. The benchmark suite covers the entire edge computing framework, including cloud servers, edge computing layers, and client-side devices. Four primary scenarios are highlighted: intensive care unit (ICU) patient monitor, surveillance camera, smart home, and autonomous vehicle. The Edge AIBench is open-sourced and aims to address the challenges and complexities of benchmarking in the evolving field of Edge AI.

6.7 Future Directions and Open Research Challenges

As the digital landscape continues to evolve, DFL and mesh networks have emerged at the forefront, showcasing a transformative potential in distributed computing and collaboration. As the merger of these two domains unfolds, it becomes imperative to foresee their trajectory, paving the way for innovations and addressing budding challenges. To harness the full potential of this confluence, it is crucial to not only understand its current state but also to envision where it might head in the future.

6.7.1 Advanced Algorithms

A pressing area of interest lies in the development of advanced algorithms tailored for DFL. Existing methodologies, while effective, have room for enhancement, especially concerning efficiency, resilience, and privacy preservation. The exploration of concepts like zero-shot learning [144] and transfer learning [145], when adapted for DFL, promises the optimization of model training with limited data or transferring knowledge across disparate domains. Envisioning further into the horizon, intriguing prospects arise, such as the integration of quantum mechanics [146] to aid FL or the application of neuromorphic computing [147], mimicking the human brain's architecture to enhance the efficacy and speed of DFL processes. These potential directions not only open doors to improved performance but also set the stage for revolutionary advancements in decentralized ML.

6.7.2 Enhanced Security Mechanisms

DFL, with its innovative approach to leveraging data from a multitude of nodes without central consolidation, inherently presents unique security challenges. As the scope of the mesh network expands and integrates more nodes, maintaining the integrity, privacy, and security of both data and models is of utmost importance.

Quantum cryptography [148] is emerging as a groundbreaking method in this arena. Utilizing the principles of quantum mechanics, it crafts encryption protocols that are considered exceptionally secure against potential breaches. It offers a heightened level of security for data transmissions across mesh networks. Another promising avenue is differential privacy. By adding carefully calibrated noise to data or queries, differential privacy ensures that individual data points within a set cannot be reverse-engineered, protecting user data while still allowing for meaningful data analysis.

6.7.3 Network Optimization

In the evolving landscape of DFL, the optimization of mesh network topologies stands out as a paramount concern. Ensuring efficient connections between a magnitude of nodes that minimize latency, maximize throughput, and ensure robustness requires meticulous design and constant adaptation. As these networks become more intricate and expansive, pinpointing and addressing potential inefficiencies and bottlenecks is of utmost importance.

One promising avenue being explored is the concept of self-organizing networks [149]. Such networks dynamically adapt and reconfigure themselves based on the real-time requirements and constraints of the system, eliminating the demand for manual intervention. They assess the network's conditions, automatically making adjustments to enhance performance, resilience, and efficiency.

Dynamic routing, another prospective solution, offers the potential to smartly route data packets based on current network conditions, traffic patterns, and even the anticipated future state of the network. By constantly evaluating and adjusting routes, this approach ensures that data travels through the most efficient path, thereby reducing latency and preventing congestion.

Finally, with the inevitable uncertainties and potential failures in any large-scale decentralized system, building in robust fault tolerance mechanisms is essential [150]. Future research will undoubtedly delve deeper into creating systems that can swiftly detect anomalies or failures and automatically reconfigure to maintain seamless operations.

6.7.4 Interoperability and Standardization

As the realm of DFL within mesh networks continues to grow, a key direction is ensuring that a myriad of devices, platforms, and protocols can seamlessly communicate and collaborate. The vision is a cohesive ecosystem where disparate systems can effortlessly exchange information, share resources, and synchronize model updates without any friction.

However, achieving this harmonious integration poses a formidable challenge. The tech world is diverse, with manufacturers, developers, and researchers often adopting varied approaches to design and functionality. This diversity, while fostering innovation, can sometimes lead to systems that operate in silos.

The challenge, therefore, is to develop universally accepted standards that strike a balance between innovation and compatibility.

A globally recognized framework can serve as a guiding light, ensuring all participants in the DFL process, regardless of their origin or design specifics, adhere to a common set of protocols and guidelines. Such standardization not only facilitates smoother operations but also ensures that the advancements in one area can be easily adopted and leveraged by others. Achieving this would mark a significant leap in the efficient and widespread application of DFL in mesh networks.

6.7.5 Energy Efficiency and Sustainability

The rise of DFL in mesh networks brings to the forefront the critical need for energy efficiency. With the proliferation of devices and nodes in these networks, it is essential to ensure that the systems are sustainable in their energy consumption. Research on the development of energy-efficient algorithms is not merely an engineering challenge but an environmental imperative. By streamlining computations and reducing overhead, these algorithms can play a pivotal role in conserving energy, thus promoting longer device lifetimes and reduced carbon footprints.

Furthermore, integrating renewable energy-powered mesh nodes offers a promising avenue toward sustainability [151]. By harnessing solar, wind, or other renewable energy sources, mesh nodes can operate independently of non-sustainable power grids, thereby reducing their carbon footprint.

Furthermore, integrating energy harvesting techniques [152] within the mesh-DFL framework provides an innovative avenue to capture and utilize ambient energy sources. Such techniques can harness energy from vibrations, thermal gradients, or even RF signals, turning otherwise wasted energy into valuable power for the system. Embracing such forward-thinking strategies can pave the way for a more sustainable future in the realm of DFL and mesh networks.

6.7.6 User-Centric Approaches

The evolution of DFL and mesh networks has illuminated the need for a paradigm shift toward more user-centric models and services. As the technology ecosystem becomes increasingly sophisticated, it is imperative that end-users remain at the core of its advancements.

Personalized FL [153] is an exemplary instance of this shift. Instead of applying a one-size-fits-all approach, personalized FL focuses on tailoring models based on individual user data, ensuring more accurate and relevant outcomes. This personalization not only enhances the efficacy of the models but also caters to the unique requirements and preferences of each user.

Furthermore, as mesh networks grow in complexity and scale, optimizing the user experience within these networks is of paramount importance. Achieving seamless connectivity, reducing latency, and ensuring uninterrupted data flow are some of the key objectives. By focusing on the user's perspective, researchers can identify pain points and streamline the network's operation to enhance overall satisfaction and usability. In essence, the future of DFL in mesh networks will be greatly influenced by how effectively it can be molded around the user, making these approaches not just desirable but essential.

6.7.7 Real-time Decentralized Federated Learning

The concept of real-time DFL presents an exciting frontier in the domain of mesh networks. While the advantages of real-time DFL, such as immediate data analysis and rapid model adaptation, are abundantly clear, its implementation is not without challenges. Mesh networks, by virtue of their decentralized structure, can face issues like network latency, data transmission delays, and node synchronization problems, all of which can impede real-time operations.

Despite these challenges, the potential applications of real-time DFL are vast and transformative. Consider autonomous vehicles: they require instantaneous data processing to make split-second driving decisions. By integrating real-time DFL, vehicles can learn from the collective experiences of other vehicles on the road, enhancing their predictive accuracy and ensuring safer navigation. Similarly, in emergency response systems, real-time DFL can enable immediate data sharing among responders, optimizing rescue operations by providing real-time insights and coordination. As the landscape of technology continues to evolve, the quest for perfecting real-time DFL will undoubtedly shape the future of responsive, intelligent systems.

6.7.8 Codesigning Hardware and Software

The evolution of DFL calls for a rethinking of the traditional boundaries between hardware and software development. Instead of considering these two essential components separately, there is a growing emphasis on a holistic approach where network devices and learning algorithms are codeveloped. By ensuring a tight integration between the two, performance can be optimized.

When hardware and software are developed collaboratively, they can take full advantage of each other's capabilities [154]. For instance, specialized hardware can be designed to accelerate specific computational tasks inherent to an algorithm. On the other hand, software can be optimized to fully utilize unique hardware features, enhancing efficiency and processing speed. This synergistic approach can potentially lead to breakthroughs in performance. As the need for real-time processing and immediate insights intensifies, the codesign strategy will be crucial in advancing the capabilities of DFL.

6.7.9 Ethical and Regulatory Considerations

As the realm of DFL continues to expand, there is a growing imperative to grapple with the ethical and regulatory challenges. The promise of FL is grounded in its capacity to process data locally, which inherently respects user privacy. However, the decentralized nature further complicates the landscape of ethical AI. How do we ensure that AI models, trained on diverse data sources, act without bias? Who truly owns the data, especially when it remains on the local devices, and how do users give or revoke consent [155]?

Furthermore, the potential misuse of decentralized federated systems, especially when they touch on sensitive domains such as healthcare or finance, is a pressing concern. The intersection of data privacy and utility is a delicate balance. Ethical considerations such as ensuring transparency, fairness, and robustness in AI models while preserving individual rights are paramount. Regulators and researchers will need to forge a path that addresses these challenges while enabling the vast potential benefits of this technology.

6.7.10 Interdisciplinary Research

DFL over mesh networks presents a landscape so intricate that its full potential can only be realized by transcending traditional disciplinary boundaries. By intertwining the field with social sciences, we can gain insights into human behavior, collective decision-making, and user preferences, offering a more nuanced understanding of how such systems will be adopted and utilized by communities.

Furthermore, as DFL grows, there is a need to ensure participation, and here economic models come into play. By developing sophisticated incentivization strategies, we can encourage more nodes to join, share, and contribute to the network, making it more robust and diverse [156]. On the frontier of technological intersection, quantum computing presents an exciting prospect [157]. Its potential to process vast amounts of data swiftly could radically transform how DFL operates over mesh networks, paving the way for faster, more efficient learning processes. Together, these interdisciplinary efforts promise richer, more comprehensive solutions for the challenges DFL faces.

6.8 Concluding Remarks

In the ever-evolving world of distributed computing and ML, the fusion of FL with mesh networks signifies a monumental stride toward a decentralized, efficient, and privacy-conscious future. Throughout this survey, we have navigated the intricate terrain of this convergence, elucidating both the challenges faced and the opportunities that lie ahead. From cutting-edge methodologies in DFL to the most recent advancements in mesh network technologies, it is evident that the synergy of these domains holds tremendous promise. The intersection of FL and mesh networks remains rich with opportunities, and we eagerly anticipate the innovations and breakthroughs the future holds.

References

1 Li, S., Ngai, E., and Voigt, T. (2023). Byzantine-robust aggregation in federated learning empowered industrial IoT. *IEEE Transactions on Industrial Informatics* 19 (2): 1165–1175.

2 Jithish, J., Alangot, B., Mahalingam, N., and Yeo, K.S. (2023). Distributed anomaly detection in smart grids: a federated learning-based approach. *IEEE Access* 11: 7157–7179.

3 Zhou, H., Zheng, Y., Huang, H. et al. (2023). Toward robust hierarchical federated learning in internet of vehicles. *IEEE Transactions on Intelligent Transportation Systems* 24 (5): 5600–5614.

4 McMahan, B., Moore, E., Ramage, D. et al. (2017). Communication-efficient learning of deep networks from decentralized data. *Proceedings of International Conference on Artificial Intelligence and Statistics (AISTATS)*, 1273–1282. Fort Lauderdal, USA.

5 He, L., Bian, A., and Jaggi, M. (2018). COLA: Decentralized linear learning. In: *Proceedings of Conference on Neural Information Processing Systems (NeurPIS)* (ed. S. Bengio, H. Wallach, H. Larochelle, et al.), 4541–4551. Montreal, Canada.

6 Savazzi, S., Nicoli, M., Bennis, M. et al. (2021). Opportunities of federated learning in connected, cooperative, and automated industrial systems. *IEEE Communications Magazine* 59 (2): 16–21.

7 Pappas, C., Chatzopoulos, D., Lalis, S., and Vavalis, M. (2021). IPLS: A framework for decentralized federated learning. *Proceedings of IFIP Networking Conference*, 1–6. Espoo and Helsinki, Finland.

8 Ye, H., Liang, L., and Li, G.Y. (2022). Decentralized federated learning with unreliable communications. *IEEE Journal of Selected Topics in Signal Processing* 16 (3): 487–500.

9 Kalra, S., Wen, J., Cresswell, J.C. et al. (2023). Decentralized federated learning through proxy model sharing. *Nature Communications* 14 (1): 2899.

10 Bruno, R., Conti, M., and Gregori, E. (2005). Mesh networks: commodity multihop ad hoc networks. *IEEE Communications Magazine* 43 (3): 123–131.

11 Li, L., Zhao, X., Geng, S. et al. (2020). Robust beamforming design for SWIPT-based multi-radio wireless mesh network with cooperative jamming. *Information* 11 (3): 138.

12 Su, X., Zhou, Y., Cui, L., and Liu, J. (2023). On model transmission strategies in federated learning with lossy communications. *IEEE Transactions on Parallel and Distributed Systems* 34 (4): 1173–1185.

13 Darroudi, S.M. and Gomez, C. (2017). Bluetooth low energy mesh networks: a survey. *Sensors* 17 (7): 1467.

14 Al Asaad, A., Gopalakrishnan, S., and Leung, V. (2009). Peer-to-peer file sharing over wireless mesh networks. *2009 IEEE Pacific Rim Conference on Communications, Computers and Signal eProcessing*, 697–702.

15 Li, T., Sahu, A.K., Talwalkar, A., and Smith, V. (2020). Federated learning: challenges, methods, and future directions. *IEEE Signal Processing Magazine* 37 (3): 50–60.

16 Nguyen, D.C., Ding, M., Pathirana, P.N. et al. (2021). Federated learning for Internet of Things: a comprehensive survey. *IEEE Communications Surveys & Tutorials* 23 (3): 1622–1658.

17 Shahdad, S.Y., Sabahath, A., and Parveez, R. (2016). Architecture, issues and challenges of wireless mesh network. *2016 International Conference on Communication and Signal Processing (ICCSP)*, 0557–0560.

18 Lim, W.Y.B., Luong, N.C., Hoang, D.T. et al. (2020). Federated learning in mobile edge networks: a comprehensive survey. *IEEE Communications Surveys & Tutorials* 22 (3): 2031–2063.

19 Jeyakarthic, M. and Subalakshmi, N. (2023). Energy saving slot allocation-based multicast routing in cloud wireless mesh network. *International Journal of Cloud Computing* 12 (2–4): 148–162.

20 Kushwah, R. (2023). A novel traffic aware reliable gateway selection in wireless mesh network. *Cluster Computing* 27 (1): 673–687.

21 Zhang, C., Xie, Y., Bai, H. et al. (2021). A survey on federated learning. *Knowledge-Based Systems* 216: 106775.

22 Beltrán, E.T.M., Pérez, M.Q., Sánchez, P.M.S. et al. (2023). Decentralized federated learning: fundamentals, state of the art, frameworks, trends, and challenges. *IEEE Communications Surveys & Tutorials* 25 (4): 2983–3013.

23 Gholami, A., Torkzaban, N., and Baras, J.S. (2022). Trusted decentralized federated learning. *2022 IEEE 19th Annual Consumer Communications & Networking Conference (CCNC)*, 1–6.

24 Roy, A.G., Siddiqui, S., Pölsterl, S. et al. (2019). BrainTorrent: A peer-to-peer environment for decentralized federated learning. *arXiv preprint arXiv:1905.06731*.

25 Aledhari, M., Razzak, R., Parizi, R.M., and Saeed, F. (2020). Federated learning: a survey on enabling technologies, protocols, and applications. *IEEE Access* 8: 140699–140725.

26 Liu, Q., Yang, B., Wang, Z. et al. (2022). Asynchronous decentralized federated learning for collaborative fault diagnosis of PV stations. *IEEE Transactions on Network Science and Engineering* 9 (3): 1680–1696.

27 Wink, T. and Nochta, Z. (2021). An approach for peer-to-peer federated learning. *2021 51st Annual IEEE/IFIP International Conference on Dependable Systems and Networks Workshops (DSN-W)*, 150–157.

28 Lian, Z. and Su, C. (2022). Decentralized federated learning for Internet of Things anomaly detection. *Proceedings of the 2022 ACM on Asia Conference on Computer and Communications Security*, 1249–1251.

29 Liu, W., Chen, L., and Zhang, W. (2022). Decentralized federated learning: balancing communication and computing costs. *IEEE Transactions on Signal and Information Processing over Networks* 8: 131–143.

30 Vogels, T., Hendrikx, H., and Jaggi, M. (2022). Beyond spectral gap: the role of the topology in decentralized learning. *Proceedings of Conference on Neural Information Processing Systems (NeurPIS)*, volume 35, 15039–15050, New Orleans, USA.

31 Duan, M., Liu, D., Ji, X. et al. (2021). FedGroup: Efficient federated learning via decomposed similarity-based clustering. *Proceedings of IEEE International Conference on Parallel & Distributed Processing with Applications, Big Data & Cloud Computing, Sustainable Computing & Communications, Social Computing & Networking (ISPA/BDCloud/SocialCom/SustainCom)*, 228–237, New York City, NY, USA.

32 Al-Abiad, M.S., Obeed, M., Hossain, M.J., and Chaaban, A. (2023). Decentralized aggregation for energy-efficient federated learning via D2D communications. *IEEE Transactions on Communications* 71 (6): 3333–3351.

33 Mendieta, M., Yang, T., Wang, P. et al. (2022). Local learning matters: rethinking data heterogeneity in federated learning. *Proceedings of the IEEE/CVF Conference on Computer Vision and Pattern Recognition*, 8397–8406.

34 Lin, F.P.-C., Hosseinalipour, S., Azam, S.S. et al. (2021). Semi-decentralized federated learning with cooperative D2D local model aggregations. *IEEE Journal on Selected Areas in Communications* 39 (12): 3851–3869.

35 Li, Y., Chen, C., Liu, N. et al. (2021). A blockchain-based decentralized federated learning framework with committee consensus. *IEEE Network* 35 (1): 234–241.

36 Conti, M. and Giordano, S. (2014). Mobile ad hoc networking: milestones, challenges, and new research directions. *IEEE Communications Magazine* 52 (1): 85–96.

37 Baran, P. (1964). On distributed communications networks. *IEEE Transactions on Communications Systems* 12 (1): 1–9.

38 Liu, Y., Tong, K.-F., Qiu, X. et al. (2017). Wireless mesh networks in IoT networks. *2017 International Workshop on Electromagnetics: Applications and Student Innovation Competition*, 183–185.

39 Parvin, J.R. (2019). An overview of wireless mesh networks. In: *Wireless Mesh Networks-Security, Architectures and Protocols*. IntechOpen.

40 Yang, Y., Wang, J., and Kravets, R. (2005). Designing routing metrics for mesh networks. *IEEE Workshop on Wireless Mesh Networks (WiMesh)*, 1–9. Santa Clara, CA, USA.

41 Liu, Q., Cheng, L., Jia, A.L., and Liu, C. (2021). Deep reinforcement learning for communication flow control in wireless mesh networks. *IEEE Network* 35 (2): 112–119.

42 Tang, J., Xue, G., and Zhang, W. (2005). Interference-aware topology control and QoS routing in multi-channel wireless mesh networks. *Proceedings of the 6th ACM International Symposium on Mobile Ad Hoc Networking and Computing*, 68–77.

43 Vaishampayan, R., Garcia-Luna-Aceves, J.J., and Obraczka, K. (2007). An adaptive redundancy protocol for mesh based multicasting. *Computer Communications* 30 (5): 1015–1028.

44 Dang, K., Shen, J.-Z., Dong, L.-D., and Xia, Y.-X. (2013). A graph route-based superframe scheduling scheme in wirelesshart mesh networks for high robustness. *Wireless Personal Communications* 71: 2431–2444.

45 Zhou, P., Fang, X., Fang, Y. et al. (2019). Beam management and self-healing for mmWave UAV mesh networks. *IEEE Transactions on Vehicular Technology* 68 (2): 1718–1732.

46 Sampaio, S., Souto, P., and Vasques, F. (2016). A review of scalability and topological stability issues in IEEE 802.11s wireless mesh networks deployments. *International Journal of Communication Systems* 29 (4): 671–693.

47 Rondón, R., Mahmood, A., Grimaldi, S., and Gidlund, M. (2020). Understanding the performance of bluetooth mesh: reliability, delay, and scalability analysis. *IEEE Internet of Things Journal* 7 (3): 2089–2101.

48 Benni, N.S. and Manvi, S.S. (2022). Channel allocation scheme to mitigate interference in 5G backhaul wireless mesh networks. *2022 International Conference on Data Science*, Agents & Artificial Intelligence (ICDSAAI), volume 01, 1–5.

49 Ebrahim, N.S. (2023). Complexity of IoT world-review of challenges and opportunities in application development. *2023 International Conference on Smart Computing and Application (ICSCA)*, 1–5.

50 Kavitha, A., Reddy, V.B., Singh, N. et al. (2022). Security in IoT mesh networks based on trust similarity. *IEEE Access* 10: 121712–121724.

51 Han, Y., Zhang, G., Meng, L. et al. (2022). Application of mesh ad-hoc network in power grid engineering. *2022 IEEE 5th Advanced Information Management, Communicates, Electronic and Automation Control Conference (IMCEC)*, volume 5, 1492–1496.

52 Yuan, X. and Chen, Y. (2022). Secure routing protocol based on dynamic reputation and load balancing in wireless mesh networks. *Journal of Cloud Computing* 11 (1): 77.

53 Yin, K., Wu, B., Zhu, R. et al. (2022). DLDP-FL: Dynamic local differential privacy federated learning method based on mesh network edge devices. *Journal of Computational Science* 63: 101789.

54 Giménez, N.L., Solé, J.M., and Freitag, F. (2023). Embedded federated learning over a LoRa mesh network. *Pervasive and Mobile Computing* 93: 101819.

55 Kairouz, P., McMahan, H.B., Avent, B. et al. (2021). Advances and open problems in federated learning. *Foundations and Trends in Machine Learning* 14 (1-2): 1–210.

56 Nouri, N.A., Aliouat, Z., Naouri, A., and Hassak, S.A. (2023). Accelerated PSO algorithm applied to clients coverage and routers connectivity in wireless mesh networks. *Journal of Ambient Intelligence and Humanized Computing* 14 (1): 207–221.

57 Zhu, J., Cao, J., Saxena, D. et al. (2023). Blockchain-empowered federated learning: challenges, solutions, and future directions. *ACM Computing Surveys* 55 (11): 1–31.

58 Shi, Y., Shen, L., Wei, K. et al. (2023). Improving the model consistency of decentralized federated learning. *arXiv preprint arXiv:2302.04083.*

59 Barolli, A., Bylykbashi, K., Qafzezi, E. et al. (2023). A comparison study of Weibull, normal and Boulevard distributions for wireless mesh networks considering different router replacement methods by a hybrid intelligent simulation system. *Journal of Ambient Intelligence and Humanized Computing* 14 (8): 10181–10194.

60 Castillo, A.C. (2023, 2023). An overview resilience in computer networks and network topologies using different metrics. In: *Future of Information and Communication Conference* (ed. K. Arai), 588–605. Springer.

61 Kurt, A., Sahin, A., Harrilal-Parchment, R., and Akkaya, K. (2023). LNMesh: Who said you need internet to send Bitcoin? Offline lightning network payments using community wireless mesh networks. *arXiv preprint arXiv:2304.14559.*

62 Chellapandi, V.P., Upadhyay, A., Hashemi, A., and Żak, S.H. (2023). On the convergence of decentralized federated learning under imperfect information sharing. *IEEE Control Systems Letters* 7: 2982–2987.

63 Gugueoth, V., Safavat, S., Shetty, S., and Rawat, D. (2023). A review of IoT security and privacy using decentralized blockchain techniques. *Computer Science Review* 50: 100585.

64 Falayi, A., Wang, Q., Liao, W., and Yu, W. (2023). Survey of distributed and decentralized IoT securities: approaches using deep learning and blockchain technology. *Future Internet* 15 (5): 178.

65 Ratnayake, H., Chen, L., and Ding, X. (2023). A review of federated learning: taxonomy, privacy and future directions. *Journal of Intelligent Information Systems* 61 (3): 923–949.

66 Hua, H., Li, Y., Wang, T. et al. (2023). Edge computing with artificial intelligence: a machine learning perspective. *ACM Computing Surveys* 55 (9): 1–35.

67 Zhu, X., Wang, J., Chen, W., and Sato, K. (2023). Model compression and privacy preserving framework for federated learning. *Future Generation Computer Systems* 140: 376–389.

68 Isik, B., Choi, K., Zheng, X. et al. (2023). Neural network compression for noisy storage devices. *ACM Transactions on Embedded Computing Systems* 22 (3): 1–29.

69 Salama, A., Stergioulis, A., Hayajneh, A.M. et al. (2023). Decentralized federated learning over slotted aloha wireless mesh networking. *IEEE Access* 11: 18326–18342.

70 Watanabe, Y., Kawamoto, Y., and Kato, N. (2023). A novel routing control method using federated learning in large-scale wireless mesh networks. *IEEE Transactions on Wireless Communications* 22 (12): 9291–9300.

71 Merling, D., Stüber, T., and Menth, M. (2023). Efficiency of BIER multicast in large networks. *IEEE Transactions on Network and Service Management* 20 (4): 4013–4027.

72 Sharma, K.K., Pokharana, A., and Sharma, S. (2023). Study and analyzing propagation models for wireless mesh networks. *2023 4th International Conference for Emerging Technology (INCET)*, 1–5. https://doi.org/10.1109/INCET57972.2023.10169921.

73 Wang, L., Zhao, X., Lu, Z. et al. (2023). Enhancing privacy preservation and trustworthiness for decentralized federated learning. *Information Sciences* 628: 449–468.

74 Xu, C., Qu, Y., Xiang, Y., and Gao, L. (2023). Asynchronous federated learning on heterogeneous devices: a survey. *Computer Science Review* 50: 100595.

75 Lu, G., Xiong, Z., Li, R. et al. (2023). DEFEAT: A decentralized federated learning against gradient attacks. *High-Confidence Computing* 3 (3): 100128.

76 Ko, S., Lee, K., Cho, H. et al. (2023). Asynchronous federated learning with directed acyclic graph-based blockchain in edge computing: overview, design, and challenges. *Expert Systems with Applications* 223: 119896.

77 Tomiyama, E., Esaki, H., and Ochiai, H. (2023). Competitive and asynchronous decentralized federated learning with blockchain smart contracts. *Proceedings of the 2023 ACM Conference on Information Technology for Social Good*, 92–99.

78 Singla, S., Sharma, P., and Sharma, P.K. (2023). Enhanced security using proxy signcryption technique for wireless mesh networks. *International Journal of System Assurance Engineering and Management* 14 (1): 474–482.

79 Yu, G., Wang, X., Sun, C. et al. (2023). IronForge: An open, secure, fair, decentralized federated learning. *arXiv preprint arXiv:2301.04006*.

80 Errounda, F.Z. and Liu, Y. (2023). Adaptive differential privacy in vertical federated learning for mobility forecasting. *Future Generation Computer Systems* 149: 531–546.

81 Zhang, H., Bosch, J., and Olsson, H.H. (2023). EdgeFL: A lightweight decentralized federated learning framework. *arXiv preprint arXiv:2309.02936*.

82 Liu, S., Wang, X., Hui, L., and Wu, W. (2023). Blockchain-based decentralized federated learning method in edge computing environment. *Applied Sciences* 13 (3): 1677.

83 Akyildiz, I.F., Wang, X., and Wang, W. (2005). Wireless mesh networks: a survey. *Computer Networks* 47 (4): 445–487.

84 Dutta, H., Bhuyan, A.K., and Biswas, S. (2023). Reinforcement learning based flow and energy management in resource-constrained wireless networks. *Computer Communications* 202: 73–86.

85 Rösler, S., Zubow, A., and Dressler, F. (2023). Opportunistic routing in LoRA-based wireless mesh networks. *2023 IEEE 24th International Symposium on a World of Wireless, Mobile and Multimedia Networks (WoWMoM)*, 364–369.

86 Alameri, I., Komarkova, J., Al-Hadhrami, T., and Alkaraawi, H. (2023). Fuzzy logic-based congestion control in AODV mesh networks. *2023 3rd International Conference on Electrical, Computer, Communications and Mechatronics Engineering (ICECCME)*, 1–10.

87 Larrakoetxea, N.G., Astobiza, J.E., López, I.P. et al. (2023). Efficient machine learning on edge computing through data compression techniques. *IEEE Access* 11: 31676–31685.

88 Prabhu, D., Alageswaran, R., and Amali, S.M.J. (2023). Multiple agent based reinforcement learning for energy efficient routing in WSN. *Wireless Networks* 29 (4): 1787–1797.

89 Xie, Z. and Song, S. (2023). FedKL: Tackling data heterogeneity in federated reinforcement learning by penalizing KL divergence. *IEEE Journal on Selected Areas in Communications* 41 (4): 1227–1242.

90 Liao, Y., Xu, Y., Xu, H. et al. (2023). Adaptive configuration for heterogeneous participants in decentralized federated learning. *IEEE INFOCOM 2023 - IEEE Conference on Computer Communications*, 1–10.

91 Wang, H., Kaplan, Z., Niu, D., and Li, B. (2020). Optimizing federated learning on Non-IID data with reinforcement learning. *IEEE INFOCOM 2020 - IEEE Conference on Computer Communications*, 1698–1707.

92 Chen, Y., Su, L., and Xu, J. (2017). Distributed statistical machine learning in adversarial settings: Byzantine gradient descent. *Proceedings of the ACM on Measurement and Analysis of Computing Systems* 1 (2): 1–25.

93 Yin, D., Chen, Y., Kannan, R., and Bartlett, P. (2018). Byzantine-robust distributed learning: towards optimal statistical rates. *Proceedings of International Conference on Machine Learning (ICML)*, 5650–5659. Stockholm, Sweden.

94 Ma, X. and Xu, D. (2023). TORR: A lightweight blockchain for decentralized federated learning. *IEEE Internet of Things Journal* 11 (1): 1028–1040.

95 Zheng, G., Jacobs, M.A., Braverman, V., and Parekh, V.S. (2023). Asynchronous decentralized federated lifelong learning for landmark localization in medical imaging. *arXiv preprint arXiv:2303.06783*.

96 Zheng, S., Wang, X., and Duan, L. (2023). Adaptive federated learning via entropy approach. *arXiv preprint arXiv:2303.14966*.

97 Fereidouni, H., Fadeitcheva, O., and Zalai, M. (2023). IoT and man-in-the-middle attacks. *arXiv preprint arXiv:2308.02479*.

98 Aziz, R., Banerjee, S., Bouzefrane, S., and Le Vinh, T. (2023). Exploring homomorphic encryption and differential privacy techniques towards secure federated learning paradigm. *Future Internet* 15 (9): 310.

99 Arora, S., Beams, A., Chatzigiannis, P. et al. (2023). Privacy-preserving financial anomaly detection via federated learning & multi-party computation. *arXiv preprint arXiv:2310.04546*.

100 Wu, J., Drew, S., Dong, F. et al. (2023). Topology-aware federated learning in edge computing: a comprehensive survey. *arXiv preprint arXiv:2302.02573*.

101 Mansour, Y., Mohri, M., Ro, J., and Suresh, A.T. (2020). Three approaches for personalization with applications to federated learning. *arXiv preprint arXiv:2002.10619*.

102 Gupta, S.K., Wani, A.R., Kumar, S. et al. (2020). Wireless mesh network security, architecture, and protocols. In: *Security and Privacy Issues in Sensor Networks and IoT* (ed. P. Ahlawat and M. Dave), 1–27. IGI Global.

103 Jin, R., Wang, B., Wei, W. et al. (2016). Detecting node failures in mobile wireless networks: a probabilistic approach. *IEEE Transactions on Mobile Computing* 15 (7): 1647–1660.

104 Perumal, S., Tabassum, M., Narayana, G. et al. (2021). ANN based novel approach to detect node failure in wireless sensor network. *Computers, Materials and Continua* 69 (2): 1447–1462.

105 Zhang, W., Lu, Q., Yu, Q. et al. (2020). Blockchain-based federated learning for device failure detection in industrial IoT. *IEEE Internet of Things Journal* 8 (7): 5926–5937.

106 Zaw, C.W., Pandey, S.R., Kim, K., and Hong, C.S. (2021). Energy-aware resource management for federated learning in multi-access edge computing systems. *IEEE Access* 9: 34938–34950.

107 Arouj, A. and Abdelmoniem, A.M. (2022). Towards energy-aware federated learning on battery-powered clients. *Proceedings of the 1st ACM Workshop on Data Privacy and Federated Learning Technologies for Mobile Edge Network*, 7–12.

108 Marfoq, O., Xu, C., Neglia, G., and Vidal, R. (2020). Throughput-optimal topology design for cross-silo federated learning. *Advances in Neural Information Processing Systems* 33: 19478–19487.

109 Wang, Z., Hu, Y., Yan, S. et al. (2022). Efficient ring-topology decentralized federated learning with deep generative models for medical data in eHealthcare systems. *Electronics* 11 (10): 1548.

110 Xing, H., Simeone, O., and Bi, S. (2020). Decentralized federated learning via SGD over wireless D2D networks. *2020 IEEE 21st International Workshop on Signal Processing Advances in Wireless Communications (SPAWC)*, 1–5.

111 Tao, Y., Zhou, J., and Yu, S. (2021). Efficient parameter aggregation in federated learning with hybrid convergecast. *2021 IEEE 18th Annual Consumer Communications & Networking Conference (CCNC)*, 1–6.

112 Lin, Y., Han, S., Mao, H. et al. (2017). Deep gradient compression: reducing the communication bandwidth for distributed training. *arXiv preprint arXiv:1712.01887*.

113 Han, P., Wang, S., and Leung, K.K. (2020). Adaptive gradient sparsification for efficient federated learning: an online learning approach. *2020 IEEE 40th International Conference on Distributed Computing Systems (ICDCS)*, 300–310. https://doi.org/10.1109/ICDCS47774.2020.00026.

114 Reisizadeh, A., Mokhtari, A., Hassani, H. et al. (2020). FedPAQ: A communication-efficient federated learning method with periodic averaging and quantization. *International Conference on Artificial Intelligence and Statistics*, 2021–2031. PMLR.

115 Wei, K., Li, J., Ding, M. et al. (2020). Federated learning with differential privacy: algorithms and performance analysis. *IEEE Transactions on Information Forensics and Security* 15: 3454–3469.

116 Zhang, C., Li, S., Xia, J. et al. (2020). BatchCrypt: Efficient homomorphic encryption for Cross-Silo federated learning. *2020 USENIX Annual Technical Conference (USENIX ATC 20)*, 493–506.

117 Sotthiwat, E., Zhen, L., Li, Z., and Zhang, C. (2021). Partially encrypted multi-party computation for federated learning. *2021 IEEE/ACM 21st International Symposium on Cluster, Cloud and Internet Computing (CCGrid)*, 828–835.

118 Bonawitz, K., Ivanov, V., Kreuter, B. et al. (2016). Practical secure aggregation for federated learning on user-held data. *arXiv preprint arXiv:1611.04482*.

119 Vepakomma, P., Gupta, O., Swedish, T., and Raskar, R. (2018). Split learning for health: distributed deep learning without sharing raw patient data. *arXiv preprint arXiv:1812.00564*.

120 Proos, D.P. and Carlsson, N. (2020). Performance comparison of messaging protocols and serialization formats for digital twins in IoV. *2020 IFIP Networking Conference (Networking)*, 10–18. IEEE.

121 Jiang, X., Shokri-Ghadikolaei, H., Fodor, G. et al. (2019). Low-latency networking: where latency lurks and how to tame it. *Proceedings of the IEEE* 107 (2): 280–306.

122 Alizadeh, M., Kabbani, A., Edsall, T. et al. (2012). Less is more: trading a little bandwidth for ultra-low latency in the data center. *9th USENIX Symposium on Networked Systems Design and Implementation (NSDI 12)*, 253–266.

123 Alotaibi, E. and Mukherjee, B. (2012). A survey on routing algorithms for wireless ad-hoc and mesh networks. *Computer Networks* 56 (2): 940–965.

124 Wu, Y., Lu, C., and Chen, Y. (2016). A survey of routing algorithm for mesh network-on-chip. *Frontiers of Computer Science* 10: 591–601.

125 Kaur, J. and Singh, H. (2023). Several routing protocols, features and limitations for wireless mesh network (WMN): a review. *ICDSMLA 2021: Proceedings of the 3rd International Conference on Data Science, Machine Learning and Applications*, 187–200. Springer.

126 Vanlalhruaia and Khan, A.K. (2023). Security challenges during handoff authentication operation for wireless mesh network. *Advances in Signal Processing, Embedded Systems and IoT: Proceedings of Seventh ICMEET-2022*, 397–405. Springer.

127 Nanda, A., Nanda, P., He, X. et al. (2020). A hybrid encryption technique for Secure-GLOR: the adaptive secure routing protocol for dynamic wireless mesh networks. *Future Generation Computer Systems* 109: 521–530.

128 Selimi, M., Kabbinale, A.R., Ali, A. et al. (2018). Towards blockchain-enabled wireless mesh networks. *Proceedings of the 1st Workshop on Cryptocurrencies and Blockchains for Distributed Systems*, 13–18.

129 Kabbinale, A.R., Dimogerontakis, E., Selimi, M. et al. (2020). Blockchain for economically sustainable wireless mesh networks. *Concurrency and Computation: Practice and Experience* 32 (12): e5349.

130 Altop, D.K., Bingöl, M.A., Levi, A., and Savaş, E. (2017). DKEM: Secure and efficient distributed key establishment protocol for wireless mesh networks. *Ad Hoc Networks* 54: 53–68.

131 Kandah, F., Zhang, W., Du, X., and Singh, Y. (2011). A secure key management scheme in wireless mesh networks. *2011 IEEE International Conference on Communications (ICC)*, 1–5.

132 Wu, M.-Y., Wu, S.-H., Chang, Y.-E. et al. (2023). Intrusion detection with radio frequency sensing based on Wi-Fi mesh network for home security. *2023 International Conference on Consumer Electronics - Taiwan (ICCE-Taiwan)*, 329–330. https://doi.org/10.1109/ICCE-Taiwan58799.2023.10226838.

133 Savazzi, S., Kianoush, S., Rampa, V., and Bennis, M. (2020). A joint decentralized federated learning and communications framework for industrial networks. *2020 IEEE 25th International Workshop on Computer Aided Modeling and Design of Communication Links and Networks (CAMAD)*, 1–7.

134 Pinyoanuntapong, P., Huff, W.H., Lee, M. et al. (2022). Toward scalable and robust AIoT via decentralized federated learning. *IEEE Internet of Things Magazine* 5 (1): 30–35.

135 Zecchin, M., Gesbert, D., and Kountouris, M. (2022). UAV-Aided decentralized learning over mesh networks. *2022 30th European Signal Processing Conference (EUSIPCO)*, 702–706.

136 Abadi, M., Agarwal, A., Barham, P. et al. (2015). TensorFlow: Large-scale machine learning on heterogeneous systems. Software available from tensorflow.org. https://www.tensorflow.org/ (accessed 1 October 2024).

137 Ziller, A., Trask, A., Lopardo, A. et al. (2021). PySyft: A library for easy federated learning. *Federated Learning Systems: Towards Next-Generation AI* 965: 111–139.

138 He, C., Li, S., So, J. et al. (2020). FedML: A research library and benchmark for federated machine learning. *arXiv preprint arXiv:2007.13518.*

139 Ludwig, H., Baracaldo, N., Thomas, G. et al. (2020). IBM federated learning: an enterprise framework white paper v0. 1. *arXiv preprint arXiv:2007.10987.*

140 Bansal, N. and Bansal, N. (2020). Microsoft azure IoT platform. In: *Designing Internet of Things Solutions with Microsoft Azure: A Survey of Secure and Smart Industrial Applications* (ed. S. Srivastava, M. Moodie, and S. Vishwakarma), 33–48. Springer.

141 Chai, D., Wang, L., Yang, L. et al. (2020). FedEval: A holistic evaluation framework for federated learning. *arXiv preprint arXiv:2011.09655.*

142 Hu, S., Li, Y., Liu, X. et al. (2022). The OARF benchmark suite: characterization and implications for federated learning systems. *ACM Transactions on Intelligent Systems and Technology (TIST)* 13 (4): 1–32.

143 Hao, T., Huang, Y., Wen, X. et al. (2019, 2019). Edge AIBench: Towards comprehensive end-to-end edge computing benchmarking. In: *Benchmarking, Measuring, and Optimizing: First BenchCouncil International Symposium, Bench 2018*, Seattle, WA, USA (10—13 December 2018), Revised Selected Papers 1 (ed. C. Zheng and J. Zhan), 23–30. Springer.

144 Romera-Paredes, B. and Torr, P. (2015). An embarrassingly simple approach to zero-shot learning. *International Conference on Machine Learning*, 2152–2161. PMLR.

145 Zhuang, F., Qi, Z., Duan, K. et al. (2021). A comprehensive survey on transfer learning. *Proceedings of the IEEE* 109 (1): 43–76.

146 Levi, A.F.J. (2023). *Applied Quantum Mechanics*. Cambridge University Press.

147 Marković, D., Mizrahi, A., Querlioz, D., and Grollier, J. (2020). Physics for neuromorphic computing. *Nature Reviews Physics* 2 (9): 499–510.

148 Pirandola, S., Andersen, U.L., Banchi, L. et al. (2020). Advances in quantum cryptography. *Advances in Optics and Photonics* 12 (4): 1012–1236.

149 Chaoub, A., Mämmelä, A., Martinez-Julia, P. et al. (2023). Hybrid self-organizing networks: evolution, standardization trends, and a 6G architecture vision. *IEEE Communications Standards Magazine* 7 (1): 14–22.

150 Zhang, W., Kandah, F., Tang, J., and Nygard, K. (2010). Interference-aware robust topology design in multi-channel wireless mesh networks. *2010 7th IEEE Consumer Communications and Networking Conference*, 1–5.

151 Shah, J.C.H. and Chothani, N.G. (2023). Protection challenges and mitigation techniques of power grid integrated to renewable energy sources: a review. *Energy Sources, Part A: Recovery, Utilization, and Environmental Effects* 45 (2): 4195–4210.

152 Van Leemput, D., Sabovic, A., Hammoud, K. et al. (2023). Energy harvesting for wireless IoT use cases: a generic feasibility model and tradeoff study. *IEEE Internet of Things Journal* 10 (17): 15025–15043. https://doi.org/10.1109/JIOT.2023.3263543.

153 Wei, K., Li, J., Ma, C. et al. (2023). Personalized federated learning with differential privacy and convergence guarantee. *IEEE Transactions on Information Forensics and Security* 18: 4488–4503.

154 Huang, J., Majumder, P., Kim, S. et al. (2021). Communication algorithm-architecture co-design for distributed deep learning. *2021 ACM/IEEE 48th Annual International Symposium on Computer Architecture (ISCA)*, 181–194.

155 Wu, L., Guo, S., Wang, J. et al. (2022). Federated unlearning: guarantee the right of clients to forget. *IEEE Network* 36 (5): 129–135.

156 Zhan, Y., Zhang, J., Hong, Z. et al. (2022). A survey of incentive mechanism design for federated learning. *IEEE Transactions on Emerging Topics in Computing* 10 (2): 1035–1044.

157 Yang, Z., Zolanvari, M., and Jain, R. (2023). A survey of important issues in quantum computing and communications. *IEEE Communications Surveys & Tutorials* 25 (2): 1059–1094.

7

Antenna Design Using Artificial Intelligence

Sotirios K. Goudos[1,4], Mohammad A. Matin[2], and George K. Karagiannidis[3,5]

[1] *Department of Physics, Aristotle University of Thessaloniki, Greece*
[2] *Department of Electrical and Computer Engineering, North South University, Dhaka, Bangladesh*
[3] *Department of Electrical and Computer Engineering, Aristotle University of Thessaloniki, Greece*
[4] *Director of the ELEDIA@AUTH lab member of the ELEDIA Research Center Network*
[5] *Head of the Wireless Communications and Information Processing (WCIP) Group*

CHAPTER MENU

7.1 Introduction

The application of artificial intelligence (AI) techniques in the antenna domain is a common practice in the literature (see Figure 7.1). A major group of AI techniques used in antenna design is the *evolutionary algorithms* (EAs). This chapter provides a classification of EAs **into mainstream and emerging** with representative algorithms of both types. *Machine learning* (ML) algorithms can also be employed in various antenna applications, such as the estimation of antenna characteristics, such as the S-parameters and gain or radiation pattern, and their use as surrogate methods. In the latter, ML algorithms are used instead of running computationally intensive simulations using full-wave simulator software. In this case, the ML algorithm runs in conjunction with an optimizer algorithm (EA or other type). Such methods could reduce the time needed to achieve a satisfactory accuracy result. The use of *knowledge representation* (KR) in the antenna domain

Artificial Intelligence for Future Networks, First Edition.
Edited by Mohammad A. Matin, Sotirios K. Goudos, and George K. Karagiannidis.

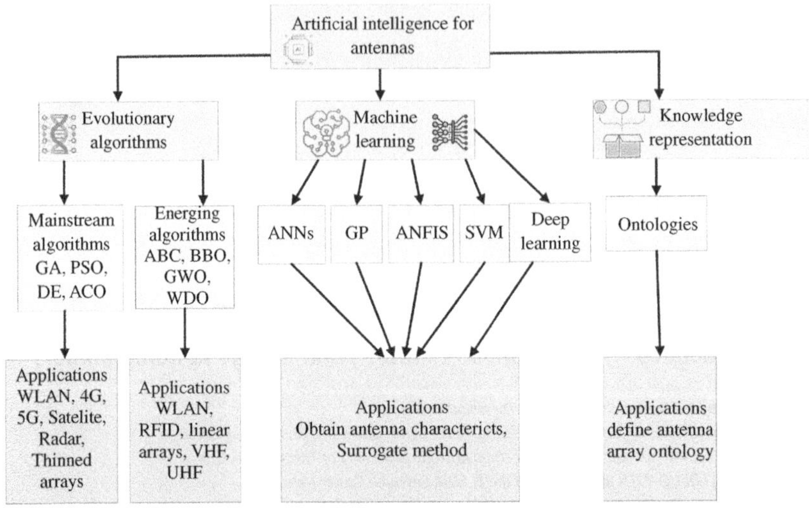

Figure 7.1 A classification of AI techniques in the antennas domain.

has been enabled by semantic web technologies such as Ontology Web Language (OWL). It should be noted that traditional antenna design methods have several limitations. For instance, the Dolph–Tschebyscheff technique is only suitable for equidistant broadside antenna arrays, and the amplitudes it produces are real numbers, which are hard to implement in practice. In addition, the rectangular patch design procedure is based on formulas that calculate the effective length and width, and thus, the dimensions for a given resonant frequency are obtained. However, this method is only applicable to rectangular patches and not to other shapes. All of these limitations are overcome by EAs, which are "black-box" optimizers and independent of the form of the objective function (i.e., the antenna). The literature contains several review papers that focus on a particular AI method or antenna type. For example, Campbell et al. [1] discusses deep learning applications for antennas, while [2] covers AI methods for reconfigurable antenna design. This chapter provides a comprehensive overview of all major AI methods and their applications to antenna design. The main contributions of this article are:

This chapter has grouped the above AI families and techniques into separate sections with the aim of facilitating their comparison and investigating their potential for future antenna applications. In Section 7.2, we present popular EAs algorithms with a brief discussion in Sections 7.2.1 and 7.2.2. Some of the most popular ML techniques used in antenna design are presented in Section 7.3. The use of semantic web technologies is presented in Section 7.4. Finally, Section 7.5 concludes the chapter.

7.2 Evolutionary Algorithms

Nature-inspired metaheuristics, such as those based on biology, are known as EAs. These algorithms imitate the behavior and evolution of biological species. Recently, physical-/chemical-based metaheuristics have also been developed. Furthermore, algorithms based on human social behavior are also being developed. All of these algorithms share the same basic principle: an initial population of vectors is evolved through iterations by applying various operators.

EAs can be divided into two main groups based on their prevalence: mainstream algorithms and emerging algorithms. Genetic algorithms (GAs) are one of the most widely used mainstream techniques. Other popular mainstream techniques include GAs, evolution strategies (ES), and evolutionary programming (EP). In addition, swarm intelligence (SI) algorithms are a special type of EA. SI is characterized by decentralized and self-organized swarm behavior and includes ant colony optimization (ACO) and particle swarm optimization (PSO). Differential evolution (DE) is a type of EA that is widely used and popular. It is a metaheuristic based on mathematical models that do not mimic any natural behavior. Antenna engineers often face a challenge when selecting the right algorithm for a given optimization problem in the antenna domain, as there are many of them.

Therefore, it is beneficial to explore new optimization strategies for antenna design and synthesis, which can be greatly improved by the use of EAs. Investigating new optimization strategies for antenna design and synthesis is worthwhile, as they can be greatly enhanced by the use of EAs. These strategies can be divided into two categories: finding the optimal geometry for a particular antenna element and optimizing the locations and excitation of an antenna array. In addition, a combination of the two issues can be considered. Antenna applications are used in a variety of wireless communication systems, such as cellular communications, satellite communications, cognitive radio applications, and wireless networks. Therefore, it is advantageous to explore new optimization strategies for antenna design and synthesis, which can be greatly improved by using EA.

7.2.1 Mainstream Algorithms

7.2.1.1 Genetic Algorithms

Genetic algorithms (GAs) are the most widely used metaheuristics in the field of antenna design [3, 4]. These algorithms typically involve crossover, mutation, and selection operators. Crossover combines two or more parent solutions to create a new child. Mutation randomly alters the current solution. Selection determines which parent vectors or chromosomes are chosen for recombination. In addition, GAs often include the feature of elitism, which means that the best solutions are preserved for the next generation. GAs can be real-coded (RGA) or binary-coded (BGA) algorithms depending on the problem type. They have been applied to a variety of antenna designs, such as the 5G base station antenna array design discussed in [5].

In [6], a standard real-coded GA is used for the optimization of multiport pixel antennas for application in multiple-input multiple-output (MIMO) communication systems. The authors design a dual-band antenna at 2.4 and 5 GHz and a single-band antenna at 5 GHz. Both antennas can be used for wireless local area network (WLAN) applications.

PSO is a SI algorithm that mimics the behavior of bird swarms when they search for food. It is a low-complexity algorithm that can be easily implemented in any programming language. To use the PSO, three control parameters must be set: the cognitive learning factor, the social learning factor, and the inertia weight. These parameters determine the balance between exploration and exploitation, the effect of the particle's memory on its best position, and the influence of the swarm's best position, respectively. PSO is also a popular algorithm for antenna design, and several variants of it have been used. Although PSO works in real space, binary extensions are also available.

7.2.1.2 Particle Swarm Optimization

PSO is also a very popular algorithm for antenna design. PSO was introduced by Kennedy and Eberhart [7]. The basic idea of PSO is to have a swarm of particles move through a search space, using information from two optimal values to modify each particle's position in each iteration. The first optimal value is the best solution (objective function value) that has been achieved so far by each particle in the swarm, called the personal best, or p_{best}. The second optimal value is the global best value obtained by any particle in the swarm, denoted g_{best}. Therefore, PSO finds these values, p_{best} and g_{best}, in each iteration.

The velocity of the ath particle in the bth dimension is updated by taking into account the global best and the personal best of the particle. This rule is essential to achieve the global optimum. The equation for the velocity update of the ath particle in the bth dimension is given by: The velocity of the ath particle in the bth dimension is adjusted based on the global best and the personal best of the particle. This rule is essential to find the global optimum. The equation for the velocity update of the ath particle in the bth dimension is expressed as:

$$q_{k+1,ba} = wq_{k,ba} + s_1 \text{rnd}_{1,ba}(p_{\text{best}_{k+1,ba}} - z_{k,ba}) + s_2 \text{rnd}_{2,ba}(g_{\text{best}_{k+1,ba}} - z_{k,ba}), \tag{7.1}$$

where $q_{k+1,ba}$ denotes the ath particle velocity in the bth dimension, $k+1$, k represent the current and the previous iteration, respectively, $z_{k,ba}$ is the ath particle position in the bth dimension, in current iteration, $\text{rnd}_{1,ba}$, $\text{rnd}_{2,ba}$ are uniformly distributed random numbers in (0, 1), w is the inertia weight, s_1, and s_2 are called the learning factors.

Several PSO variants have also been used. Although PSO inherently works in real spaces, there are also binary extensions. Various velocity update rules have been suggested in the literature, the most popular being the one proposed by Clerc [8]. This rule introduces a new parameter, the constriction factor (KF), which replaces the inertia weight. The advantage of this constriction factor is that it encourages convergence when the swarm ceases to move. The velocity update rule for this case is expressed as follows:

$$q_{k+1,ba} = \text{KF}\left[q_{k,ba} + w_1\text{rand}_{1,ba}(p_{\text{best}k+1,ba} - z_{k,ba}) + w_2\text{rand}_{2,ba}(g_{\text{best}k+1,ba} - z_{k,ba})\right], \tag{7.2}$$

where KF is the constriction factor. The definition is given by

$$\text{KF} = \frac{2}{\left|2 - \vartheta - \sqrt{\vartheta^2 - 4\vartheta}\right|}, \tag{7.3}$$

where

$$\vartheta = \begin{cases} w_1 + w_2 & \text{if } w_1 + w_2 > 4 \\ 1 & \text{if } w_1 + w_2 \leq 4. \end{cases}$$

It is typical to set both w_1 and w_2 to 2.05 when using the PSO algorithm, also known as Constriction Factor PSO (CFPSO) due to its constriction factor parameter.

The authors in [9] apply a modified binary PSO (MBPSO) to design a planar wideband antenna array for millimeter Wave operation. In [10], a variant of PSO with velocity mutation (PSOvm) variant is utilized for log-periodic antenna (LPA) wideband antenna working between 790 and 6000MHz. Thus, the antenna covers the 2G, 3G, 4G, and WLAN bands. The operating frequency band also makes the antenna suitable for cognitive radio applications. The optimization problem is formulated as a single objective function that consists of a weighted sum of four different optimization goals.

7.2.1.3 Differential Evolution

Differential evolution is a mathematical design that does not depend on any particular biological phenomenon [11]. It uses three operators, mutation, crossover, and selection, to evolve vectors. Various DE strategies have been developed by modifying one or more of these operators, and the most suitable one is determined depending on the type of problem. The DE algorithm has two control parameters, MP for mutation and CO for crossover, and is based on a greedy selection approach. We can express the tth mutant vector $\overline{W}_{g,t}$ for each target vector $\overline{Y}_{g,m}$ in the g iteration in the following way:

$$\begin{aligned}&\text{DE/rand/1/bin}\\&\overline{W}_{g,t} = \overline{Y}_{g,r_1} + \text{MP} \times (\overline{Y}_{g,r_2} - \overline{Y}_{g,r_3}), r_1 \neq r_2 \neq r_3\\&\text{DE/best/1/bin}\\&\overline{W}_{g,t} = \overline{Y}_{g,\text{best}} + \text{MP} \times (\overline{Y}_{g,r_1} - \overline{Y}_{g,r_2}), r_1 \neq r_2,\end{aligned} \tag{7.4}$$

where the randomly chosen individuals r_1, r_2, and r_3 are used to create a trial vector $\overline{R}_{g,t}$ with the crossover operator, and the mutation control parameter or mutation scale factor, which is denoted as MP.

$$W_{g,kt} = \begin{cases} R_{g,kt}, & \text{if} \quad \text{rand}_{k[0,1)} \leq \text{CR} \quad \text{or} \quad k = rn(t) \\ Y_{g,kt}, & \text{if} \quad \text{rand}_{k[0,1)} > \text{CO} \quad \text{and} \quad k \neq rn(t). \end{cases} \tag{7.5}$$

The DE algorithms employ a greedy selection operator for minimization problems. For each k from 1 to D, a uniformly distributed random number $\text{rand}_{k[0,1)}$ is generated in the interval $[0, 1)$, and an index rn(m) is randomly chosen from the set $(1, 2, \ldots, D)$. In addition, another DE control parameter, the crossover constant, is taken from the interval $[0, 1]$.

$$\overline{Y}_{g+1,t} = \begin{cases} \overline{R}_{g,t}, \text{if } f(\overline{R}_{g,t}) < f(\overline{Y}_{g,t}) \\ \overline{Y}_{g,t}, \text{ otherwise,} \end{cases} \tag{7.6}$$

where $f(\overline{R}_{g,t}), f(\overline{Y}_{g,t})$ are the objective function values of the trial and the old vector, respectively. Therefore, the new trial vector $\overline{R}_{g,t}$ replaces the old vector $\overline{Y}_{g,t}$ if it is better than the previous one. Otherwise, the old vector remains in the population.

DE is a popular choice for antenna design, especially for antenna array design, because of its widespread use. The authors of [12] present a DE algorithm that incorporates a novel encoding technique. Their aim is to create large planar arrays with unequal spacing and a minimum distance between elements. Arrays of this type can be utilized in radar, sonar, and satellite communication systems.

7.2.1.4 Ant Colony Optimization

ACO is an algorithm inspired by the ants' communication system [13–15]. In nature, ants use chemical pheromones to communicate with each other. By sensing and following trails with high concentrations of pheromones, ants can find the shortest paths between their nests and food sources. ACO mimics this behavior by modeling it mathematically. One of its main benefits is that it can be used to solve combinatorial optimization problems, which are common in the antenna domain, such as the design of linear subarrays or thinned array design. The authors in [16] use a multiobjective lazy ant colony optimization (MOLACO) algorithm for a 3D frequency selective structures (FSSs) design operating in C-band.

Algorithm	Features	Optimization Problem	Application
GA [6]	Mimics evolution of chromosomes. Crossover and mutation operators	Single-objective optimization of pixel antennas for application in MIMO systems	2.4–5 GHz WLAN
PSO [9]	Models behavior of bird swarms	Single-objective optimization of planar Wideband Antenna Array	5G
DE [12]	Mutation, crossover, and selection operators Multiple strategies exist	Single-objective optimization of planar arrays under constraints.	Radars, sonars, and Satellite comm.
ACO [16]	Mimics behavior of ants in search of food.	Multiobjective optimization of 3D FSS	Satellite comm. C-band

Algorithm	Advantages	Disadvantages
GAs	Popular (available in multiple codes)	Extensive effort is needed to fine-tune the operators.
	Considerable adaptability	Unguided mutation operator
	Parallelizing is simple.	
	Capacity to seek a vast and extensive solution space	
PSO	Simple to implement	Numerous tuning parameters for optimal operation
	Scaling insensitivity of design variables	Potentially confined in high-dimensional local optima
	Effective worldwide inquiry	low convergence rate
DE	Very few control parameters are present in DE	Problem-solving challenges may arise when confronted with nonseparable functions.
	Compared to other EAs, more simple and straightforward to implement	Potentially stuck in local optima
	Limited spatial complexity	
	Guided operator for mutation	
ACO	Extremely effective when solving discrete optimization issues	Unsuitable for all types of problems
	Easily parallelizable, Easy to apply to dynamic applications (e.g., adapts to changes such as new distances)	The local search ability of ant colony algorithm is weak, local convergence, slow convergence speed, and other problems are easy to occur
		Long computation time and stagnation phenomenon are easy to appear in the whole process.
BBO	Highly effective when applied to combinatorial optimization issues.	BBO is poor in exploiting the solutions
	Computational complexity similar to other Eas	No provision is made to identify the best individuals from each generation.
	Distinctive, straightforward, and uncomplicated to implement	
	Excellent exploration skills	
	Widely available source code	
GWO	Straightforward to implement	May get trapped into local optima
	Parameter-free no tuning required	Low-exploitation ability
	Widely available source code	
WDO	Effectively addresses antenna design challenges	Numerous tuning control parameters
	Fast convergence	Lack of testing across multiple application domains
	Widely available source code	

7.2.2 Emerging Algorithms

7.2.2.1 Biogeography-Based Optimization

Biogeography-based optimization (BBO) is a mathematical model that has been developed based on the biological domain of biogeography. Predicts animal migration patterns and can be used to estimate the number of species in a habitat. In BBO, the habitat is represented by a vector of unknown variables, called suitability index variables (SIVs). Each vector is assigned a habitat suitability index (HSI) value which is used to measure habitat fitness. BBO utilizes both migration and mutation operators, which is a unique feature that separates it from other algorithms. This feature makes BBO a good candidate for solving combinatorial optimization problems. To use BBO, two control parameters must be set: modification probability and mutation probability. The authors of [17] employed BBO to design a Yagi–Uda antenna. The cost function was a single-objective function that utilized a weighted-sum approach.

7.2.2.2 Grey Wolf Optimizer

Grey Wolf Optimizer (GWO) is an EA based on mathematical models that mimic the hierarchy and hunting strategies of grey wolves. A key feature of GWO is that it maintains information about the search space during the iteration process. In addition, GWO does not require the setting of any control parameters. The algorithm divides the population into different categories and the optimization process, which is modeled after the social behavior of a wolf pack, is driven by these population groups. In [18], the authors use different versions of the GWO algorithm to design sparse linear arrays, Yagi–Uda antennas, and thinned planar arrays.

7.2.2.3 Wind-Driven Optimization

Wind-driven optimization (WDO) is a novel physics-based algorithm that predicts atmospheric motion. This algorithm is based on Newton's second law of motion and has several control parameters that can be adjusted, such as the friction coefficient, air temperature, universal gas constant, Coriolis force contribution parameter, gravitational constant, and maximum allowable velocity. The authors in [19] explain how to tune these parameters and demonstrate the efficiency of the WDO by applying it to various antenna design problems, including a linear antenna array, an E-shaped patch antenna for Wi-Fi operation at 5 GHz, and artificial magnetic conducting (AMC) ground planes.

7.2.2.4 Salp Swarm Algorithm

The Salp Swarm Algorithm (SSA) mathematically models the swarming behavior of salps when they navigate and search for food in the ocean [20]. In SSA, each salp is represented by a D-dimensional solution vector. Moreover, the salp (solution vector) that obtains the best value of the objective function in each iteration is considered the food position. In SSA, the behavior of salps is represented by dividing the population into two categories. The salps at the front of the chain are referred to as leaders, while the remaining salps are known as followers. This implies that SSA labels the salps (solution vectors) with the highest objective function values as leaders and those with the lowest objective function values as followers. Hence, SSA splits the population into two halves of the same size $\frac{NP}{2}$, where NP stands for the population size.

The SSA approach has been used to address the issue of joint power allocation and user association in nonorthogonal multiple access networks [21]. Furthermore, SSA was used for the design of MIMO antennas for 5G communication systems in [22].

7.2.2.5 Artificial Bee Colony (ABC)

Artificial Bee Colony ABC belongs to the class of SI algorithms. The basic idea is to model the behavior of honey bees in food foraging. The terminology used in the ABC algorithm is the position of a food source and the food source. The position of the food source is the vector of the decision variables of the potential solution, and the food source is the fitness value of the objective function. There are three types of bees in ABC that use different formulations to find the best food position; employed bees, onlooker bees, and scout bees. ABC is a simple-to-implement algorithm and requires only one parameter to set (except the maximum number of iterations and the population size); the limit parameter that represents the maximum number of trials to improve a food source (solution). The authors of [23] use ABC for the design of UHF radio frequency identification (RFID) tags. To this end, they express the objective function using an exact penalty method.

7.2.2.6 Harmony Search (HS)

Harmony search (HS) is an EA that takes inspiration from the process of creating music. A harmony in HS terminology is a potential solution vector that is composed of pitches (decision variables). As with other EA, HS has control parameters, such as harmony memory size (HMS), which is analogous to the population size in other algorithms, and harmony memory consideration rate (HMCR), which determines how the pitches will be chosen. The pitch adjusting rate (PAR) is another control parameter that indicates the probability of adjusting the original value of the pitches selected from the harmony memory (HM). The authors in [24] use the HS algorithm to synthesize sparse linear matrices, the formulation being based on a single objective function with weights.

7.2.2.7 Shuffled Frog-Leaping Algorithm

The shuffled frog-leaping algorithm (SFLA) [25] is a nature-inspired metaheuristic based on the behavior of frogs. In this algorithm, the virtual population is made up of individuals (frogs) who each carry a single meme that is composed of one or more memotypes. The population is divided into several parallel communities (memeplexes). Within each memeplex, the frogs undergo a memetic evolution, which involves changing their memes to improve their performance. After a certain number of steps, a shuffling mix process is applied between memeplexes. This process of memetic evolution and shuffling is repeated until certain conditions are met. An example of a population distributed in four memeplexes is shown in Figure 7.2. SFLA could be applied to antenna design problems and especially discrete-valued problems such as thinned arrays and subarrays.

A brief description of the main advantages and disadvantages of some of the algorithms presented in this section are presented in table in Section 7.2.1.4.

7.2.3 Antenna Optimization Using Evolutionary Algorithms

7.2.3.1 Problem Formulation

We take into account a linear array of elements P arranged symmetrically along the x-axis (Figure 7.3). When P is an even number, there is no zero element. The x–z plane's array factor can be expressed as

$$\mathrm{AF}(\vartheta, \overline{x}, \overline{\phi}, \overline{A}) = \sum_{p=1}^{P} A_p e^{j\left(\frac{2\pi}{\lambda} x_p \sin\vartheta + \phi_p\right)} \qquad (7.7)$$

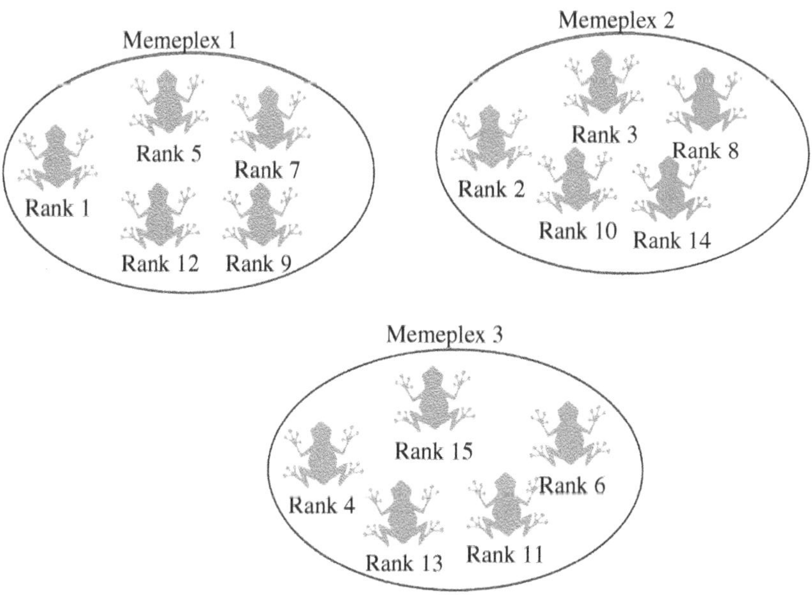

Figure 7.2 A population of 15 frogs distributed in 3 memeplexes.

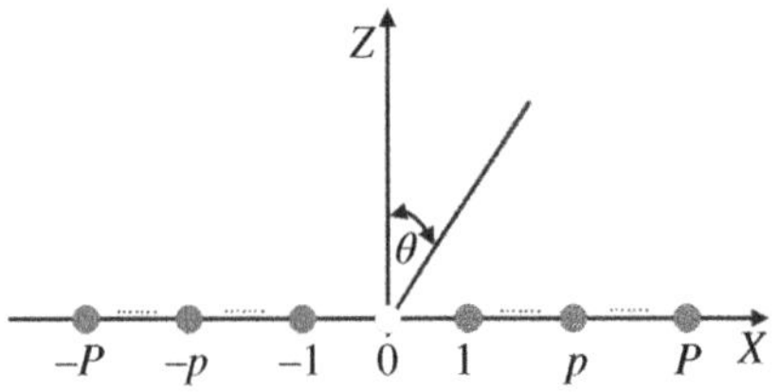

Figure 7.3 Symmetrically placed linear array geometry.

The wavelength λ and the steering angle ϑ are given, the latter measured in relation to the positive direction of the z axis. The position, amplitude, and phase of the pth element are represented by x_p, A_p, and ϕ_p, respectively, and the corresponding vectors are $\bar{x}, \bar{A}$, and $\bar{\phi}$. For a symmetrically excited array, see Eq. (7.7) becomes

$$\mathrm{AF}(\vartheta, \bar{x}, \bar{\phi}, \bar{A}) = 2\sum_{p=1}^{W} A_p \cos\left[\frac{2\pi}{\lambda} x_p \sin\vartheta + \phi_p\right] + A_0, \tag{7.8}$$

where $W = \left\lfloor \frac{P}{2} \right\rfloor$ is the largest integer less than or equal to $P/2$. For an even number of elements $A_0 = 0$, and for an odd number of elements, we set $A_0 = 1$. The position of elements in this array satisfies the following constraints:

$$\begin{aligned} & x_1 \geq \tfrac{v_{\min}}{2}, x_1 \leq \tfrac{v_{\max}}{2}, && \text{when} \quad P = 2W \\ & |x_0 - x_1| \geq v_{\min}, |x_0 - x_1| \leq v_{\max}, && \text{when} \quad P = 2W + 1 \\ & \left.\begin{array}{l} |x_k - x_{k+1}| \geq v_{\min} \\ |x_k - x_{k+1}| \leq v_{\max} \end{array}\right\}, 1 \leq k \leq W - 1, \end{aligned} \tag{7.9}$$

where $v_{\min}$ and $v_{\max}$ are the minimum and maximum distances between two adjacent elements, respectively.

The goal of optimization is to identify an array with the least number of elements that conforms to a desired pattern. This requirement to conform to a desired pattern can be expressed by the mean square error (MSE) as follows:

$$F(\bar{x}, \bar{\phi}, \bar{A}) = \frac{1}{S}\sum_{s=1}^{S}\left[|\mathrm{AF}_d(\vartheta_s)| - |\mathrm{AF}(\vartheta_s, \bar{x}, \bar{\phi}, \bar{A})|\right]^2, \tag{7.10}$$

where $\mathrm{AF}_d(\vartheta) = \sum_{p=1}^{P'} A'_p e^{j\left[\frac{2\pi}{\lambda} x'_p \sin\vartheta + \phi'_p\right]}$, $P' > P$ is the desired pattern, P' is the number of elements of the desired pattern, K is the number of pattern samples taken, and x'_p, I'_p, and ϕ'_p are the position, amplitude, and the phase of the pth element of the desired pattern, respectively. The objective is to identify the locations, phases, and magnitudes of the elements that minimize Eq. (7.10). The optimization process uses the distances between two neighboring elements as unknown position variables instead of absolute positions x_p. Equation (7.10) can also be used as a performance metric of how closely a desired pattern is matched.

7.2.3.2 Numerical Results

This section looks into the creation of a pattern with fewer elements that has been taken from the literature. We assume that the minimum and maximum distances between elements are 0.5 and 1 wavelengths, respectively. We optimize the element positions and amplitudes (assuming that the element phases are zero) using a design scheme. The objective function in (7.10) is calculated using 901 pattern samples from 0 to 90° with steps of 0.1 degrees. The linear-array synthesis design case is executed 20 times, and the best results are reported. The size of the population and the number of generations are set at 100 and 2000, respectively. The number of unknown variables is $2W$. The algorithms applied are the BBO [26], the PSO algorithm with constriction factor [8], the Whale Optimization Algorithm (WOA) [27], the SSA [20], and the GWO [18].

The example of linear array synthesis can be found in [28]. This design was achieved using PSO to optimize the placement of elements to reduce sidelobe levels and control nulls. The pattern from [28] is a 32-element array with nonuniform spacing and uniform excitation. We try to design the same pattern using a reduced number of elements for two cases with 20 and 22 elements, respectively. The desired pattern has a maximum sidelobe level of −18.73 dB, a beamwidth of 8.6°, and a null level of −62.13 dB in the direction of 9°.

We tested the algorithms for the two scenarios with 20 and 22 elements, respectively.

Tables 7.1 and 7.2 list the comparative results for both cases. We observe that MSE was reduced by a factor of 10 when two additional elements were added.

Table 7.1 Comparative results of all algorithms for $P = 20$.

Algorithm	Best	Worst	Mean	Median	Std. Dev
GWO	0.002910	0.013128	0.008973	0.008953	0.003904
WOA	0.008765	0.019948	0.014891	0.015141	0.003682
SSA	0.005163	0.014619	0.011972	0.013277	0.002992
BBO	0.008127	0.014571	0.012711	0.013422	0.002109
PSO	0.006670	0.015910	0.013737	0.014244	0.002638

Table 7.2 Comparative results of all algorithms for $P = 20$.

Algorithm	Best	Worst	Mean	Median	Std. Dev
GWO	0.000761	0.01398	0.006246	0.006512	0.004667
WOA	0.010794	0.019892	0.015055	0.014391	0.002840
SSA	0.005748	0.012752	0.008189	0.007977	0.002201
BBO	0.004190	0.013887	0.009880	0.010013	0.003332
PSO	0.000927	0.014261	0.010642	0.011539	0.004037

We notice that for 12 elements the algorithms obtain different results. The GWO performance is better than the other algorithms in this case. The SSA obtains the second-best result. However, the numerical results are quite close. This suggests that any of these algorithms can be employed to tackle this kind of issue. It is possible to achieve the same result when considering the 22-element case. Again, GWO outperformed the other algorithms. PSO in this case obtains the second best result.

The boxplot shown in Figure 7.4 indicates that BBO and PSO have the least variation in their outcomes when $P = 20$. Although GWO achieved the best result, it had the highest standard deviation values. High standard deviation values are also evident for the WOA and SSA algorithms. The boxplot in Figure 7.5 for $P = 22$ reveals that all algorithms have a wide range of values, except SSA.

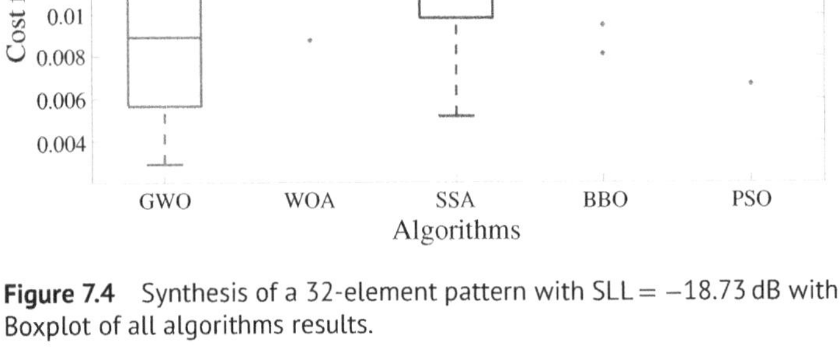

Figure 7.4 Synthesis of a 32-element pattern with SLL = −18.73 dB with $P = 20$. Boxplot of all algorithms results.

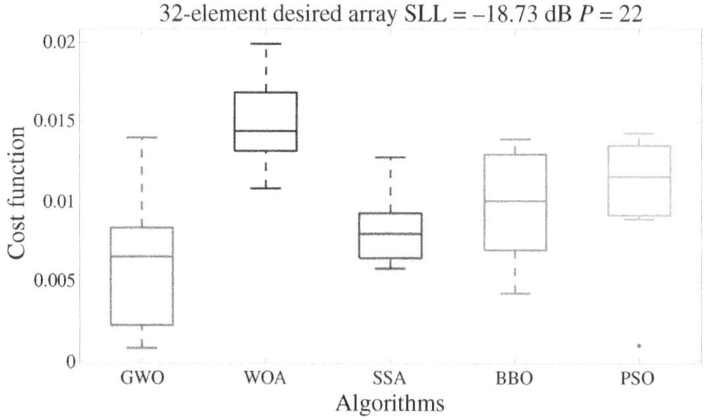

Figure 7.5 Synthesis of a 32-element pattern with SLL = −18.73 dB with $P = 22$. Boxplot of all algorithms results.

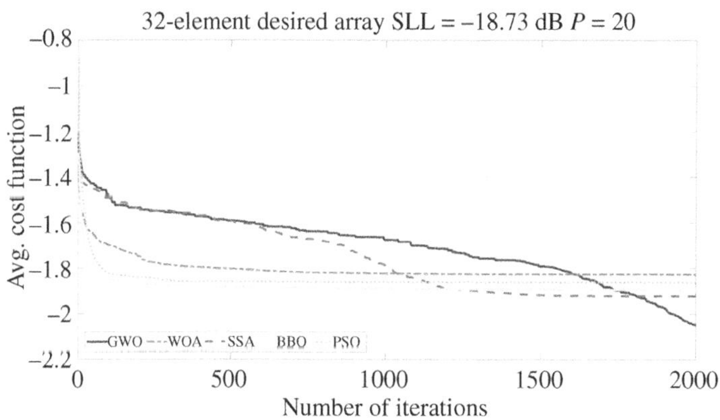

Figure 7.6 Synthesis of a 32-element pattern with SLL = −18.73 dB with $P = 20$. Convergence rate graph.

Figures 7.6 and 7.7 show the convergence rate graphs, indicating that GWO and WOA have a similar convergence rate, which is faster than the other algorithms.

The characteristics of the radiation pattern for the desired and reconstructed patterns are shown in Table 7.3. It is possible to create the same pattern with a 20-element array by using amplitude-position synthesis. The pattern consisting of 22 elements is nearly identical to what is desired. The peak sidelobe level (SLL) of all the arrays found by GWO closely matches the peak SLL of [28]. The GWO

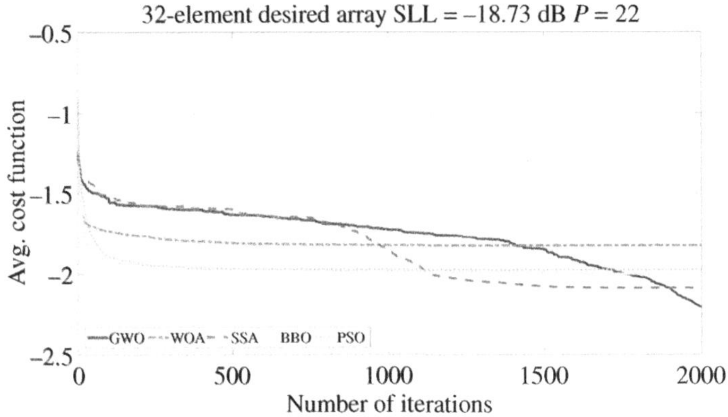

Figure 7.7 Synthesis of a 32-element pattern with SLL = −18.73 dB with $P = 22$. Convergence rate graph.

Table 7.3 Radiation pattern features for the pattern.

Feature	Desired [28]	GWO $P = 20$	GWO $P = 22$
SLL (dB)	−18.73	−18.70	−18.73
BW (deg)	8.6	8.8	8.6
MSE	N/A	1.158×10^{-7}	4.682×10^{-6}

Source: Adapted from Khodier and Christodoulou [28].

pattern with $P = 20$ was slightly different from the desired one, with a peak sidelobe level of −18.70 dB. On the other hand, the GWO pattern with $P = 22$ closely matched the desired pattern, with an SLL of −18.73 dB. The 22-element array precisely matches the beamwidth of the desired array, while the other array closely resembles the beamwidth of the desired arrays, differing only by 2.3%. The MSE value is satisfactory for the $P = 20$ case. The array obtained with 22 elements improves the MSE by an order of magnitude. To achieve lower MSE values, the number of elements must be increased by two.

The illustration in Figure 7.8 shows the desired pattern and the two reconstructed patterns. It is obvious that for $P = 20$ the pattern obtained is different from the desired pattern for angles greater than 80°. When the number of elements of the array increases, the reconstructed pattern is almost the same as the desired pattern.

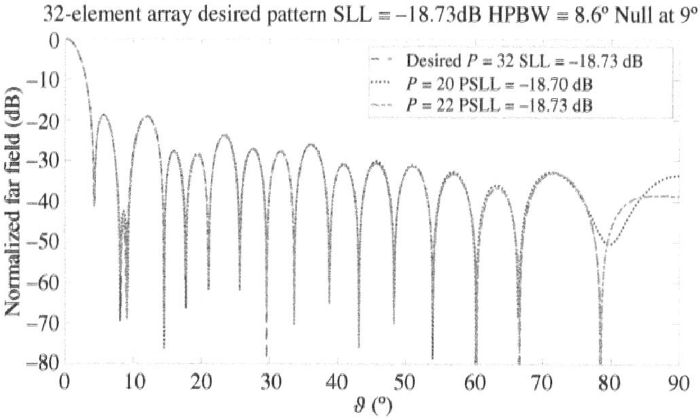

Figure 7.8 Synthesis of a 32-element pattern with SLL = −18.73 dB with $P = 20$ and $P = 22$. Radiation patterns of the best results found.

Table 7.4 Positions and amplitudes of the best reconstructed patterns found by GWO for the 32-element array design, for $P = 20$ and $P = 22$.

$P = 20$			$P = 22$		
p	x_p	A_p	p	x_p	A_p
1	0.445	1.000	1	0.417	1.000
2	1.294	1.000	2	1.210	1.000
3	2.044	0.958	3	1.913	0.948
4	2.689	1.000	4	2.514	0.992
5	3.429	1.000	5	3.161	1.000
6	4.271	0.896	6	3.884	0.895
7	5.156	0.782	7	4.630	0.777
8	6.037	0.666	8	5.399	0.715
9	6.946	0.567	9	6.187	0.626
10	7.719	0.569	10	7.033	0.58
			11	7.751	0.575

The best amplitudes and positions found by GWO are reported in Table 7.4 The last element in [28] is placed at 7.755λ. Table 7.4 shows that for both reconstructed patterns, the total size of the array is smaller than that in [28].

The results of the amplitudes and positions are shown in Table 7.4. It is evident that the element positions calculated in both cases are smaller than the total length of the desired array pattern.

7.3 Machine Learning

ML algorithms, which can learn from data, have been applied to a variety of antenna design problems. These algorithms learn from experience E when used to complete a task T with a performance metric P, and their performance P in task T increases with experience E. ML algorithms can be divided into supervised and unsupervised, depending on the type of experience they gain during the learning process. Therefore, several supervised ML algorithms have been used in antenna problems. A comprehensive review of ML techniques in electromagnetics can be found in [29]. Generally, three main types of antenna problems have been addressed by ML algorithms, namely antenna synthesis, antenna analysis, and direction-of-arrival estimation (DoA).

7.3.1 Artificial Neural Networks (ANNs)

Artificial neural networks (ANNs) are the most commonly used base learners in the antenna domain. The main benefit of neural networks is their ability to identify the relationship between the physical parameters of a problem and the outcomes they produce through a complex physical process. Therefore, an ANN can effectively determine the function that describes this physical process, which has no closed formulation. ANNs are especially useful in applications such as data classification, prediction approximation, and signal processing. Multilayer perceptron (MLP) networks are composed of an input layer with a small or large number of nodes, one or more hidden layers, and an output layer with one (more commonly) or more nodes. The activation functions of the nodes in the last hidden layer are usually linear, while the activation functions of the nodes in the preceding hidden layers are nonlinear. A MLP network is composed of an input layer with N_I nodes, N_H hidden layers, and an output layer with N_O nodes. The activation functions of the last hidden layer's nodes are linear, while the activation functions of the nodes in the first $N_H - 1$ hidden layers are nonlinear. The number of neurons in the mth hidden layer is denoted by N_m. In [30], an ANN model was used to simultaneously obtain antenna characteristics such as S_{11} value, gain, and radiation pattern from three different branches, in combination with a support vector machine (SVM).

7.3.2 Support Vector Machines

SVMs are a type of ML algorithm that is often used in antenna-related problems. They can be used for both classification and regression tasks. One of the benefits of SVM is that it uses convex quadratic programming, which is a global optimization technique. The theoretical basis of SVMs is statistical learning theory. According to [2], SVM was used to address the issue of antenna array element failure. In [31], an SVM was used to determine the resonant frequency of the UHF patch antennas. This was also compared with an adaptive neuro-fuzzy inference system (ANFIS) using the same data, and the authors concluded that both ML algorithms can be efficiently applied to this problem.

7.3.3 Gaussian Process (GP)

Recently, GP has been used as the primary ML algorithm for antenna-related issues. GP is thought to be comparable to ANNs, but it is simpler to implement and requires fewer parameters to be trained. In [32], GP regression was used to model the input characteristics of a planar antenna. This approach allowed a reduction of 80% in computational time, with the antenna geometry being the input parameters and the S_{11} value being the output. The authors in [32] use GP regression for modeling the input characteristics of planar antennas. The authors using GP can have reduced computational time by 80%. The input parameters are the antenna geometry, and the output is S_{11} value.

7.3.4 Deep Learning (DL)

Deep learning (DL) is a form of ANN that uses multiple hidden layers to process data. In recent years, DL has become increasingly popular, particularly in the field of image recognition. However, its use has been extended to a variety of other domains, including antenna design. A recent study [33] proposed a surrogate modeling technique based on a pyramidal deep regression network for the design of different types of patch antenna. In [33] the authors present a surrogate modeling technique using a pyramidal deep regression network [33]. The authors design different patch antenna types with this framework.

7.3.5 ANFIS

ANFIS is a type of Fuzzy Inference System (FIS) that combines the benefits of both ANNs and FIS. The primary purpose of ANFIS is to optimize the parameters of the corresponding FIS by utilizing a learning algorithm and input–output data sets. ANFIS has been widely used to address antenna problems, such as predicting the performance of pyramidal and conical corrugated horn antennas. The problem of predicting the performance of pyramidal and conical corrugated horn antennas was addressed in [34]. Again, the input parameters are the geometrical variables of the antenna, and the output parameters are the S_{11} value and the radiation pattern.

Table in Section 7.3.5 provides an overview of the advantages and disadvantages of ML methods used for antenna design.

Algorithm	Advantages	Disadvantages
ANN	Method generic for modeling any form of data	Extensively computational
	Noise resistance in training data	Easily overfitted
	Potential use of various training algorithms	Structure definition lacks a norm.
	Models linear and nonlinear functions	Does not "understand" the problem
SVM	Highly efficient in spaces with a high dimension	Kernel function selection is a challenging endeavor
	Memory effectiveness	Poor performance with noise in data
	Manages nonlinear data with effectiveness	Large data sets exhibit subpar performance
GP	Capture the model's uncertainty directly	Highly intensive in terms of computation
	Obtain nonlinear and smooth models	
DL	Work very well with image detection tasks	Highly intensive in terms of computation
	Same DL structure can be applied to different data types	Vast training data is necessary
	Very flexible for future problem adaptation	No rule for suitable DL type selection
	Able to effectively utilize GPUs effectively and parallel	
	A multitude of open source frameworks	
ANFIS	Rapid learning	Type and number of membership functions
	Capacity to adapt	The location of a membership function
	Models linear and nonlinear functions	Curse of dimensionality
	Inserts rules to the model based on prior knowledge	

7.3.6 Surrogate Modeling

Antenna issues require a general modeling approach as a supervised regression task. Consequently, a variety of supervised ML techniques have been used to address antenna problems. A comprehensive overview of supervised ML

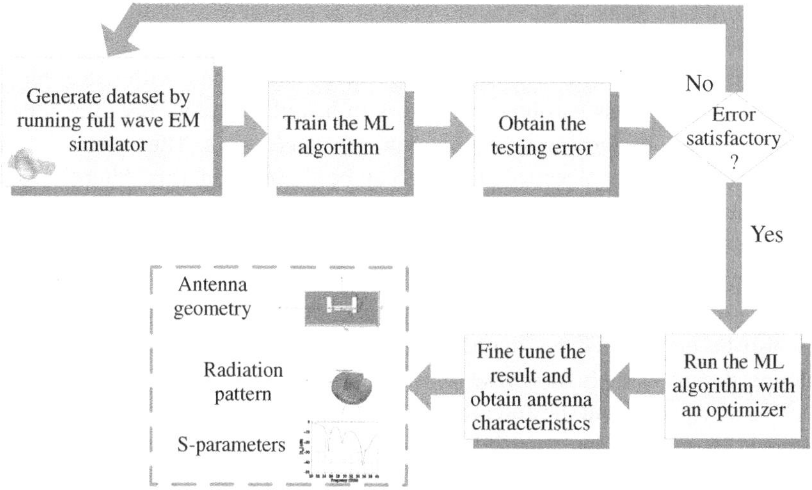

Figure 7.9 Use of an ML algorithm in antenna synthesis.

approaches in electromagnetics is provided in [29]. Generally, three major antenna problem types are tackled by ML algorithms: antenna synthesis, antenna analysis, and direction-of-arrival estimation.

An ML algorithm is often used as a substitute approach for full-wave analysis in antenna synthesis. As full-wave electromagnetic (EM) solvers for modern antennas are time-consuming and require a lot of computing power, the use of a surrogate method is a popular choice in antenna research. Generally, the process involves the following steps (see Figure 7.9):

- *Step 1:* The antenna geometry and/or frequency of operation are altered to create the dataset. A full-wave EM simulator is then used to generate the output, which can be S parameters, gain, or radiation patterns at one or more frequencies.
- *Step 2:* Train the ML algorithm using about 80% of the obtained data set.
- *Step 3:* To get the testing error, use the last 20% of the data set as input for the ML algorithm. If the testing error is not satisfactory, go back to step one, create more data, or adjust the ML algorithm.
- *Step 4:* Instead of using a full-wave EM simulator, employ an ML algorithm in combination with an optimization technique, such as an EA.
- *Step 5:* With the best-obtained design, refine the most successful design by re-executing the full-wave EM simulator with the most successful geometrical parameters.

In antenna analysis, a training dataset is acquired by executing full-wave simulations or actual antenna measurements. The ML algorithm is then trained with these data and used to forecast certain antenna characteristics, such as the radiation pattern, scattering responses, S parameters, mutual coupling characteristics, and resonant frequencies.

Note that more intricate surrogate models can be constructed with two or more outputs. These models can be combined with a multiobjective EA to generate various antenna design cases as points on the Pareto front. The most suitable case that meets all the objective functions can be chosen from this. Alternatively, constraint functions can be defined to identify only the Pareto front points within a particular area of interest. In addition, the surrogate model can be used to perform a sensitivity analysis, in which the input parameters are varied within a certain tolerance to determine the tolerance level of the output. ML models can also be employed for antenna array failure correction, which is another antenna design case. In this case, one or more elements of an array are not working as expected and the antenna array must adjust the remaining operational elements to achieve the same initial radiation pattern. This can be seen as a binary classification problem. Different radiation patterns can be used as training data, which can be classified as either functioning correctly or not based on the element excitations. The ML algorithm is initially trained with DoA data and then used to identify the optimal excitation of the antenna array to reduce interference and decrease SLL. This ML model functions as an adaptive beamformer, which directs the main lobe of the antenna array toward the desired signal, places nulls in the path of the interference signals, and suppresses the SLL.

7.3.6.1 Surrogate Modeling Example

An example case for 5G cellular communications is presented in this section using surrogate modeling. The antenna engineer is faced with novel challenges in the 5G era of cellular communications.

The design of such antennas that cover a wide range of frequencies in the millimeter wave range is a requirement for 5G, as specified by 3GPP. The frequency range of the 5G NR FR2 is 24.25 to 52.6 GHz [35]. EAs are a highly effective optimization technique that can be utilized for complex antenna design projects [36]. The proposed antenna is a modified E-shaped antenna [37] that incorporates an additional third slot. Figure 7.10 shows the antenna geometry. The antenna is mounted on a Taconic substrate with a dielectric constant of $\epsilon_r = 2.21$, a $\tan\delta$ of 0.001, and a thickness of 1.58 mm. The antenna requires the adjustment of fourteen distinct geometric parameters. It is clear that the construction of this design cannot be accomplished without the use of an optimization technique. However, optimizing this geometry requires a very intensive computational load. Thus, surrogate modeling can be applied to reduce computational time.

Therefore, the optimization goal is to design such an antenna for wide-band operation using a surrogate model. The selected optimization algorithm is GWO.

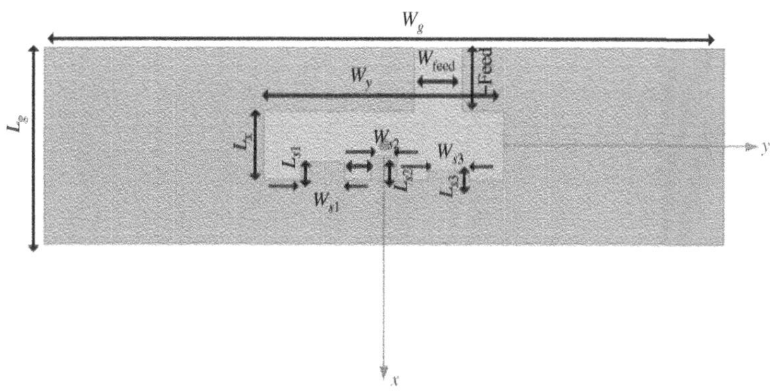

Figure 7.10 Geometry of the proposed modified E-shaped patch antenna.

We first obtain a data set using the HFSS commercial EM full-wave solver. Unlike the training set, the test set is selected at random from the entire data set.

Then, we apply the Adaboost algorithm and build a ML model. AdaBoost (Adaptive Boosting) is a well-known boosting ML algorithm for building an ensemble model.

The fundamental concept behind AdaBoost is to train a new instance with a focus on instances that were previously misclassified [38]. Before beginning the training, weights are allocated to each instance. The weights are adjusted iteratively so that instances that are misclassified receive greater weight than instances that are correctly classified. This algorithm operates on the fundamental principle of iteratively instructing numerous weak learners using altered instances of the data. Adaboost models require the adjustment of a reduced number of hyperparameters, albeit at the expense of diminished flexibility compared to gradient-boosting models.

The surrogate model has 14 input variables (which correspond to the geometric parameters of the antenna) and a single output that represents the antenna S_{11} maximum value over the selected frequency range.

We use the mean absolute error (MAE) as a performance measure for the model. This is given by

$$\text{MAE} = \frac{1}{Q_{\text{test}}} \sum_{s=1}^{Q_{\text{test}}} \left| d_s - y_{o,s} \right|, \tag{7.11}$$

where d_s is the sth input pattern and $y_{o,s}$ is the output when the sth input data pattern is entered as input. The error measurement metric values in this case reflect the differences between the predicted S_{11} values and the corresponding values of the test set data. Model performance in terms of MAE is shown in Figure 7.11.

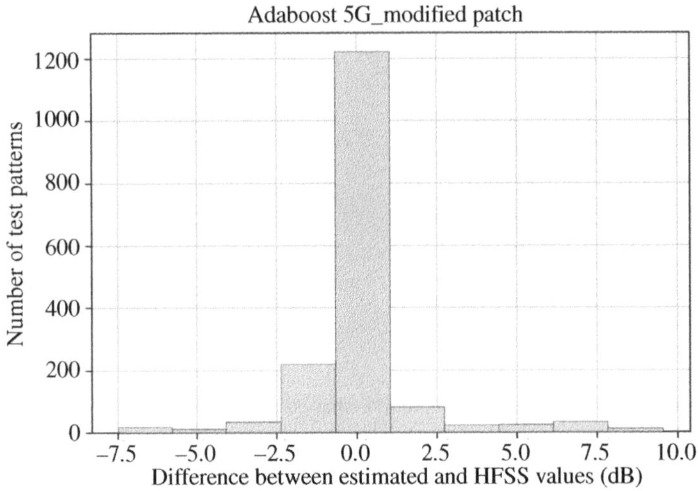

Figure 7.11 Statistical distribution of the difference between estimated and real values.

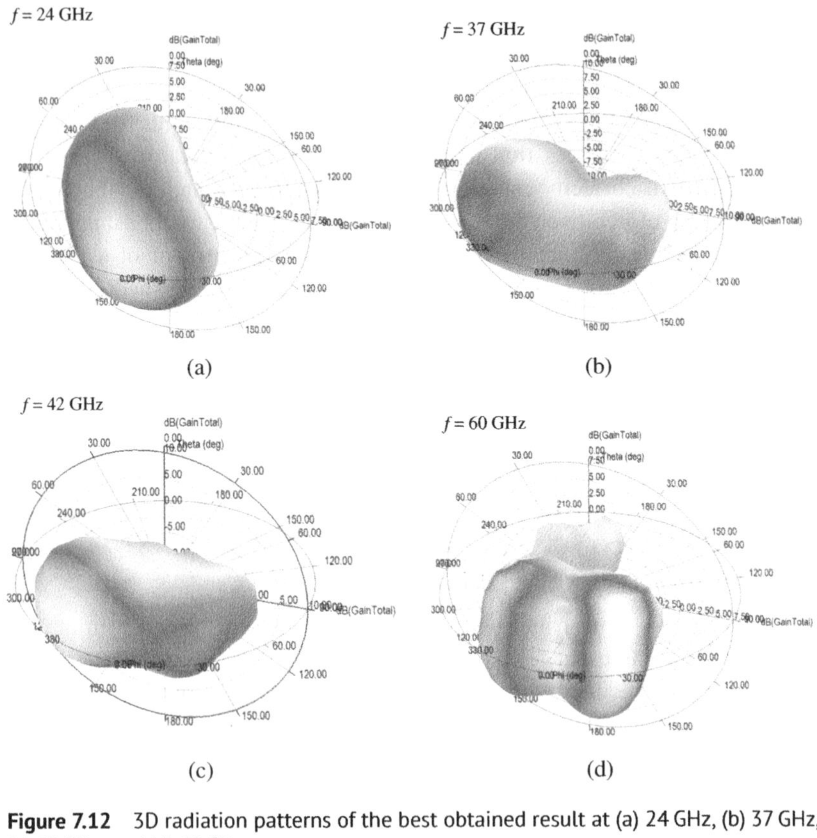

Figure 7.12 3D radiation patterns of the best obtained result at (a) 24 GHz, (b) 37 GHz, (c) 42 GHz, and (d) 60 GHz.

We applied the GWO algorithm using a population size of 20 and a total number of 2000 iterations. We use a wrapper program in Python in conjunction with the ML model. Formulating the objective function as follows

$$F(\overline{w}) = -G + \Sigma \times \max\left(0, S_{11}^{\max}(\overline{w}) - M_{\mathrm{dB}}\right). \tag{7.12}$$

In this instance, $\overline{w}$ represents the vector of the antenna's geometric parameters, and S_{11}^{maxl} denotes the maximal value of S_{11} at the solution frequencies. $S_{11}^{\mathrm{maxval}} = \max\left(S_{11}^{24\,\mathrm{GHz}}, S_{11}^{37\,\mathrm{GHz}}, S_{11}^{42\,\mathrm{GHz}}, S_{11}^{60\,\mathrm{GHz}}\right)$. The symbol M_{dB} represents the acceptable S_{11} limit for the algorithm. In this particular instance, we have set $M_{\mathrm{dB}} = -10\,\mathrm{dB}$. A penalty factor, denoted Σ, is applied to the objective function value when the constraints are not met.

The 3D radiation patterns of the most successful result are shown in Figure 7.12. In addition, the surface current distribution for various frequencies is displayed in Figure 7.13. Figure 7.14 also illustrates the frequency response of the optimal design that was acquired. It is possible to observe that the proposed antenna exhibits a broad frequency range, spanning from the 5G FR2 region to 65 GHz. Therefore, the proposed antenna is compatible with the operation of 5G and IEEE802.11ad.

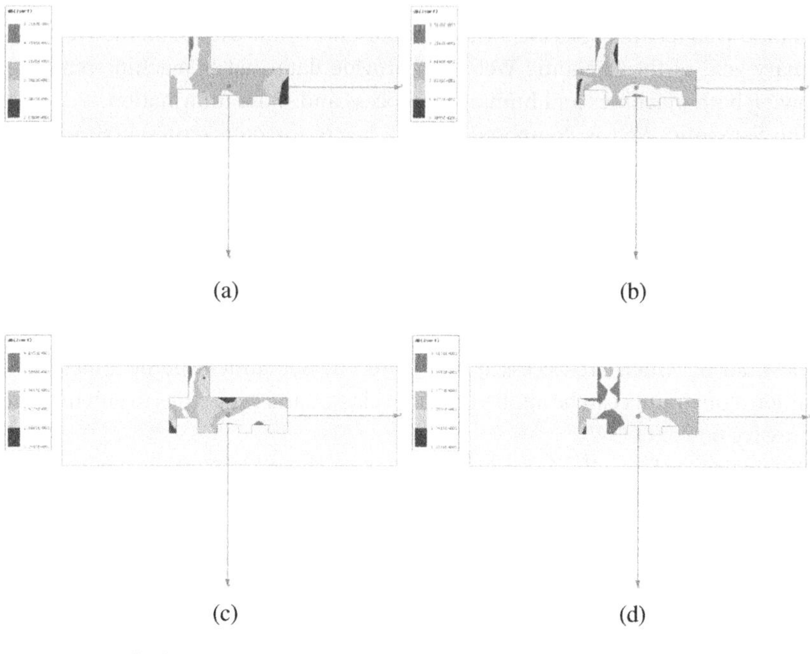

Figure 7.13 Surface current distribution of the best obtained result at (a) 24 GHz,(b) 37 GHz, (c) 42 GHz, and (d) 60 GHz.

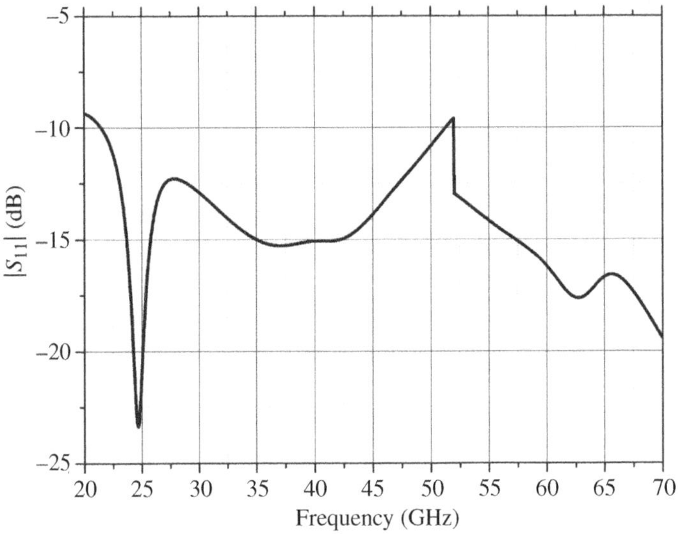

Figure 7.14 Reflection coefficient (S_{11} magnitude) versus frequency of the best obtained patch antenna for 5G operation.

7.4 Knowledge Representation

Semantic Web technologies are a way of representing knowledge on the Web. The primary goal of the Semantic Web is to provide data that is machine-readable, allowing both machines and humans to process and share information.

The Semantic Web is composed of four layers for data representation from the bottom up: Extensible Markup Language (XML), Resource Description Framework (RDF), Ontology, and Logic. Web Ontology Language (OWL) is the primary language for constructing ontologies on the Web and is a standard of the Web Ontology Working Group of the World Wide Web Consortium (W3C) Web Ontology Working Group. Ontologies in OWL are created by defining classes (which have instances or individuals) and properties (which have data values), all of which are accessed and stored as Semantic Web documents. In addition, constraints can be applied to these classes and properties to enhance the expressive power of OWL.

Ontologies are formalized vocabularies that are shared by a group of people and typically focus on a single domain. They provide a description of how each term in the ontology relates to other concepts by outlining the definitions of each term.

In 2012, the W3C established OWL2, which added new features and increased the expressiveness of the language. OWL2 can be viewed as a RDF graph. It also defines three sublanguages, or profiles, in OWL terminology: OWL 2 EL, OWL 2

QL, and OWL 2 RL. Each profile makes different compromises with OWL's expressive capacity in exchange for various computational and/or implementation benefits. OWL 2 EL is an ideal choice for reasoning tasks that require polynomial-time algorithms; it is especially useful for applications that involve very large ontologies. OWL 2 QL enables the answer of conjunctive queries in LogSpace with the aid of traditional relational database technology. It is especially suitable for scenarios where lightweight ontologies are used to organize a large number of individuals, and it is beneficial or necessary to access the data directly through relational queries. It is particularly well-suited for cases where lightweight ontologies are used to organize a large number of class instances and it is advantageous or necessary to directly access the data using relational queries. OWL 2 RL is particularly advantageous for scenarios where it is advantageous or necessary to take action on data in the form of RDF triples and where relatively lightweight ontologies are used to manage a large number of individuals. Furthermore, OWL 2 RL enables the use of rule-augmented database systems that operate directly with RDF triples to execute polynomial-time reasoning algorithms. The utilization of knowledge representation (KR) in the antenna field is a novel concept. In [39], the authors presented an example of an OWL DL ontology for antennas, which focused on the concept of antenna array. This utilization of ontologies in the antenna domain could be the initial step toward the machine-aided design of antennas.

7.5 Conclusion

This book chapter provides an overview of the main AI families used in the antenna domain. We have divided the techniques into three categories: EAs, ML, and KR. We review the most common evolutionary algorithms and the newer algorithms that have been applied to fewer antenna design cases. In addition, we discuss and present the most commonly used ML methods in the antenna domain. We also consider KR using the OWL language. Finally, we briefly outline potential future research directions, focusing on additional algorithms and ML methods that can be applied to antenna design.

References

1 Campbell, S.D., Jenkins, R.P., O'Connor, P.J., and Werner, D. (2021). The explosion of artificial intelligence in antennas and propagation: how deep learning is advancing our state of the art. *IEEE Antennas and Propagation Magazine* 63 (3): 16–27. https://doi.org/10.1109/MAP.2020.3021433.

2 Zardi, F., Nayeri, P., Rocca, P., and Haupt, R. (2021). Artificial intelligence for adaptive and reconfigurable antenna arrays: a review. *IEEE Antennas*

and Propagation Magazine 63 (3): 28–38. https://doi.org/10.1109/MAP.2020.3036097.

3 Goldberg, D.E. (1989). *Genetic Algorithms in Search, Optimization and Machine Learning*. New York: Addison Wesley.

4 Holland, J.H. (1975). *Adaptation in Natural and Artificial Systems*. Ann Arbor, MI: The University of Michigan Press.

5 Oliveri, G., Gottardi, G., Robol, F. et al. (2017). Codesign of unconventional array architectures and antenna elements for 5G base stations. *IEEE Transactions on Antennas and Propagation* 65 (12): 6752–6767. https://doi.org/10.1109/TAP.2017.2738065.

6 Soltani, S., Lotfi, P., and Murch, R.D. (2018). Design and optimization of multiport pixel antennas. *IEEE Transactions on Antennas and Propagation* 66 (4): 2049–2054. https://doi.org/10.1109/TAP.2018.2800759.

7 Kennedy, J. and Eberhart, R. (1995). Particle swarm optimization. *Proceedings of ICNN'95 - International Conference on Neural Networks*, volume 4, 1942–1948. https://doi.org/10.1109/ICNN.1995.488968.

8 Clerc, M. (1999). The swarm and the queen: towards a deterministic and adaptive particle swarm optimization. *Proceedings of the 1999 Congress on Evolutionary Computation. CEC 99*, volume 3, 1951–1957.

9 Dai, X., Li, X., and Luk, K.-M. (2021). A planar wideband millimeter-wave antenna array with low sidelobe using ';±1' excitations. *IEEE Transactions on Antennas and Propagation* 69 (10): 6999–7004. https://doi.org/10.1109/TAP.2021.3076217.

10 Zaharis, Z.D., Gravas, I.P., Yioultsis, T.V. et al. (2017). Exponential log-periodic antenna design using improved particle swarm optimization with velocity mutation. *IEEE Transactions on Magnetics* 53 (6): 1–4. https://doi.org/10.1109/TMAG.2017.2660061.

11 Storn, R. and Price, K. (1997). Differential evolution—a simple and efficient heuristic for global optimization over continuous spaces. *Journal of Global Optimization* 11 (4): 341–359.

12 Liu, F., Liu, Y., Han, F. et al. (2020). Synthesis of large unequally spaced planar arrays utilizing differential evolution with new encoding mechanism and Cauchy mutation. *IEEE Transactions on Antennas and Propagation* 68 (6): 4406–4416. https://doi.org/10.1109/TAP.2020.2969741.

13 Dorigo, M. and Gambardella, L.M. (1997). Ant colonies for the travelling salesman problem. *Biosystems* 43 (2): 73–81.

14 Dorigo, M., Maniezzo, V., and Colorni, A. (1996). Ant system: optimization by a colony of cooperating agents. *IEEE Transactions on Systems, Man, and Cybernetics, Part B: Cybernetics* 26 (1): 29–41.

15 Dorigo, M. and Stutzle, T. (2004). *Ant Colony Optimization*. Cambridge, MA: The MIT Press.

16 Zhu, D.Z., Gregory, M.D., Werner, P.L., and Werner, D.H. (2018). Fabrication and characterization of multiband polarization independent 3-D-printed frequency selective structures with ultrawide fields of view. *IEEE Transactions on Antennas and Propagation* 66 (11): 6096–6105. https://doi.org/10.1109/TAP.2018.2866507.

17 Singh, U., Kumar, H., and Kamal, T.S. (2010). Design of Yagi-Uda antenna using biogeography based optimization. *IEEE Transactions on Antennas and Propagation* 58 (10): 3375–3379.

18 Li, X. and Guo, Y.-X. (2021). The Gray Wolf Optimizer for Antenna Optimization Designs: continuous, binary, single-objective, and multiobjective implementations. *IEEE Antennas and Propagation Magazine* 64 (6): 29–40. https://doi.org/10.1109/MAP.2021.3127798.

19 Bayraktar, Z., Komurcu, M., Bossard, J.A., and Werner, D.H. (2013). The wind driven optimization technique and its application in electromagnetics. *IEEE Transactions on Antennas and Propagation* 61 (5): 2745–2757.

20 Mirjalili, S., Gandomi, A.H., Mirjalili, S.Z. et al. (2017). Salp Swarm Algorithm: a bio-inspired optimizer for engineering design problems. *Advances in Engineering Software* 114: 163–191. https://doi.org/10.1016/j.advengsoft.2017.07.002.

21 Goudos, S.K. (2019). Joint power allocation and user association in non-orthogonal multiple access networks: an evolutionary approach. *Physical Communication* 37: https://doi.org/10.1016/j.phycom.2019.100841.

22 Boursianis, A.D., Goudos, S.K., Yioultsis, T.V. et al. (2020). MIMO antenna design for 5G communication systems using Salp swarm algorithm. *2020 International Workshop on Antenna Technology (iWAT)*, 1–3. https://doi.org/10.1109/iWAT48004.2020.1570618331.

23 Goudos, S.K., Siakavara, K., and Sahalos, J.N. (2014). Novel spiral antenna design using artificial bee colony optimization for UHF RFID applications. *IEEE Antennas and Wireless Propagation Letters* 13: 528–531.

24 Yang, S.-H. and Kiang, J.-F. (2015). Optimization of sparse linear arrays using harmony search algorithms. *IEEE Transactions on Antennas and Propagation* 63 (11): 4732–4738.

25 Eusuff, M., Lansey, K., and Pasha, F. (2006). Shuffled frog-leaping algorithm: a memetic meta-heuristic for discrete optimization. *Engineering Optimization* 38 (2): 129–154. https://doi.org/10.1080/03052150500384759.

26 Simon, D. (2008). Biogeography-based optimization. *IEEE Transactions on Evolutionary Computation* 12 (6): 702–713.

27 Mirjalili, S. and Lewis, A. (2016). The whale optimization algorithm. *Advances in Engineering Software* 95: 51–67. https://doi.org/10.1016/j.advengsoft.2016.01.008.

28 Khodier, M.M. and Christodoulou, C.G. (2005). Linear array geometry synthesis with minimum sidelobe level and null control using particle swarm optimization. *IEEE Transactions on Antennas and Propagation* 53 (8): 2674–2679. https://doi.org/10.1109/TAP.2005.851762.

29 Massa, A., Oliveri, G., Salucci, M. et al. (2018). Learning-by-examples techniques as applied to electromagnetics. *Journal of Electromagnetic Waves and Applications* 32 (4): 516–541.

30 Xiao, L.Y., Shao, W., Jin, F.L., and Wang, B.Z. (2018). Multiparameter modeling with ANN for antenna design. *IEEE Transactions on Antennas and Propagation* 66 (7): 3718–3723.

31 Kayabasi, A. and Akdagli, A. (2015). Predicting the resonant frequency of e-shaped compact microstrip antennas by using ANFIS and SVM. *Wireless Personal Communications* 82 (3): 1893–1906. https://doi.org/10.1007/s11277-015-2321-6.

32 Jacobs, J.P. and Koziel, S. (2014). Two-stage framework for efficient Gaussian process modeling of antenna input characteristics. *IEEE Transactions on Antennas and Propagation* 62 (2): 706–713.

33 Koziel, S., Çalık, N., Mahouti, P., and Belen, M.A. (2022). Accurate modeling of antenna structures by means of domain confinement and pyramidal deep neural networks. *IEEE Transactions on Antennas and Propagation* 70 (3): 2174–2188. https://doi.org/10.1109/TAP.2021.3111299.

34 Pujara, D., Modi, A., Pisharody, N., and Mehta, J. (2014). Predicting the performance of pyramidal and corrugated horn antennas using ANFIS. *IEEE Antennas and Wireless Propagation Letters* 13: 293–296.

35 3GPP (2021). 5G-NR specifications series. https://www.3gpp.org/DynaReport/38-series.htm/ (accessed 1 February 2022).

36 Goudos, S.K. (2021). *Emerging Evolutionary Algorithms for Antennas and Wireless Communications*, Electromagnetic Waves. Institution of Engineering and Technology. ISBN: 9781785615528.

37 Goudos, S.K., Boursianis, A., Mohamed, A.W. et al. (2022). Wideband antenna design for 5G mmWave applications using enhanced adaptive differential evolution. *2022 IEEE International Symposium on Antennas and Propagation and USNC-URSI Radio Science Meeting (AP-S/URSI)*, 63–64. https://doi.org/10.1109/AP-S/USNC-URSI47032.2022.9886187.

38 Sagi, O. and Rokach, L. (2018). *Ensemble Learning: A Survey*, Data Mining and Knowledge Discovery. Wiley Interdisciplinary Reviews.

39 Esposito, A., Tarricone, L., and Vallone, L. (2007). Knowledge modeling in electromagnetic applications [em programmer's notebook]. *IEEE Antennas and Propagation Magazine* 49 (5): 130–137.

8

AI-Driven Approaches for Solving Electromagnetic Inverse Problems

Marco Salucci[1], *Maokun Li*[2,3], *and Andrea Massa*[1,3,4,5]

[1] ELEDIA Research Center (ELEDIA@UniTN – University of Trento), DICAM – Department of Civil, Environmental, and Mechanical Engineering, Trento, Italy
[2] Department of Electronic Engineering, Beijing National Research Center for Information Science and Technology (BNRist), Tsinghua University, Beijing, China
[3] ELEDIA Research Center (ELEDIA@TSINGHUA – Tsinghua University), Beijing, China
[4] ELEDIA Research Center (ELEDIA@UESTC – UESTC), School of Electronic Science and Engineering, Chengdu, China
[5] School of Electrical Engineering, Tel Aviv University, Tel Aviv, Israel

CHAPTER MENU

8.1 Introduction

Nowadays, we are witnessing an exceptional growth of interest toward artificial intelligence (*AI*) in many research areas including electromagnetics (*EM*) [1–4]. As a matter of fact, the number of *AI*-related publications in *EM* problems has been exponentially growing during the last few years due to an unprecedented technological development of the so-called deep learning (*DL*) paradigm [1]. Among the several *EM* applications, *AI* currently represents a disruptive framework with the potential of revolutionizing the solution of many inverse problems (*IP*s). For instance, *AI*-based methods have been recently proposed to solve highly nonlinear/ill-posed *EM* imaging and inverse scattering (*IS*) problems with real-time capabilities, high accuracy, and robustness [5–9].

Artificial Intelligence for Future Networks, First Edition.
Edited by Mohammad A. Matin, Sotirios K. Goudos, and George K. Karagiannidis.

In such a context, the development of *AI* techniques to solve *EM–IP*s has its roots in pioneer works based on simple ("shallow") artificial neural network (*ANN*) architectures dating back to the early 1990s [10–14]. Afterward, significant advances have been achieved thanks to the development of other effective learning-by-examples (*LBE*) techniques (i.e., a branch of machine learning methods often referred to as "supervised learning" ones) with high generalization capabilities, such as those based on the "kernel trick" (e.g., support vector machines, *SVM*s [3]). Moreover, significant efforts have been devoted at improving the capabilities of such prediction techniques, while trying to minimize the amount of training data necessary for building them. More specifically, "three-step *LBE*" formulations have been proposed effectively exploiting dimensionality reduction techniques and smart sampling strategies to generate low-cardinality and highly informative training sets, alleviating the overall cost of the off-line learning stage and improving the robustness against the presence of noise [15–18].

Besides, the system-by-design (*SbD*) has recently emerged as a powerful *AI*-driven paradigm for the computationally efficient solution of complex *EM* problems [19]. In the context of *EM–IPs*, the *SbD* aims at exploiting physics/domain knowledge to enable a profitable exploitation of *LBE* techniques and their "interactive" integration with global optimization tools in order to efficiently process the measured data and yield accurate reconstructions of the imaged domain with very high time savings with respect to standard strategies [20].

Regardless of the wide and rapid spreading of *AI* methods for solving EM-IPs, the "no-free lunch theorems" [21, 22] clearly state that there is no algorithm universally performing better than others in any type of problem. Therefore, it is paramount to understand the main features and underlying hypotheses of each algorithm in order to select and properly customize the most suitable one for the problem at hand, avoiding a "black-box" strategy unavoidably leading to sub-optimal/unacceptable results.

The goal of this chapter is to provide an overview of the existing *AI*-driven techniques for solving different *EM–IP*s. Toward this end, Section 8.2 briefly recalls the main definitions and theory of *EM* imaging/*IS* problems. Afterward, a survey of different *AI* solution frameworks such as (i) three-step *LBE*, (ii) *SbD*, and (iii) *DL* is given in Section 8.3. Then, some of the most recent and appealing applications of such methodologies are outlined in Section 8.4. Eventually (Section 8.5), some conclusions and final remarks are drawn.

8.2 Mathematical Formulation

*EM–IP*s can be generally decomposed into a set of cascaded subtasks, aimed at gradually recovering, in a noninvasive way, complementary information on an inaccessible domain $\mathcal{D}$ starting from its probed "EM signature." Such sub-tasks

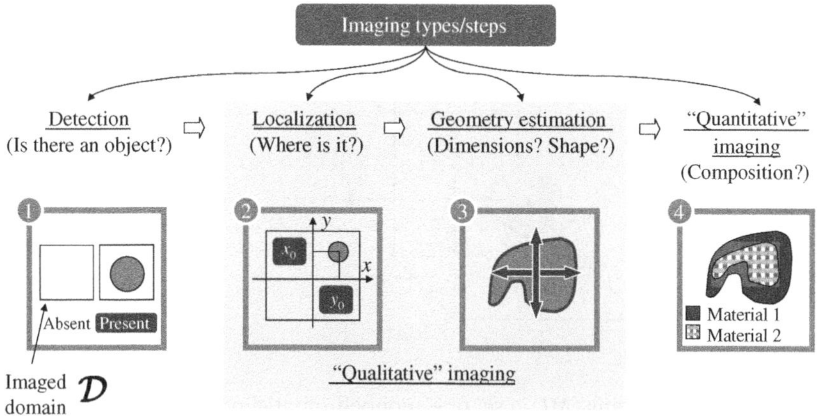

Figure 8.1 Classification of *EM* imaging types/steps.

can be defined as follows (Figure 8.1):

1) *Detection*: Refers to the identification of the presence of an unknown target within $\mathcal{D}$;
2) *Localization*: Aims at recovering an estimation of the position of the detected object;
3) *Geometry estimation*: Addresses the reconstruction of the size and shape of the localized target. Together with the previous step, this task is often referred to as "qualitative" imaging;
4) *"Quantitative" imaging*: Refers to the retrieval of the *EM* composition of the unknown scatterer (e.g., permittivity and conductivity).

The solution of each task requires the inversion of the underlying *EM* forward problem (*FP*), which can be briefly denoted as the process of generating the measured data $\underline{d} = \left\{d_k;\, k = 1, \dots, K\right\}$ starting from a set of descriptors $\underline{p} = \left\{p_q;\, q = 1, \dots, Q\right\}$ physically modeling, in a suitably chosen representation basis, the unknown solution

$$\underline{d} = \Phi\left\{\underline{p}\right\}, \tag{8.1}$$

$\Phi\{\,.\,\}$ being the known forward operator. For the sake of notation simplicity and without loss of generality, let us consider in the following the mathematical formulation of the *FP* in the microwave regime under free-space conditions as a representative case of nonlinear *EM* imaging problems. The extension to other modalities (e.g., eddy current testing, *ECT*, or electrical impedance tomography, *EIT*) and scenarios (e.g., subsurface, through-the-wall) only requires a suitable reformulation of the Green's operators and of the primary/equivalent sources according to problem-specific customizations of the Maxwell's equations and of the involved boundary conditions.

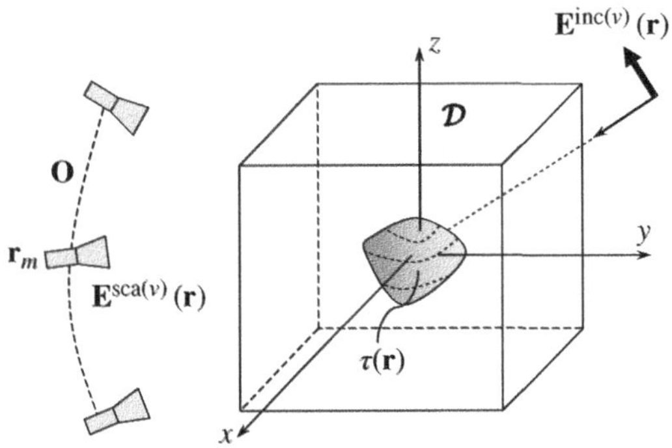

Figure 8.2 Geometry of the *3D MI* problem.

In microwave imaging (*MI*), a set of V monochromatic incident electric fields at frequency f with associated complex phasors [1]

$$\mathbf{E}^{\mathrm{inc}(v)}(\mathbf{r}) = \sum_{\xi\in\{x,y,z\}} E_{\xi}^{\mathrm{inc}(v)}(\mathbf{r})\,\mathbf{u}_{\xi};\quad v = 1,\ldots,V, \tag{8.2}$$

$\mathbf{u}_{\xi}$ being the unit vector along the ξth ($\xi \in \{x,y,z\}$) Cartesian axis, is used to illuminate $\mathcal{D}$ in order to retrieve its *EM* properties (Figure 8.2). These latter are mathematically modeled in terms of the so-called contrast function [23]

$$\tau(\mathbf{r}) = \frac{\tilde{\varepsilon}(\mathbf{r})}{\tilde{\varepsilon}_b} - 1, \tag{8.3}$$

where $\mathbf{r} = (x, y, z)$ is the position vector and

$$\tilde{\varepsilon}(\mathbf{r}) = \varepsilon_0\varepsilon(\mathbf{r}) - j\frac{\sigma(\mathbf{r})}{2\pi f} \tag{8.4}$$

is the complex permittivity in $\mathbf{r} \in \mathcal{D}$, $\varepsilon(\mathbf{r})$ and $\sigma(\mathbf{r})$ being the relative permittivity and the conductivity, respectively, while $\varepsilon_0 = 8.85 \times 10^{-12}$ [F/m] denotes the permittivity of vacuum. Similarly, $\tilde{\varepsilon}_b = \varepsilon_0\varepsilon_b - j\frac{\sigma_b}{2\pi f}$ is the complex permittivity of the homogeneous, isotropic, and nonmagnetic background medium of relative permittivity ε_b and conductivity σ_b. Under such hypotheses, the induced *EM* scattering phenomena are governed by the following *state*

$$\mathbf{E}^{\mathrm{inc}(v)}(\mathbf{r}) = \mathbf{E}^{\mathrm{tot}(v)}(\mathbf{r}) - \iiint_{\mathcal{D}} \mathbf{G}(\mathbf{r}|\mathbf{r}')\cdot\mathbf{J}^{(v)}(\mathbf{r}')\,d\mathbf{r}';\quad \mathbf{r}\in\mathcal{D} \tag{8.5}$$

and *data*

$$\mathbf{E}^{\mathrm{sca}(v)}(\mathbf{r}) = \iiint_{\mathcal{D}} \mathbf{G}(\mathbf{r}|\mathbf{r}')\cdot\mathbf{J}^{(v)}(\mathbf{r}')\,d\mathbf{r}';\quad \mathbf{r}\in\mathcal{O}, \tag{8.6}$$

1 The time-domain representation is obtained as $\mathbf{E}^{\mathrm{inc}(v)}(\mathbf{r},t) = \Re\left\{E_{\xi}^{\mathrm{inc}(v)}(\mathbf{r})\exp(-j2\pi f)\right\}$ ($v = 1,\ldots,V$), $\Re\{.\}$ being the real part.

equations, where $\mathbf{E}^{\mathrm{tot}(v)}(\mathbf{r})$ and

$$\mathbf{E}^{\mathrm{sca}(v)}(\mathbf{r}) \triangleq \left(\mathbf{E}^{\mathrm{tot}(v)}(\mathbf{r}) - \mathbf{E}^{\mathrm{inc}(v)}(\mathbf{r})\right); \quad v = 1, \dots, V \tag{8.7}$$

are the vth ($v = 1, \dots, V$) total and scattered field, respectively. Moreover, $\mathbf{G}\left(\mathbf{r}|\,\mathbf{r}'\right)$ is the free-space dyadic Green's function [23] and

$$\mathbf{J}^{(v)}(\mathbf{r}) = \tau(\mathbf{r})\,\mathbf{E}^{\mathrm{tot}(v)}(\mathbf{r}); \quad v = 1, \dots, V \tag{8.8}$$

is the equivalent current induced within $\mathcal{D}$ ($\mathbf{J}^{(v)}(\mathbf{r}) = \sum_{\xi\in\{x,y,z\}} J_{\xi}^{(v)}(\mathbf{r})\,\mathbf{u}_{\xi}$, $v = 1, \dots, V$). Finally, $\mathcal{O}$ denotes an external observation domain not intersecting $\mathcal{D}$ where the scattered field is derived according to (8.7) over a set of M field probing locations $\left\{\mathbf{r}_m \in \mathcal{O};\ m = 1, \dots, M\right\}$.

To numerically deal with the solution of the *MI* scattering problem, a set of N three-dimensional (3D) pulse basis functions

$$\Upsilon^{(n)}(\mathbf{r}) = \begin{cases} 1, & \text{if } \mathbf{r} \in \mathcal{D}^{(n)}, \\ 0, & \text{otherwise,} \end{cases} \quad n = 1, \dots, N \tag{8.9}$$

is used to discretize $\mathcal{D}$, $\mathcal{D}^{(n)}$ being the nth voxel having barycenter in $\mathbf{r}_n$ ($\mathcal{D} = \sum_{n=1}^{N} \mathcal{D}^{(n)}$). Accordingly, both the contrast and the equivalent currents can be expressed in discrete form as follows

$$\tau(\mathbf{r}) = \sum_{n=1}^{N} \tau^{(n)}\Upsilon^{(n)}(\mathbf{r}), \tag{8.10}$$

$$\mathbf{J}^{(v)}(\mathbf{r}) = \sum_{n=1}^{N} \sum_{\xi\in\{x,y,z\}} J_{\xi}^{(v,n)}\Upsilon^{(n)}(\mathbf{r})\,\mathbf{u}_{\xi}; \quad v = 1, \dots, V, \tag{8.11}$$

respectively, where $\tau^{(n)} = \tau\left(\mathbf{r}_n\right)$ and $J_{\xi}^{(v,n)} = J_{\xi}^{(v)}\left(\mathbf{r}_n\right)$, $n = 1, \dots, N$. Under such assumptions, the state and data equations can be rewritten in compact matrix form as

$$\begin{bmatrix} \underline{E}_x^{\mathrm{inc}(v)} \\ \underline{E}_y^{\mathrm{inc}(v)} \\ \underline{E}_z^{\mathrm{inc}(v)} \end{bmatrix} = \begin{bmatrix} \underline{E}_x^{\mathrm{tot}(v)} \\ \underline{E}_y^{\mathrm{tot}(v)} \\ \underline{E}_z^{\mathrm{tot}(v)} \end{bmatrix} - \underline{\underline{G}}_{\mathcal{D}} \times \begin{bmatrix} \underline{J}_x^{(v)} \\ \underline{J}_y^{(v)} \\ \underline{J}_z^{(v)} \end{bmatrix}, \tag{8.12}$$

$$\begin{bmatrix} \underline{E}_x^{\mathrm{sca}(v)} \\ \underline{E}_y^{\mathrm{sca}(v)} \\ \underline{E}_z^{\mathrm{sca}(v)} \end{bmatrix} = \underline{\underline{G}}_{\mathcal{O}} \times \begin{bmatrix} \underline{J}_x^{(v)} \\ \underline{J}_y^{(v)} \\ \underline{J}_z^{(v)} \end{bmatrix}, \tag{8.13}$$

where $\underline{\underline{G}}_{\mathcal{D}} \in \mathbb{C}^{3N\times 3N}$ and $\underline{\underline{G}}_{\mathcal{O}} \in \mathbb{C}^{3M\times 3N}$ are the internal and the external Green's matrices [23], respectively, $\underline{E}_{\xi}^{\mathrm{inc}(v)} = \left[E_{\xi}^{\mathrm{inc}(v,n)};\ n = 1, \dots, N\right]^{\mathrm{T}}$, $E_{\xi}^{\mathrm{inc}(v,n)} =$

$E_\xi^{\mathrm{inc}(v)}\left(\mathbf{r}_n\right)$, $\underline{E}_\xi^{\mathrm{tot}(v)} = \left[E_\xi^{\mathrm{tot}(v,n)};\, n=1,\dots,N\right]^{\mathrm{T}}$, $E_\xi^{\mathrm{tot}(v,n)} = E_\xi^{\mathrm{tot}(v)}\left(\mathbf{r}_n\right)$, $\underline{J}_\xi^{(v)} = \left[J_\xi^{(v,n)};\, n=1,\dots,N\right]^{\mathrm{T}}$, and $\underline{E}_\xi^{\mathrm{sca}(v)} = \left[E_\xi^{\mathrm{sca}(v,m)};\, m=1,\dots,M\right]^{\mathrm{T}}$, $E_\xi^{\mathrm{sca}(v,m)} = E_\xi^{\mathrm{sca}(v)}\left(\mathbf{r}_m\right)$, for $\xi \in \{x, y, z\}$ and $v = 1, \dots, V$, $.^{\mathrm{T}}$ being the transpose operator.

The above formulation can be simplified when dealing with two-dimensional (2D) *MI* problems under the hypothesis of invariant cylindrical scattering properties along the z-axis and considering a transverse-magnetic (*TM*) polarization of the electric field (i.e., $\mathbf{E}^{(v)}(\mathbf{r}) = E_z^{(v)}(\mathbf{r})\,\mathbf{u}_z$). In such cases, the state and data equations assume the following scalar forms

$$E_z^{\mathrm{inc}(v)}(\mathbf{r}) = E_z^{\mathrm{tot}(v)}(\mathbf{r}) - \iint_D g\left(\mathbf{r}|\,\mathbf{r}'\right) J_z^{(v)}\left(\mathbf{r}'\right) d\mathbf{r}'; \quad \mathbf{r} \in D, \tag{8.14}$$

$$E_z^{\mathrm{sca}(v)}(\mathbf{r}) = \iint_D g\left(\mathbf{r}|\,\mathbf{r}'\right) J_z^{(v)}\left(\mathbf{r}'\right) d\mathbf{r}'; \quad \mathbf{r} \in \mathcal{O}, \tag{8.15}$$

respectively, where $g\left(\mathbf{r}|\,\mathbf{r}'\right)$ is the 2D Green's function [23]. Moreover, after discretizing D into N pixels, the scattering phenomena can be numerically modeled by means of the following matrix equations

$$\underline{E}_z^{\mathrm{inc}(v)} = \underline{E}_z^{\mathrm{tot}(v)} - \underline{\underline{g}}_D \times \underline{J}_z^{(v)}, \tag{8.16}$$

$$\underline{E}_z^{\mathrm{sca}(v)} = \underline{\underline{g}}_{\mathcal{O}} \times \underline{J}_z^{(v)}, \tag{8.17}$$

where $\underline{\underline{g}}_D \in \mathbb{C}^{N\times N}$ and $\underline{\underline{g}}_{\mathcal{O}} \in \mathbb{C}^{M\times N}$ are the discrete Green's operators for the 2D *MI* scenario [23].

8.3 *AI*-Based *EM–IP* Solution Strategies

The goal of the *EM–IP* is to retrieve a guess $\underline{\tilde{p}}$ of the physical descriptors of the unknown target within D starting from the measured data $\underline{d}$, by solving

$$\underline{\tilde{p}} = \Phi^{-1}\left\{\underline{d}\right\}. \tag{8.18}$$

$\Phi^{-1}\{.\}$ being the *unknown* inverse operator. Although there exist several ways to classify the existing state-of-the-art approaches to robustly and reliably solve (8.18) through *AI*-based strategies, a convenient classification for the forthcoming discussion is between "direct inversion schemes" (*DIS*s) and "iterative inversion schemes" (*IIS*s).

DIS techniques aim at directly inverting the available data (e.g., the scattered field or equivalent quantities, such as the scattering parameters measured by a vector network analyzer [24] or the received signal strength indicator, *RSSI*, collected by a wireless network [25]) to recover $\underline{\tilde{p}}$, by learning a fast predictor $\tilde{\Psi}\{.\}$ of the unknown inverse relationship between data and unknowns (i.e., $\tilde{\Psi}\{\underline{d}\} = \tilde{\Phi}^{-1}\{\underline{d}\}$). Toward this end, according to a supervised learning (*SL*) mechanism, a training set of *S* input/output (*I*/*O*) pairs

$$\mathbb{T} = \left\{ \left(\underline{d}^{(s)}, \underline{p}^{(s)} \right);\quad s = 1, \ldots, S \right\} \tag{8.19}$$

is generated offline during the so-called "training phase." Then, the information embedded within $\mathbb{T}$ is exploited to build $\tilde{\Psi}\{.\}$, which is successively used in the online "testing phase" to predict the unknown solution $\tilde{\underline{p}}$ corresponding to a previously unseen data sample $\underline{d}$ collected in the observation domain [3].

Otherwise, *IIS* methods generate a sequence of solutions $\tilde{\underline{p}}_i$, $i = 1, \ldots, I$, I being the number of iterations needed to converge to the global optimum, starting from a properly initialized guess $\tilde{\underline{p}}_0$. Differently from *DIS* approaches, *IIS* ones often rely on the generation of a fast surrogate model (*SM*) of the data mismatch cost function. For instance, in 2D *MI* problems a predictor $\tilde{\Psi}\{.\}$ can be built to yield a computationally fast evaluation of [20]

$$\Xi\left(\underline{p}\right) = \frac{\sum_{v=1}^{V}\sum_{m=1}^{M}\left|E_z^{\mathrm{sca}(v,m)} - \tilde{E}_z^{\mathrm{sca}(v,m)}\left(\underline{p}\right)\right|^2}{\sum_{v=1}^{V}\sum_{m=1}^{M}\left|E_z^{\mathrm{sca}(v,m)}\right|^2}, \tag{8.20}$$

(i.e., $\tilde{\Psi}\left\{\underline{p}\right\} = \tilde{\Xi}\left(\underline{p}\right)$), where $\tilde{E}_z^{\mathrm{sca}(v,m)}\left(\underline{p}\right)$ is the reconstructed scattered field due to the guess solution represented by $\underline{p}$. In such cases, the training set is built as

$$\mathbb{T} = \left\{ \left[\underline{p}^{(s)}, \Xi\left(\underline{p}^{(s)}\right)\right];\quad s = 1, \ldots, S \right\}, \tag{8.21}$$

and the predictor is used to bypass the accurate but time-consuming full-wave solver to accelerate the minimization of (8.20) by means of either deterministic or stochastic optimization algorithms.

In Sections 8.3.1–8.3.3, a survey of the most widespread *AI*-based techniques for solving *EM–IPs* is given.

8.3.1 3-Step Learning-by-Examples (*LBE*) Framework

LBE methods are *SL* techniques and therefore, as previously mentioned, they are typically formulated as two-step approaches. The first step is the so-called "training phase," which is performed offline in order to build a computationally fast predictor $\tilde{\Psi}\{.\}$ starting from the information collected within a training set $\mathbb{T}$ of *S I*/*O* pairs. The second step is the "testing phase," which represents the online operation, in which the trained predictor inputs new input samples to efficiently estimate the corresponding output [3].

More, in general, *LBE* techniques can be also formulated in terms of the following three steps [15–18]:

- *Step 1*: Reduction of the input space dimensionality;
- *Step 2*: Exhaustive sampling of the input space;
- *Step 3*: Definition of the *LBE* prediction method.

The first step is aimed at counteracting the so-called "*curse of dimensionality*" (*CoD*) [26], which refers to the exponential growth of the required training samples (S) when the input space dimensionality increases. This is a crucial issue, for instance, in *DIS* methods where the measured data $\underline{d}$ represents the input space of the predictor $\tilde{\Psi}\{.\}$. Indeed, in many *EM* imaging problems, the number of "measured features" K directly coming from the observation domain $\mathcal{O}$ is very large [16], and as a consequence, a huge amount of training samples would be required to sufficiently explore such an input space and build an accurate *SM* of the underlying *I/O* relationship. When this happens, a possible solution to reduce S and, therefore, the cost/complexity of the overall training phase is to reduce the dimensionality of the input space from K to $H \ll K$ "reduced features" carrying the same amount of information on the output. More in detail, the generation of low-cardinality/highly informative training sets can be performed according to the following procedure (Figure 8.3):

1) *Parameters space sampling*: Sample the Q-dimensional parameters space to generate S known training configurations of the target, $\left\{\underline{p}^{(s)}; s = 1, \ldots, S\right\}$;
2) *FP solution*: Apply the forward operator $\Phi\{.\}$ to each generated sth ($s = 1, \ldots, S$) sample to compute/measure the corresponding K-dimensional data vectors $\left[\underline{d}^{(s)} = \Phi\left\{\underline{p}^{(s)}\right\}; s = 1, \ldots, S\right]$;

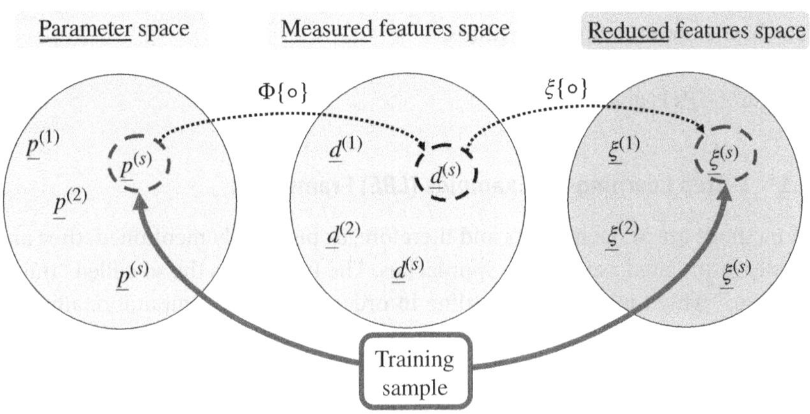

Figure 8.3 Schematic representation of the training set generation exploiting dimensionality reduction techniques to mitigate the negative effects of the curse of dimensionality.

3) *Dimensionality reduction*: Apply a mapping operator $\zeta\{.\}$ to each sth ($s = 1, \dots, S$) measured features vector $\underline{d}^{(s)}$ to derive the corresponding set of H-dimensional reduced features, $\underline{\xi}^{(s)} = \left\{\xi_h^{(s)};\ h = 1, \dots, H\right\} = \zeta\left\{\underline{d}^{(s)}\right\}$;
4) *Training set construction*: Build the training set exploiting the reduced features ($\underline{\xi}^{(s)}$, $s = 1, \dots, S$) as a lower-dimensionality substitute of the measured ones ($\underline{d}^{(s)}$, $s = 1, \dots, S$), by letting

$$\mathbb{T} = \left\{\left(\underline{\xi}^{(s)}, \underline{p}^{(s)}\right);\quad s = 1, \dots, S\right\}. \tag{8.22}$$

Several techniques can be exploited to implement the mapping operator $\zeta\{.\}$ for dimensionality reduction. These account for both (i) feature selection and (ii) feature extraction methods [5]. As suggested by their name, the techniques belonging to class (i) aim at selecting a small subset of the K-measured features that have been found to be the most informative ones on the predicted output. Unfortunately, such a selection requires some kind of sensitivity analysis to discover which features carry the largest amount of information, making them often computationally cumbersome. Moreover, in several contexts, it is often hard to identify specific features that are more related than others to the imaged target, since such information is generally equally spread over the complete set of measured features. To overcome such a limitation, methods belonging to the family (ii) are often preferred. As a matter of fact, feature extraction methods generate a *new* set of H features by processing together all the original ones without performing any selection. Such an "extraction" process can be either linear or nonlinear depending on the definition of the mapping operator $\zeta\{.\}$. For instance, the principal component analysis (*PCA*) [27] is a widespread linear technique which transforms the measured features according to the following linear mapping

$$\underline{\xi}^{(s)} = \underline{d}^{(s)} \times \underline{\underline{W}}, \tag{8.23}$$

where $\underline{\underline{W}}$ is a $(K \times H)$ matrix whose columns are the eigenvectors associated with the H largest eigenvalues of the covariance matrix of $\left\{\underline{d}^{(s)};\ s = 1, \dots, S\right\}$. Another well-known linear feature extraction technique is the partial least squares (*PLS*) [16, 28], which derives, differently from the *PCA*, the transformation matrix $\underline{\underline{W}}$ in a supervised way (i.e., by jointly processing input, $\underline{d}$, and output, $\underline{p}$, samples). As a matter of fact, while the *PCA* is aimed at identifying orthogonal directions of maximum variance in the measured features space, the *PLS* seeks the directions that are most informative on the predicted output [16, 28]. Nonlinear variants of both the *PCA* and the *PLS* have also been proposed based, for instance, on the "kernel trick," which allows the handling of highly nonlinear *I/O* relationships [15].

Concerning the second *LBE* step, it is aimed at selecting the S *I/O* samples forming the training database. Toward this end, both single-shot and adaptive sampling techniques can be exploited. The former are simple strategies that perform a sampling of the input space in a noniterative way, such as the uniform grid sampling and the uniform random sampling. Although computationally efficient, such techniques have their own drawbacks. For instance, uniform grid sampling is suitable only for low-dimensionality spaces, since the number of generated samples exponentially increases with the number of variables. Differently, uniform random sampling suffers from undersampling and oversampling. Such issues are partially mitigated by the Latin hypercube sampling (*LHS*) technique [29], which exhibits the so-called "input space filling" property by systematically evenly covering the entire space. On the other hand, adaptive sampling techniques perform an iterative sampling of the input space by looking for the optimal trade-off between "exploration" and "exploitation" [30]. More in detail, the term "exploration" refers to the need for exploring regions of the input space where a low density of samples has been detected. Differently, "exploitation" indicates the need for additional samples where the *I/O* relationship has been found, by analyzing previously collected samples, rapidly changing and/or highly nonlinear. Among the several techniques belonging to this group, it is worth mentioning the LOLA-Voronoi [31] and the output space filling (*OSF*) [15–18].

Finally, the third *LBE* step is aimed at building a fast predictor $\tilde{\Psi}\{.\}$ from the information collected inside the training set. The choice of the most suitable technique starts from the definition of the task to be solved. As a matter of fact, *EM-IPs* can be formulated either as classification or regression problems. More in detail, indicating with $\underline{\chi}$ the generic set of input variables, [2] classification problems are aimed at retrieving a discrete-valued/integer label $\tilde{\Psi}\left\{\underline{\chi}\right\} = \ell\left\{\underline{\chi}\right\}$ among a predefined set of C classes (i.e., $\ell\left\{\underline{\chi}\right\} \in \{\ell_c \in \mathbb{Z};\, c = 1, \dots, C\}$). One of the most widespread methods for solving classification tasks in *EM* imaging is *SVM* [32]. When dealing with binary classification tasks (i.e., $C = 2$ and $\ell_1 = -1$, $\ell_2 = +1$) the *SVM* predictor takes the following form

$$\tilde{\Psi}\left\{\underline{\chi}\right\} = \mathrm{sgn}\left\{\sum_{s=1}^{S} \alpha_s \ell\left(\underline{\chi}^{(s)}\right) \mathcal{K}\left(\underline{\chi}^{(s)}, \underline{\chi}\right) + \delta\right\}, \tag{8.24}$$

where sgn $\{.\}$ is the sign function, α_s $(s = 1, \dots, S)$ are Lagrange's multipliers, δ is a constant bias term, and $\mathcal{K}\left(\underline{\chi}^{(s)}, \underline{\chi}\right)$ is the so-called "kernel function" expressing the scalar product between $\underline{\chi}^{(s)}$ and $\underline{\chi}$ within a higher dimensionality space where the samples belonging to both classes become linearly separable. When dealing with multi-class classification problems (i.e., $C > 2$), the "one-against-one"

2 Depending on the problem at hand and according to the previous definitions, $\underline{\chi}$ could represent either the original measured features (i.e., $\underline{\chi} \leftarrow \underline{d}$) or the extracted features (i.e., $\underline{\chi} \leftarrow \underline{\xi}$) in *DIS* methods. Otherwise, in *IIS* methods, it could coincide with the set of solution descriptors (i.e., $\underline{\chi} \leftarrow \underline{p}$).

strategy is often adopted by training a binary *SVM* for each pair of classes and then selecting the class that receives the most votes among all classifiers [33].

Differently, when dealing with regression tasks, the goal is to estimate a continuous-valued output (i.e., $\tilde{\Psi}\left\{\underline{\chi}\right\} \in \mathbb{R}$). Toward this end, support vector regression (*SVR*) [34] can be exploited as a natural extension of *SVMs* defining $\tilde{\Psi}\{.\}$ as follows

$$\tilde{\Psi}\left\{\underline{\chi}\right\} = \sum_{s=1}^{S} \left(-\beta_s + \beta_s^*\right) \mathcal{K}\left(\underline{\chi}^{(s)}, \underline{\chi}\right) + \gamma, \tag{8.25}$$

where $\left(\beta_s, \beta_s^*\right)$ $(s = 1, \dots, S)$ are Lagrange's multipliers and γ is a fixed bias.

Another popular regression technique is the ordinary kriging (*OK*) [35], according to which

$$\tilde{\Psi}\left\{\underline{\chi}\right\} = \mu + \sum_{a=1}^{S}\sum_{b=1}^{S} \rho\left(\underline{\chi}, \underline{\chi}^{(a)}\right) \underline{\underline{U}}\Big|_{a,b} \left(\Psi^{(a)} - \mu\right), \tag{8.26}$$

where

$$\mu = \frac{\sum_{a=1}^{S} \sum_{b=1}^{S} \underline{\underline{U}}\Big|_{a,b} \Psi^{(b)}}{\sum_{a=1}^{S} \sum_{b=1}^{S} \underline{\underline{U}}\Big|_{a,b}}, \tag{8.27}$$

$\Psi^{(a/b)} = \Psi\left\{\underline{\chi}^{(a/b)}\right\}$, and $\underline{\underline{U}} = \underline{\underline{R}}^{-1}$, $\underline{\underline{R}}$ being the $(S \times S)$ training correlation matrix with entries equal to

$$\underline{\underline{R}}\Big|_{a,b} = C\left(\underline{\chi}^{(a)}, \underline{\chi}^{(b)}\right); \quad a, b = 1, \dots, S. \tag{8.28}$$

Moreover, $C\left(\underline{\chi}^{(a)}, \underline{\chi}^{(b)}\right)$ expresses the correlation between samples $\underline{\chi}^{(a)}$ and $\underline{\chi}^{(b)}$, while

$$\rho\left(\underline{\chi}, \underline{\chi}^{(a)}\right) = C\left(\underline{\chi}, \underline{\chi}^{(a)}\right); \quad a = 1, \dots, S \tag{8.29}$$

is the correlation between each training sample and the test sample $\underline{\chi}$.

8.3.2 System-by-Design (*SbD*) Framework

The *SbD* is a powerful framework for the optimization-driven solution of complex *EM* problems. Although it has been introduced to address several *EM* synthesis problems, such as those arising in the design of elementary radiators, wide angle impedance matching layers, reflectarrays, and meta-materials [19], it recently proved its capabilities in solving *EM–IPs* [20].

According to such a paradigm, the problem at hand is decomposed into a set of subtasks solved by properly interconnected "functional blocks" defined under the shared goal of an effective, robust, and computationally efficient exploitation of stochastic global optimization techniques such as evolutionary algorithms (*EAs*) [19]. Toward this goal, the customization of the *SbD* to *EM* imaging problems involves (i) a suitable reformulation of the problem at hand as a global optimization one, whose solution is described by a minimum-dimensionality set of descriptors yielding a "smart" modeling of the class of imaged targets, and (ii) the integration of *EAs* with computationally efficient *SMs* of the data mismatch cost function generated by means of *LBE* techniques, in order to speed up the evaluation of each trial guess solution. In other words, the *SbD* is aimed at overcoming the limitation of deterministic optimization algorithms, which suffer of local minima unless properly initialized in the "attraction basin" of the global optimum, while solving the IP with a comparable computational efficiency.

More specifically, the *SbD* for solving 2D *MI* problems can be implemented by defining the following functional blocks (Figure 8.4) [20]:

1) *Problem formulation* (*PF*): Formulates the imaging problem at hand to enable an effective, and computationally-efficient exploitation of *EAs* by coding the unknowns into a minimum-dimension but highly-flexible set of descriptors $\underline{p}$. Moreover, it defines a suitable cost function measuring the "quality" of the solution described by $\underline{p}$;
2) *Data computation* (*DC*): Maps $\underline{p}$ into a pixel-based representation of the unknown equivalent currents within $\mathcal{D}$ in order to compute the corresponding scattered field distribution in $\mathcal{O}$;
3) *Cost function evaluation* (*CFE*): Efficiently evaluates the cost function by means of a computationally fast *SM* of the accurate (but time-consuming) full-wave solver;
4) *Solution space exploration* (*SSE*): Explores the solution space described by $\underline{p}$ by leveraging on the "hill-climbing" features of *EA*-based optimization methods (e.g., the Particle Swarm Optimization, *PSO* [36]) and on its smart interaction with the predictor to yield a fast and reliable convergence toward the global optimum.

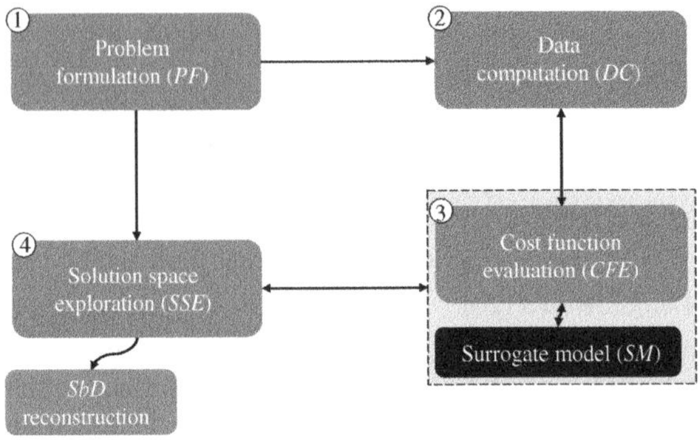

Figure 8.4 Block scheme of the *SbD*-driven solution framework for *EM* imaging problems.

Clearly, the choice, selection, implementation, and integration of each *SbD* functional block is driven by the *EM–IP* at hand. For instance, exploiting the *a-priori*/available information on the sought targets is fundamental to determine the minimum-cardinality set of unknowns defining the solution space. As a matter of fact, the identification of the most suitable model for the imaged objects depends on the specific applicative scenario and plays a fundamental role in incorporating physics-driven (PD) *LBE* strategies not requiring huge amounts of training data nor suffering from the negative effects of the *CoD*. Furthermore, it enables a profitable exploitation of *EA*s with limited population sizes, whose selection always complies with the well-known no-free-lunch theorem for optimization [21].

8.3.3 Deep Learning (*DL*) Framework

The term deep learning (*DL*) refers to a specific class of *AI* algorithms and, specifically, of machine learning (*ML*) methods. As a matter of fact, it identifies a subset of *ANN*s such as convolutional neural networks (*CNN*s), generative adversarial neural networks (*GAN*s), and variational autoencoders (*VAE*s) that are called "deep" since comprising at least three hidden layers (i.e., layers that do not coincide neither with the input nor the output of the network) [2, 4]. In the last few years, we have witnessed an exponential growth of research in this promising field, since *DL* proved unprecedented capabilities in terms of accuracy, resolution, and computational efficiency in solving many complex *EM* problems [1, 2, 4].

Concerning *EM—IP*s, *DL* has been extensively exploited to develop effective *DIS* and *IIS* methodologies, demonstrating a huge potential in overcoming traditional limitations of "shallow" *ANN*s and other *LBE* methods since capable of performing pixel-based inversions almost in real time [5, 6].

In this framework, it is worth noticing that there has been a progressive shift during the last years from purely "data-driven" (*DD*) *DL* approaches to more sophisticated *PD* methods [9]. On the one hand, *DD* strategies rely exclusively on the information gathered from the training data to learn all the underlying *EM* physics. Therefore, such methods lack physical information and are generally providing suboptimal inversion performance since barely exploited as "black-boxes." One the other hand, *PD* methods aim at integrating, in different ways and at different stages of the learning process, the available physics and *prior* knowledge on the solution and on the scenario at hand. As a consequence, such additional information provides not only an effective way to mitigate the cost of the training phase but also helps in regularizing the *IP*, retrieving more robust and reliable guesses of the *EM* properties of $\mathcal{D}$. To provide more insights on the recent trends in this emerging field, it is possible to classify existing *PD–DL* approaches depending on how/where the physics is included in the inversion process. More specifically, the following classification can be adopted [9]:

1) Learning after physics processing;
2) Learning with physics loss;
3) Learning with physics models.

The methods belonging to the first group ("learning after physics processing") are the most straightforward and easy to implement since they simply exploit *DL* as an image processing technique. For instance, in *MI*, they often aim at building a pixel-based predictor $\tilde{\Psi}\{.\}$ capable of removing artifacts and improving the resolution of a preliminary low-resolution/coarse guess of the contrast, $\underline{\tau}_{guess}$, obtained by means of fast noniterative techniques such as backpropagation [37]

$$\tilde{\underline{\tau}} = \tilde{\Psi}\left\{\underline{\tau}_{guess}\right\}. \tag{8.30}$$

The *EM* physics is therefore introduced only *before* entering the learning process in order to produce $\underline{\tau}_{guess}$ from the knowledge of the collected scattered field and the Green's operator for the scenario at hand. However, the training phase does not follow *EM* wave physics since it is only aimed at enhancing $\underline{\tau}_{guess}$, without considering any quantification of the data mismatch. In other words, the training phase is aimed at minimizing a cost function accounting only for the discrepancy between the retrieved ($\tilde{\underline{\tau}}$) and actual ($\underline{\tau}_{act}$) contrast over a set of S known examples $\mathbb{T} = \left\{\left(\tilde{\underline{\tau}}^{(s)}, \underline{\tau}_{act}^{(s)}\right); s = 1, \dots, S\right\}$, i.e.,

$$\mathcal{L}\left(\underline{\Omega}\right) \propto \frac{1}{S}\sum_{s=1}^{S}\left\|\tilde{\underline{\tau}}^{(s)} - \underline{\tau}_{act}^{(s)}\right\|^2, \tag{8.31}$$

where $\underline{\Omega}$ denotes the set of trainable hyper-parameters of the network ($\underline{\Omega}^{opt} = \arg\left\{\min_{\underline{\Omega}}\mathcal{L}\left(\underline{\Omega}\right)\right\}$ being the optimal setup derived from $\mathbb{T}$) [9].

Differently, the methods of the second class ("learning with physics loss") aim to introduce physics knowledge directly into the learning process. Toward this end, the minimized cost function is extended to include an additional term quantifying the mismatch of the retrieved data ($\Phi\left\{\tilde{\underline{\tau}}^{(s)}\right\}$) with the measured one ($\underline{d}^{(s)}$),

$$\mathcal{L}\left(\underline{\Omega}\right) \propto \frac{1}{S}\sum_{s=1}^{S}\left(w_1 \times \left\|\tilde{\underline{\tau}}^{(s)} - \underline{\tau}_{act}^{(s)}\right\|^2 + w_2 \times \left\|\Phi\left\{\tilde{\underline{\tau}}^{(s)}\right\} - \underline{d}^{(s)}\right\|^2\right), \tag{8.32}$$

where $w_{1/2}$ are real weights. In this way, the *DL* technique is somehow "forced" to obey the *EM* physics, resulting in higher accuracy and reduced ill-posedness. However, it is worth noticing that the physics still cannot guide the online testing phase, since once the network is trained, new inversions are produced without minimizing the above cost function [9].

Finally, the methods of the third class ("learning with physics models") are aimed at embedding physics operators (e.g., the Green's function) and knowledge directly within the network architecture. In such cases, the imaging task relies mainly on physics computations, significantly reducing the overall number of hyperparameters to be trained and yielding better generalization capabilities. However, the "price to pay" is a more complex implementation, since the network architecture needs to be carefully tailored to the specific *EM–IP* at hand [9, 38].

8.4 Applications

In the following, some of the most recent and appealing *AI*-based methods applied to *EM–IP*s are surveyed.

8.4.1 Microwave Imaging of Free-Space and Buried Objects

Pioneer works exploiting shallow *ANN* architectures such as multilayer perceptron [39, 40] and radial basis function networks [41] have been among the first attempts to solve free-space *MI* problems with almost real-time capabilities. However, such methods suffer from typical problems of classical *ANN*s, such as over-fitting (causing the overall prediction accuracy to be highly training-dependent) and local minima occurrence [42].

To partially mitigate such drawbacks, an *SVR*-based inversion technique has been proposed in [42] to deal with the detection of 2D buried objects starting from aspect-limited scattered data collected above a planar air–soil interface. According to such a methodology, implementing a *DIS* procedure, a predictor is trained to estimate a set of $Q = 5$ parameters $\underline{p} = \{x_0, y_0, \wp, \varepsilon_r, \sigma\}$ completely describing a cylindrical buried target, (x_0, y_0) being its barycenter, $\wp$ its radius, and (ε_r, σ) homogeneous permittivity and conductivity values, respectively, starting from the scattered field collected in M probing locations (i.e., $\underline{d} = \{\underline{E}_z^{sca(v)};\ v = 1, \dots, V\}$). Extensions of the approach have been successively proposed in [43, 44].

More recently, an *SbD* methodology has been proposed in [20] to address the computationally efficient yet reliable solution of fully nonlinear 2D *TM* free-space *IS* problems. Such an inversion method allows an effective exploration of the multimodal landscape of the data mismatch cost function with the same time cost of a standard deterministic algorithm. Such a result has been yielded thanks to the introduction of a smart and flexible minimum-dimensionality coding of complex-shaped targets by means of a spline-based modeling of their external contour. On the one hand, thanks to such an approach, a more "favorable" operative condition has been guaranteed to the underlying *PSO*-based optimization strategy. On the other hand, the *CoD* has been alleviated, improving the prediction accuracy of an *OK*-based *SM* of the data mismatch cost function. Concerning this latter, the confidence level of the outputted predictions has been exploited to set up of a "collaborative framework" between the optimizer and the *SM*. As a matter of fact, such additional information enables an effective exploration of the solution space, which is adaptively sampled at "promising" locations to enhance the accuracy of the *SM*, while accelerating the converge toward the global optimum. Such an *SbD*-based methodology is, therefore, a reliable tool for solving highly nonlinear *MI* problems, breaking the widely diffused idea that solving *EM–IPs* with *EAs* is computationally unaffordable [20].

Among the first attempts to solve free-space *IS* problems within the *DL* framework, it is worth mentioning those illustrated in [37], where three *CNN*-based strategies have been developed to perform pixel-wise inversions with high computational efficiency. More specifically, it has been verified that a *DIS* approach directly inverting the scattered field provides suboptimal performance due to the very high complexity of the learning task. To mitigate such a limitation, more advanced strategies have been proposed based on the exploitation of noniterative reconstruction methods to provide a preliminary guess of the contrast and exploit the *CNN* method as an image processing tool according to a "learning after physics processing" scheme [37] (Section 8.3.3).

Later, significant efforts have been spent in developing more sophisticated physics-embedded *DL* methods [9]. For instance, Shan et al. [45] introduced a "neural Born iterative method" inversion scheme implemented with either supervised or unsupervised learning strategies. Differently, a novel *DL* framework for solving *MI* problems has been proposed in [46] based on the effective combination of the Supervised Descent Algorithm with a *CNN*-based prediction technique. Such a method has a low computational complexity and can realize imaging in real time with high accuracy, as demonstrated by several numerical examples. Furthermore, an innovative inversion algorithm has been proposed in [47] exploiting a *VAE* neural network. More in detail, the dimensionality reduction capabilities of the *VAE* have been exploited to compress the unknowns of the *EM–IP* into a small number of "latent variables," where a Gauss–Newton optimization algorithm has been employed to efficiently minimize the data mismatch cost function, exhibiting remarkable reconstruction capabilities with both numerical and experimental data.

8.4.2 Biomedical Imaging

In recent years, several *AI*-based methodologies have been proposed to solve biomedical imaging problems arising in several medical contexts. For instance, the use of *SVM/SVR LBE* strategies has been widely investigated for the real-time detection, localization, and identification of brain strokes [18, 24, 48, 49]. In this

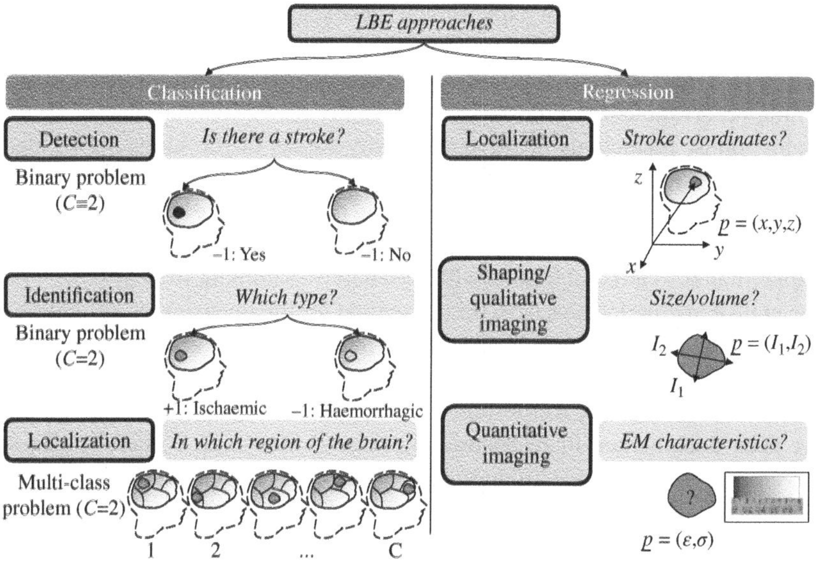

Figure 8.5 Possible *LBE*-based strategies for brain stroke imaging exploiting both classification and regression schemes.

context, it is worth mentioning that *AI* techniques have been implemented to solve different problems. For instance, with reference to Figure 8.5, binary classification strategies have been exploited to efficiently address both "detection" (i.e., recognizing the presence/absence of a stroke within the patient's head [49]) and "identification" (i.e., distinguishing between ischaemic and hemorrhagic strokes, which is of paramount importance to decide the treatment for the patient [50]). Otherwise, multiclass classification problems are suitable for formulating the "localization" task as the retrieval of which brain region (among a predefined set of $C > 2$ possibilities) has been affected by the stroke [50]. Differently, the same task can also be solved by means of regression strategies by predicting, for instance, the (x, y, z) coordinates of the stroke barycenter [24]. Moreover, regression techniques can be exploited to recover continuous-valued stroke descriptors such as its size/volume (i.e., solving a "shaping/qualitative imaging" task – Figure 8.5) or even its *EM* composition (i.e., addressing a "quantitative imaging" task).

Changing frequency range and diagnosis modality, an innovative three-step *LBE* methodology has been proposed in [17, 18] for the inversion of *EIT* data for the real-time estimation of the human lung conductivities starting from boundary voltages collected around the chest. The proposed method exploits the *PLS* algorithm to reduce the dimensionality of the input/data space by projecting the

EIT measurements into a low-dimensional space spanned by a limited number of highly informative extracted features. Moreover, it leverages an adaptive sampling strategy to select the minimum number of non-redundant/most-informative *I/O* training observations by uniformly exploring the *PLS*-generated feature space. Finally, robust predictions of the conductivity values of the lungs are yielded thanks to an *SVR* predictor. The reported numerical validation has verified that the *PLS* can effectively reduce the dimensionality of the input space by more than 98%, enabling robust conductivity predictions exploiting a limited number of training samples thanks to its combination with the adaptive sampling strategy [18].

Several *DL* methodologies have been proposed for the inversion of *EIT* data, as well [8, 51–53]. For instance, Wei et al. [51] introduced a *CNN*-based inversion method to recover an image of the internal impedance of the body or subject from external surface electrode measurements. Remarkable reconstruction capabilities have been documented in reconstructing targets with sharp corners or edges, assessing an enhanced resolution with respect to traditional *EIT* systems, also when dealing with experimental data.

Within the same applicative context, Zhang et al. [53] developed the *SDL–EIT* inversion algorithm. Such a methodology approximates the mapping from measured data to the conductivity image by a series of descent directions learned from the training set. It has been implemented in both 2D and 3D cases, and its capabilities have been successfully evaluated using both synthetic and measured thoracic data. The *SDL–EIT* method has the advantage of flexibly integrating *prior* information into the inversion through the generation of suitable training data. Moreover, it outperforms a traditional imaging technique based on the Gauss—Newton inversion method in terms of both accuracy and time efficiency.

8.4.3 Non-destructive Testing and Evaluation (*NDT/NDE*)

Nowadays, non-destructive testing and evaluation (*NDT/NDE*) is a very active field of research aimed at non-invasively inspecting devices, components, and complex structures to check their structural/functional integrity. Depending on the application, *NDT/NDE* analyses deal with the detection and characterization of anomalies, defects, or cracks inside the structure under test, which may negatively impact its working and/or the safety of the operators [15, 16]. Therefore, having a reliable and robust tool for efficiently detecting anomalies is of paramount importance to provide a decision-support basis for successive and more accurate (often invasive) investigations.

In this context, effective three-step *LBE* methodologies have been recently proposed for the real-time inversion of *ECT* measurements collected over planar [16] or cylindrical [15] metallic structures. Such techniques effectively combine the *PLS* feature extraction [16] as well as its nonlinear version (i.e., the kernel orthogonal *PLS*, KOPLS [15]) with a customized version of the *OSF* adaptive sampling strategy for generating optimal training sets, while relying on the *SVR* technique for making fast and accurate predictions of descriptors defining the position and dimension of non-visible cracks. Representative numerical and experimental test cases have been reported and discussed, assessing the high reliability and robustness of the proposed *AI* schemes as well as their superior performance in comparison with state-of-the-art approaches.

8.4.4 Wireless Detection, Localization, and Tracking of Targets

DIS methodologies based on *LBE* strategies have also been proposed for the *EM* passive localization and tracking of moving targets in wireless network-infrastructured environments [25, 54, 55].

In this context, we have recently witnessed a rapid diffusion of wireless sensor networks (*WSN*s) thanks to the increasing availability of low-power and pervasive devices integrating sensing, processing, and wireless transmission of information. Due to their dense distribution and low power consumption, which ensure a long lifetime, *WSN*s represent a suitable technology for location and tracking purposes. In such a context, an *LBE* inversion procedure for the localization and tracking of passive (i.e., noncollaborative) targets starting from the measurements of the *RSSI* taken by the nodes of a *WSN* has been presented in [25]. Since the transmission of information among the wireless nodes is allowed by radio-frequency signals, any target/person lying within the environment interacts with such *EM* radiation, becoming *de facto* an equivalent current source. Therefore, the arising *EM* signature measured by the *WSN* nodes can be profitably processed by reformulating the problem at hand as an inverse source one and efficiently solved by means of an *SVM* classification algorithm. Experimental results have assessed the effectiveness and reliability of the proposed approach in dealing with the tracking of single and multiple human beings both in indoor and outdoor environments [25].

More recently, the estimation of the presence, position, and posture of non-cooperative targets has been addressed in [56] through a multiresolution *LBE* strategy for the real-time inversion of the measured channel state information (*CSI*) of a single *IEEE 802.11ac WiFi* link. More in detail, a series of binary *SVM* classifiers have been trained to estimate multiple location-based features of the target (i.e., presence detection, localization, and posture recognition). The reported experimental results have pointed out a robust wireless sensing, with failure rates lower than 2.5% [56].

8.5 Conclusions

In this chapter, a review of the recent applications of *AI* techniques for solving *EM–IPs* is provided. Such problems are inherently complex because of their high nonlinearity and ill-posedness. Throughout the last decades, there has been a remarkable advancement in the development of effective, fast, and reliable *AI*-driven methodologies, which proved to overcome, at least in specific operative scenarios, traditional techniques in terms of both accuracy and computational efficiency. However, there are still many unsolved challenges that will clearly lead the research efforts in the next years. Among them, it is worth mentioning the current main limitation of *DL* methods, which is the need for extremely large amounts of training samples to learn thousands of unknown parameters. As discussed in this chapter, one valid recipe for coping with such an issue is the development of new and effective strategies to embed *a-priori* information and physical knowledge directly within the inversion methodology.

Acknowledgments

This work benefited from the networking activities carried out within the Project "Hub Life Science – Advanced Diagnosis (HLS-AD), PNRR PNC-E3-2022-23683266 PNC-HLS-DA, INNOVA" funded by the Italian Ministry of Health under the National Complementary Plan Innovative Health Ecosystem (CUP: E63C22003780001 – Unique Investment Code: PNC-E.3), the Project "AURORA – Smart Materials for Ubiquitous Energy Harvesting, Storage, and Delivery in Next Generation Sustainable Environments" funded by the Italian Ministry for Universities and Research within the PRIN-PNRR 2022 Program (CUP: E53D23014760001), the Project "Telecommunications of the Future" (PE00000001 – program "RESTART"), funded by the European Union – Next Generation EU under the Italian National Recovery and Resilience Plan (NRRP), Mission 4, Component 2, Investment 1.3 (CUP: D43C22003080001), the Project "Telecommunications of the Future" (PE00000001 – program "RESTART"), funded by the European Union – Next Generation EU under the Italian National Recovery and Resilience Plan (NRRP), Mission 4, Component 2, Investment 1.3 (CUP: J33C22002880001), the Project "Telecommunications of the Future" (PE00000001 – program "RESTART"), funded by the European Union – Next Generation EU under the Italian National Recovery and Resilience Plan (NRRP), Mission 4, Component 2, Investment 1.3 (CUP: J33C22002880001), the Project "Telecommunications of the Future" (PE00000001 – program "RESTART"), funded by the European Union – Next Generation EU under the Italian National Recovery and Resilience Plan (NRRP), Mission 4, Component 2, Investment 1.3 (CUP: B53C22003970001), and the Project DICAM-EXC (Grant L232/2016) funded by the Italian Ministry of Education, Universities and Research (MUR) within the "Departments of Excellence 2023–2027" Program (CUP: E63C22003880001). A. Massa wishes to thank E. Vico for her never-ending inspiration, support, guidance, and help.

References

1 Campbell, S.D., Jenkins, R.P., O'Connor, P.J., and Werner, D. (2021). The explosion of artificial intelligence in antennas and propagation: how deep learning is advancing our state of the art. *IEEE Antennas and Propagation Magazine* 63 (3): 16–27.

2 Massa, A., Marcantonio, D., Chen, X. et al. (2019). DNNs as applied to electromagnetics, antennas, and propagation–a review. *IEEE Antennas and Wireless Propagation Letters* 18 (11): 2225–2229.

3 Massa, A., Oliveri, G., Salucci, M. et al. (2018). Learning-by-examples techniques as applied to electromagnetics. *Journal of Electromagnetic Waves and Applications* 32 (4): 516–541.

4 Salucci, M. and Li, M. (ed.) (2023). *Applications of Deep Learning in Electromagnetics - Teaching Maxwell's Equations to Machines.* London: IET.

5 Salucci, M., Arrebola, M., Shan, T., and Li, M. (2022). Artificial intelligence: new frontiers in real-time inverse scattering and electromagnetic imaging. *IEEE Transactions on Antennas and Propagation* 70 (8): 6349–6364.

6 Arrebola, M., Li, M., and Salucci, M. (2022). Guest editorial artificial intelligence: new frontiers in real-time inverse scattering and electromagnetic imaging. *IEEE Transactions on Antennas and Propagation* 70 (8): 6131–6134.

7 Li, M., Guo, R., Zhang, K. et al. (2021). Machine learning in electromagnetics with applications to biomedical imaging: a review. *IEEE Antennas and Propagation Magazine* 63 (3): 39–51.

8 Chen, X., Wei, Z., Li, M., and Rocca, P. (2020). A review of deep learning approaches for inverse scattering problems (invited review). *Progress In Electromagnetics Research* 167: 67–81.

9 Guo, R., Huang, T., Li, M. et al. (2023). Physics-embedded machine learning for electromagnetic data imaging: examining three types of data-driven imaging methods. *IEEE Signal Processing Magazine* 40 (2): 18–31.

10 Low, T.S. and Chao, B. (1992). The use of finite elements and neural networks for the solution of inverse electromagnetic problems. *IEEE Transactions on Magnetics* 28: 2811–2813.

11 Hoole, S.R.H. (1993). Artificial neural networks in the solution of inverse electromagnetic field problems. *IEEE Transactions on Magnetics* 29 (2): 1931–1934.

12 Chkrabarti, S., Bindal, N., and Theagharajan, K. (1995). Robust target classifier using artificial neural networks. *IEEE Transactions on Neural Networks* 6: 760–766.

13 Azimi-Sadjandi, M.R., Pole, D.E., Sheedvash, S. et al. (1992). Detection and classification of buried dielectric anomalies using a separated aperture sensor and a neural network discriminator. *IEEE Transactions on Instrumentation and Measurement* 41: 137–143.

14 Azimi-Sadjandi, M.R. and Stricker, S.A. (1994). Detection and classification of buried dielectric anomalies using neural networks-further results. *IEEE Transactions on Instrumentation and Measurement* 43: 34–39.

15 Salucci, M., Anselmi, N., Oliveri, G. et al. (2019). A nonlinear kernel-based adaptive learning-by-examples method for robust NDE-NDT of conductive tubes. *Journal of Electromagnetic Waves and Applications* 33 (6): 669–696.

16 Salucci, M., Anselmi, N., Oliveri, G. et al. (2016). Real-time NDT-NDE through an innovative adaptive partial least squares SVR inversion approach. *IEEE Transactions on Geoscience and Remote Sensing* 54 (11): 6818–6832.

17 Salucci, M. and Oliveri, G. (2019). Robust real-time inversion of electrical impedance tomography data for human lung ventilation monitoring. *Microwave and Optical Technology Letters* 61 (1): 5–8.

18 Salucci, M., Oliveri, G., and Massa, A. (2019). Real-time electrical impedance tomography of the human chest by means of a learning-by-examples method. *IEEE Journal of Electromagnetics, RF and Microwaves in Medicine and Biology* 3 (2): 88–96.

19 Massa, A. and Salucci, M. (2022). On the design of complex *EM* devices and systems through the system-by-design paradigm: a framework for dealing with the computational complexity. *IEEE Transactions on Antennas and Propagation* 70 (2): 1328–1343.

20 Salucci, M., Poli, L., Rocca, P., and Massa, A. (2022). Learned global optimization for inverse scattering problems: matching global search with computational efficiency. *IEEE Transactions on Antennas and Propagation* 70 (8): 6240–6255.

21 Wolpert, D.H. and Macready, W.G. (1997). No free lunch theorems for optimization. *IEEE Transactions on Evolutionary Computation* 1 (1): 67–82.

22 Wolpert, D.H. (2002). The supervised learning no-free-lunch theorems. In: *Soft Computing and Industry* (eds. Roy, R., Köppen, M., Ovaska, S., Furuhashi, T., and Hoffmann, F.). London: Springer. https://doi.org/10.1007/978-1-4471-0123-9_3.

23 Chen, X. (2018). *Computational Methods for Electromagnetic Inverse Scattering*. Hoboken, NJ: Wiley.

24 Salucci, M., Polo, A., and Vrba, J. (2021). Multi-step learning-by-examples strategy for real-time brain stroke microwave scattering data inversion. *Electronics* 10 (1): 1–17.

25 Viani, F., Rocca, P., Benedetti, M. et al. (2010). Electromagnetic passive localization and tracking of moving targets in a WSN-infrastructured environment. *Inverse Problems* 26: 1–15.

26 Bellman, R.E. (1961). *Adaptive Control Processes*. Princeton, NJ: Princeton University Press.

27 Shlens, J. (2014). A tutorial on principal component analysis. *arXiv preprint arXiv:1404.1100*.

28 Wold, S., Sjostrom, M., and Eriksson, L. (2001). PLS-regression: a basic tool of chemometrics. *Chemometrics and Intelligent Laboratory Systems* 58 (2): 109–130.

29 McKay, M.D., Beckman, R.J., and Conover, W.J. (1979). A comparison of three methods for selecting values of input variables in the analysis of output from a computer code. *Technometrics* 21 (2): 239.

30 Liu, H., Ong, Y.-S., and Cai, J. (2017). A survey of adaptive sampling for global metamodeling in support of simulation-based complex engineering design. *Structural and Multidisciplinary Optimization* 57 (1): 393–416.

31 Li, B., Salucci, M., Tang, W., and Rocca, P. (2020). Reliable field strength prediction through an adaptive total-variation CS technique. *IEEE Antennas and Wireless Propagation Letters* 19 (9): 1566–1570.

32 Vapnik, V. (2013). *The Nature of Statistical Learning Theory*. Springer Science & Business Media.

33 Mathur, A. and Foody, G.M. (2008). Multiclass and binary SVM classification: implications for training and classification users. *IEEE Geoscience and Remote Sensing Letters* 5 (2): 241–245.

34 Smola, A.J. and Scholkopf, B. (2004). A tutorial on support vector regression. *Statistics and Computing* 14 (3): 199–222.

35 Forrester, A., Sobester, A., and Keane, A. (2008). *Engineering Design Via Surrogate Modelling: A Practical Guide*. Wiley.

36 Rocca, P., Benedetti, M., Donelli, M. et al. (2009). Evolutionary optimization as applied to inverse scattering problems. *Inverse Problems* 25 (12): 123003.

37 Wei, Z. and Chen, X. (2019). Deep-learning schemes for full-wave nonlinear inverse scattering problems. *IEEE Transactions on Geoscience and Remote Sensing* 57 (4): 1849–1860.

38 Liu, J., Zhou, H., Ouyang, T. et al. (2022). Physical model-inspired deep unrolling network for solving nonlinear inverse scattering problems. *IEEE Transactions on Antennas and Propagation* 70 (2): 1236–1249.

39 Caorsi, S. and Gamba, P. (1999). Electromagnetic detection of dielectric cylinders by a neural network approach. *IEEE Transactions on Geoscience and Remote Sensing* 37: 820–827.

40 Mydur, R. and Michalski, K.A. (2001). A neural-network approach to the electromagnetic imaging of elliptic conducting cylinders. *Microwave and Optical Technology Letters* 28: 303–306.

41 Rekanos, I.T. (2001). On-line inverse scattering of conducting cylinders using radial basis-functions neural networks. *Microwave and Optical Technology Letters* 28: 378–380.

42 Bermani, E., Boni, A., Caorsi, S., and Massa, A. (2003). An innovative real-time technique for buried object detection. *IEEE Transactions on Geoscience and Remote Sensing* 41 (4): 927–931.

43 Massa, A., Boni, A., and Donelli, M. (2005). A classification approach based on SVM for electromagnetic sub-surface sensing. *IEEE Transactions on Geoscience and Remote Sensing* 43 (9): 2084–2093.

44 Bermani, E., Boni, A., Caorsi, S. et al. (2004). A multi-source strategy based on a learning-by-examples technique for buried object detection. *Progress in Electromagnetics Research* 48: 185–200.

45 Shan, T., Lin, Z., Song, X. et al. (2023). Neural Born iterative method for solving inverse scattering problems: 2D cases. *IEEE Transactions on Antennas and Propagation* 71 (1): 818–829.

46 Yao, H.M., Guo, R., Li, M. et al. (2022). Enhanced supervised descent learning technique for electromagnetic inverse scattering problems by the deep convolutional neural networks. *IEEE Transactions on Antennas and Propagation* 70 (8): 6195–6206.

47 Guo, R., Lin, Z., Li, M. et al. (2022). A nonlinear model compression scheme based on variational autoencoder for microwave data inversion. *IEEE Transactions on Antennas and Propagation* 70 (11): 11059–11069.

48 Pokorny, T., Vrba, J., Fiser, O. et al. (2023). On the role of training data for SVM-based microwave brain stroke detection and classification. *Sensors* 23 (4): 1–21.

49 Salucci, M., Vrba, J., Merunka, I., and Massa, A. (2017). Real-time brain stroke detection through a learning-by-examples technique–an experimental assessment. *Microwave and Optical Technology Letters* 59 (11): 2796–2799.

50 Salucci, M., Gelmini, A., Vrba, J. et al. (2019). Instantaneous brain stroke classification and localization from real scattering data. *Microwave and Optical Technology Letters* 61 (3): 805–808.

51 Wei, Z., Liu, D., and Chen, X. (2019). Dominant-current deep learning scheme for electrical impedance tomography. *IEEE Transactions on Biomedical Engineering* 66 (9): 2546–2555.

52 Lin, Z., Guo, R., Zhang, K. et al. (2022). Feature-based inversion using variational autoencoder for electrical impedance tomography. *IEEE Transactions on Instrumentation and Measurement* 71: 1–12. Art no. 4504712.

53 Zhang, K., Guo, R., Li, M. et al. (2021). Supervised descent learning for thoracic electrical impedance tomography. *IEEE Transactions on Biomedical Engineering* 68 (4): 1360–1369.

54 Viani, F., Rocca, P., Oliveri, G. et al. (2011). Localization, tracking and imaging of targets in wireless sensor network: an invited review. *Radio Science* 46 (05): 1–12.

55 Viani, F., Robol, F., Polo, A. et al. (2013). Wireless architectures for heterogeneous sensing in smart home applications: concepts and real implementations. *Proceedings of the IEEE* 101 (11): 2381–2396.

56 Viani, F., Migliore, M.D., Polo, A. et al. (2018). Iterative classification strategy for multi-resolution wireless sensing of passive targets. *Electronics Letters* 54 (2): 101–103.

9

RA-Based RIS-1 Design Using Support Vector Machines to Enhance mmWave 5G Coverage

Álvaro F. Vaquero[1], *Eduardo Martinez-de-Rioja*[2], *Jesús A. López-Fernández*[3], *and Manuel Arrebola*[3]

[1]*Department of Mathematics, Universidad de Oviedo, Oviedo, Asturias, Spain*
[2]*Department of Signal Theory and Communications and Telematic Systems and Computing, Universidad Rey Juan Carlos, Fuenlabrada, Madrid, Spain*
[3]*Department of Electrical Engineering, Group of Signal Theory and Communications, Universidad de Oviedo, Gijon, Asturias, Spain*

CHAPTER MENU

9.1 Introduction

Wireless technologies have offered attractive opportunities for the development of several applications. Traditionally, these applications have focused on radio frequency identification systems [1], remote sensing [2], or wireless power transfer [3]. Nowadays, wireless communications are the cornerstone of most communication systems. The current peak of wireless technology development is being reached with the latest research in communication systems for the deployment of the fifth generation (5G) of mobile communications [4, 5].

In this context, the deployment of 5G networks is divided into two main groups, depending on the operational frequency of the radio access interface. The sub-6 GHz band, also known as frequency range 1 (FR1), is defined for the systems that operate at frequencies below 6 GHz [4]. The second group is associated with those systems operating at much higher frequencies, particularly in the millimeter wave (mmWave) band [6]. One of the main challenges to address

Artificial Intelligence for Future Networks, First Edition.
Edited by Mohammad A. Matin, Sotirios K. Goudos, and George K. Karagiannidis.

in mmWave networks, not only with 5G but also with future generations like *beyond* 5G or the next sixth generation (6G), is to use this part of the spectrum to increase the bandwidth and data rate in wireless communications. In particular, these systems are planned to operate from around 24 GHz to hundreds of GHz. In 6G, it is estimated that short-range communications around 200–300 GHz will be developed [7]. Focusing on current 5G, the main band is defined in the K_a-band, with central frequencies around 27–28 GHz in terrestrial communications and 39 GHz in some cases of satellite communications. This range is defined as frequency range 2 (FR2). Other higher bands around 60 and 100 GHz are also being explored for 5G and *beyond* 5G. The primary goal of exploring higher frequency bands is to achieve higher-speed wireless access in cellular networks [6].

The development of communication systems in mmWave bands is associated with some problems that are not present in FR1 or previous mobile generations. Primarily, propagation in mmWave bands is characterized by high propagation and penetration losses, making signals much more sensitive to blockage by obstacles and physical barriers [8, 9]. Therefore, the presence of these barriers, which can be buildings (in outdoor scenarios) or walls, partitions, glass, etc. (in indoor scenarios), can result in the generation of "blind zones" that are areas with poor coverage, i.e., very low levels of the received signal, within regions that are supposed to offer high coverage. Blind zones disrupt the link between the user and the base station (BS) and, thus, they should be avoided. To minimize the appearance and extension of blind zones and, thus, to extend coverage in outdoor scenarios, one approach is to deploy more BSs [10]. However, this solution comes with the main drawback of increasing network costs as well as energy consumption. Another alternative is the use of radio frequency boosters (RF-boosters) to expand cellular coverage areas in urban environments by using a greater number of smaller coverage cells (pico- or femto-cells). Still, this alternative would significantly increase network costs, once again leading to higher energy consumption and the potential for increased interference between systems. Moreover, the complexity of the network equipment and the aforementioned factors would be further heightened if RF-boosters were to be combined with other 5G technologies, such as massive multiple-input and multiple-output (MIMO) and/or ultradense networks (both are primary solutions of 5G) [11, 12]. Therefore, it is highly desirable to find alternative solutions that allow the efficient deployment of FR2 communications using low-cost and energy-efficient systems.

Reflective intelligent surfaces (RISs) [13, 14] represent a low-cost and energy-efficient alternative for mitigating blind zones in mmWave networks. These structures typically consist of planar surfaces (although, recently, curved and conformal surfaces are being explored) made up of a quasiperiodic array of reflective elements. Those elements introduce the appropriate phase and/or amplitude adjustment in the incident wave (coming from the BS), which results in a reflected beam that meets the user link's requirements. Broadly speaking, RIS can redirect part of the power radiated by the BS toward the direction where the blind zone is located, ensuring that the user-BS link remains unaffected [15, 16]. The huge potential of RIS to improve the quality of wireless communications makes them a key element in the development of smart radio environments [15, 17].

RIS can be classified based on their capabilities using the nomenclature RIS-n, where n defines the number of capabilities the intelligent surface possesses [18]. Hence, RIS-1 defines reflective surfaces; RIS-2 adds beam reconfigurability with regard to RIS-1; RIS-3 includes amplification of the signal in addition to the capacities of RIS-2; and finally, RIS-4 is also able to regenerate the incident signal. One of the main advantages of RIS-1 is that they are based on panels that can be deployed at a low cost, and have minimal visual impact (RIS-1 can be installed on walls, ceilings, etc.). In contrast with RIS-2 to RIS-4, which rely on active elements and RF electronics that need to be powered, RIS-1 does not consume any energy. In the technical literature, RIS-1 is also called smart electromagnetic skin (SES) [19], due to their ability to provide a smart reflection of the incident signal. On the other hand, reconfigurable RISs allow to implement adaptive beamforming methods [20, 21], at the cost of a more complex and expensive fabrication, and higher power consumption.

In this chapter, we will focus on RIS-1 or SES, which not only reflects the incident signal in the specular direction (as reflectors or aluminum plates that are very cost-effective) but also shapes the amplitude and/or phase of the radiated field, which notably improves the performance of the network. In recent years, most of the proposed RIS-1/SES have been based on quasiperiodic surfaces, such as reflectarray (RA) or reflective metasurfaces. In this line, RA technology [22] has great potential to be applied to the design of passive RIS to improve wireless connectivity in mmWave 5G networks, as RAs have shown their capability to produce high-gain collimated beams [23] or contoured beams with a specific shaping [24, 25] by a proper design of the phase shifts introduced in the incident field by the printed elements.

9.1.1 RA-Based Reflective Intelligent Surface

RA antennas are classified as Spatially Fed Array (SFA) antennas, see Figure 9.1. Therefore, they do not require feeding or beamforming networks, so they achieve a significant reduction in ohmic losses, fabrication complexity, and cost with respect to phased array antennas [22]. Compared to other conventional approaches (such as metallic reflectors), RAs offer additional capabilities to operate at multiple frequencies [26, 27], to generate independent beams in each polarization [28, 29], and to provide beam reconfigurability when switching devices (PIN diodes,

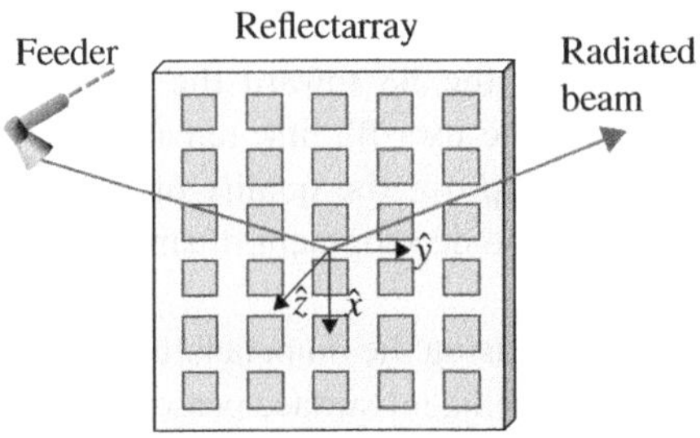

Figure 9.1 Schematic representation of a printed RA.

varactor diodes, etc.) are integrated into their cells [30] (expanding RA-based RIS from RIS-1 to RIS-4). In addition to their flat surface and lightweight, printed RAs are characterized by their simple and low-cost manufacturing process, which employs the same techniques used for conventional printed circuits.

In the last years, RA antennas have been proposed for several 5G applications at mmWave bands. These applications include steerable-beam antennas [31, 32] and fixed-beam antennas [33, 34] for BSs and point-to-point communications, terminal antennas for connected vehicles [35], and passive RIS to improve mmWave coverage [36–39]. The use of passive RA-based RIS (RIS-1 or SES) is an attractive and energetically efficient solution to the problem of blind zones in mmWave networks, since they can be deployed at reduced cost in both indoor and outdoor scenarios, as shown in Figure 9.2, installed on the walls of buildings or in the ceilings of rooms with very low visual impact [39, 40]. This RIS-1 will concentrate the power received from the BS and will redirect it to a specific coverage region, enhancing wireless access for the users located in that area.

Compared to other works on RAs for 5G applications [31–35], the design of RA-based RIS-1 involves some particular aspects that must be carefully addressed. First, it would be desirable to produce a shaped beam (e.g., a fan beam or flat-top beam) instead of a pencil beam, so that the beam shape matches the coverage specifications of the application scenario. Second, the RIS panel will be illuminated

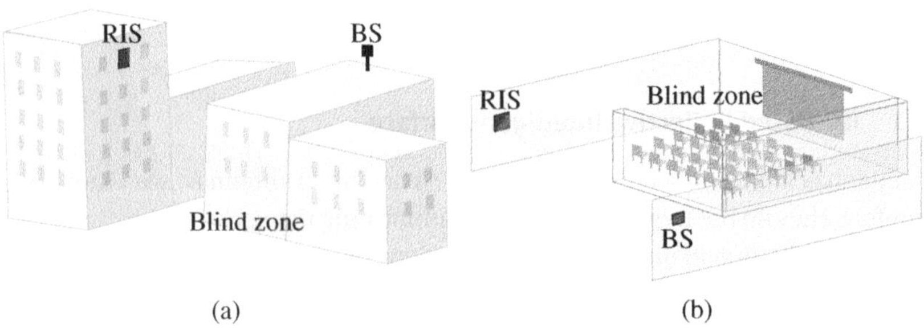

Figure 9.2 Application scenarios for RA-based RIS illuminated from a nearby BS: (a) outdoor scenario and (b) indoor scenario.

from the BS antenna, which can be located at a significant distance from the RIS (e.g., between 10 and 100 m), and depending on the geometry of the scenario, the incidence angles from the BS on the RIS may be considerable (e.g., up to 50°). This represents a major difference with respect to previous works [31–35] with centered or slightly offset close feeds, which lead to lower incidence angles. This factor must be considered in the design of the RIS, since the cells must be able to provide a satisfactory performance under strong oblique incidence. The last challenge is related to the large number of cells that form the RIS, which may increase the complexity of the RIS analysis and design processes.

The first implementations of RA-based RIS-1/SES for 5G [36, 37] showed limited capabilities, since they were only able to change the direction of the reflected signal, without providing an optimal shaping of the beam. Moreover, the RIS operated in a single frequency band (27–28 GHz). More recently, other authors have proposed the design of RIS-1 [38, 39] that is able to generate a reflected beam with a specific shape and pointing, which are selected to provide a proper illumination of the blind zone. The shaped coverage in [38, 39] was accomplished by applying a phase-only synthesis (POS) technique based on the Intersection Approach algorithm [41, 42], but other beamforming techniques are also possible [19, 43]. The RA-based RIS reported in [38, 39] presents a stable radiation performance within a 1-GHz bandwidth at the 28-GHz band, which is enough to cover a pair of standard 5G channels, each channel occupies around 400 MHz [44].

9.1.2 Considerations of RA-Based RIS Design

The fact that RIS-1/SES can be based on RA panels or reflective metasurfaces enables the application of design techniques from both antenna types to RIS. Therefore, a RIS can be designed by adjusting the phase shifts introduced by each of the RIS elements. The phase shift can be achieved using different techniques depending on the reflected beam sought. In general, two types are commonly used: analytical techniques based on geometric optics and synthesis or optimization techniques. The type of beam sought will depend on the scenario where the RIS is installed to provide coverage. For example, if the RIS is being deployed in an indoor scenario, it is common to seek a reflected beam with specific near field (NF) features. On the other hand, in an urban environment, the beam specifications will be typically defined in the far field (FF). Although this criterion is not absolute and depends on the aperture size and frequency, it covers most cases. For simple cases, like designing RIS with an NF-focused beam or a high-gain beam in the FF for point-to-point communications, geometric optics-based designs provide analytical solutions for phase-shift calculations with great results. However, for designing shaped patterns, whether in the NF or FF, analytical techniques may not be accurate enough, and optimization techniques are required.

One of the most popular techniques in the literature for obtaining shaped patterns is POS that models the RIS element solely as a phase shifter on the copolar component, ignoring the effect over the cross polar, which greatly simplifies the antenna design. The output of POS is the phase-shift value that each element should introduce, thus, it is necessary to design the elements accordingly in a succeeding stage. This technique is based on the assumption that the copolar coefficient directly controls the shape of the radiation pattern. Despite the mentioned simplification, POS has provided excellent results, producing highly complex radiation patterns in RA and transmitarray antennas, where it has been particularly applied in satellite and terrestrial communications [13, 45]. It is worth noting that this technique models the RIS elements solely as an ideal phase shifter, making it independent of the topology of the physical element used for the design. On the other hand, it does not provide control over cross-polar components, bandwidth, angle of incidence, or more complex configurations such as dual-polarization or multiple feed systems.

In the last years, machine learning (ML) techniques have been successfully used to accelerate RA analysis, design, and optimization. These techniques can be used to produce surrogate models that characterize the unit cell. In a further step, the ML technique can be applied to the design and optimization of RAs by integrating the surrogate models in an analysis routine that is called hundreds or even thousands of times to search the cell features that provide the adequate phase shift to each RA element. To deliver surrogate models, a full-wave electromagnetic technique that accurately characterizes the cell is used to generate samples of the parameters that portray its electromagnetic behavior (e.g., the reflection coefficient matrix) considering several values of different input variables, among others: the frequency, the angle of incidence, and/or the geometrical features. Subsequently, the selected ML technique is trained with those samples to yield a regression function that, ideally, matches samples of the same parameters not used during the training process. Artificial neural networks (ANNs) were the first technique of this type used to characterize the phase response of RA unit cells [46–48], which is the main parameter to accomplish antenna designs meeting specific copolar requirements. ANNs have also been devoted to the characterization of the cross-polarization coefficients for dual-polarized cells [49]. Among the ML-based cell characterizations presented in the last years, it is worth mentioning the one in [50] where the whole matrix of reflection coefficients is accurately modeled by applying support vector machines (SVMs) to a regression problem, support vector regression (SVR), reducing the time cost of reflection coefficients computation in more than three orders of magnitude. This technique was afterward used to speed up the design and the cross-polar optimization of a dual-polarization shaped-beam RA for space applications in [51] revealing precise and efficient results.

9.2 RIS-1 Unit-Cell Characterization Using SVR

9.2.1 Passive Unit Cell for RIS-1 Design

One of the key aspects in the design of RA-based RIS is choosing an appropriate configuration for the RA cell since it has a great impact on the performance and capabilities of the RIS. In this section, we present a brief review of the main types of unit-cell configurations that have been used for RA-based RIS.

RA cells formed by a single resonant element (e.g., the rectangular patch used in [36], see Figure 9.3a) typically show a limited phase range of the reflection coefficient (less than 360°) with a steep phase response. Single-layer multiresonant cells, such as those based in groups of coplanar parallel dipoles [38, 39], see Figure 9.3b, are able to provide a greater range of phase variation with a smoother phase response, which leads to a broadband performance of the RA panel. Furthermore, the RA cells used in Martinez-de-Rioja et al. [38, 39] provide a robust performance under large incidence angles, which is a crucial feature for some deployment scenarios, as mentioned above. RA cells based on stacked rectangular patches [52] can provide a similar phase-shift performance to single-layer multiresonant cells, but they require several layers of printed elements and dielectrics, which results in more complex manufacturing processes and larger weight, volume, and fabrication costs.

Concerning the polarization of the electromagnetic waves, the RA cells used in phase-shift [38, 39] enable independent control of the phase response in each linear polarization (LP), thanks to the uncoupling between the orthogonal sets of dipoles, as shown in [53]. The passive RIS reported in [38, 39] operates in dual-LP, producing the same radiation pattern in both LPs. In this case, the beam requirements are equal for both polarizations, ensuring users' connectivity in the coverage zone, but the same unit cells can be used to design a RIS with different beam requirements in each polarization. This is not the case of other single-layer RA cells with twofold symmetry (see Figure 9.3c), as those proposed in [37, 54], since they do not allow to introduce a different phase shift in each LP.

A dual-layer version of the dipole-based RA cell was proposed in [55] to design a dual-band passive RIS for mmWave 5G applications. The dual-layer configuration allows independent control of the phase shift introduced by the RA elements at each frequency band, so the RIS can be designed to produce a different coverage at each frequency band, i.e., with different requirements of beam pointing and beamwidth. Moreover, the dual-band RIS can be illuminated simultaneously from different BS, each operating at a different frequency, as in [55]. The dual-band RIS presented in [55] has been designed to produce two pencil beams, but shaped beams can also be generated by applying suitable beamforming techniques, as in [38, 39].

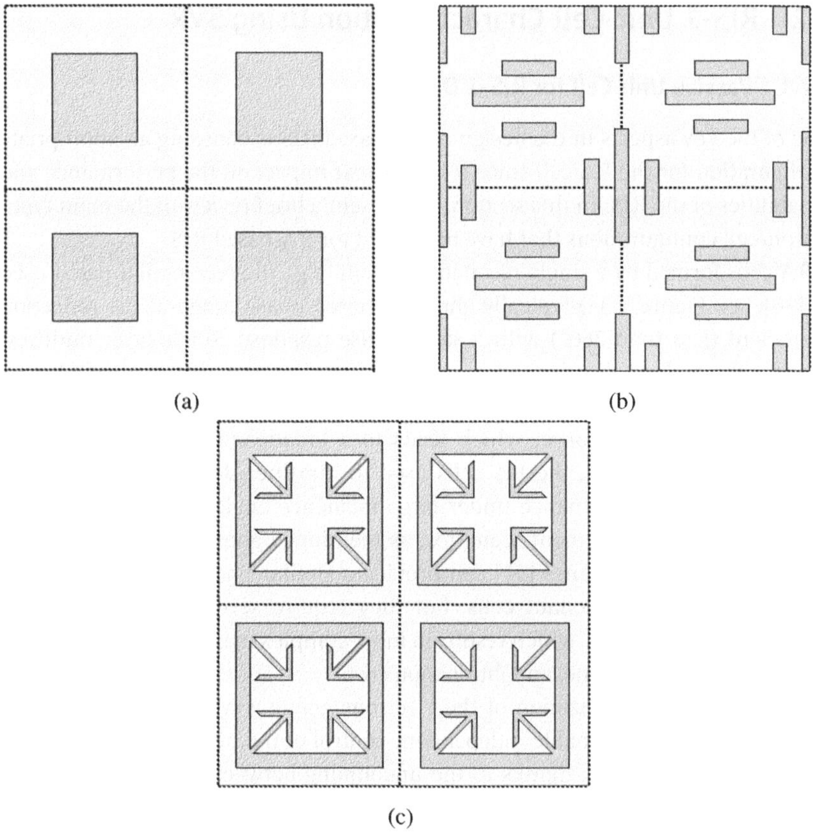

Figure 9.3 RA cells used for passive RIS: (a) rectangular patches. Source: Adapted from Kitayama et al. [36], (b) orthogonal groups of parallel dipoles. Source: [38, 39], (c) multiresonant element with twofold symmetry. Source: Adapted from Lu et al. [37].

Outside the frequency band for which they have been designed, conventional RA-based RIS behave as a flat metallic panel, i.e., as the ground plane of the RA, which produces a specular reflection of impinging signals. The reflections of the out-of-band impinging signals can be useful in some particular scenarios, but not in others, since they can produce blocking or interferences in the links of other communication systems. An interesting design option is to replace the ground plane of the RA with a frequency selective surface (FSS) [56], which will exhibit a similar performance to a stop-band filter: in the operating bands of the RIS, the FSS reflects the incident signals acting as the ground plane of the RA, while outside the RIS operating bands, the impinging signals go through it without reflection [56].

9.2.2 SVR-Based Models of RA Unit Cells

In this section, we are going to describe how to apply the theory of SVR to obtain surrogate models of RA unit cells. We will try to make this section as general as it is possible, though we will rely on the RA cell represented in Figure 9.4 to exemplify the theory and to illustrate some of the techniques reviewed here.

In brief, our goal is to model the reflection coefficient matrix, **R**, of a RA unit cell. This matrix describes the relationship between the tangential components of the incident, $\vec{E}_{\text{inc}}$, and reflected, $\vec{E}_{\text{ref}}$, electric fields at each element of the RA. Noting (x_m, y_n) as the coordinates of the (m, n)th element, or cell, center over the RA surface, we may write:

$$\vec{E}_{\text{ref}}\left(x_m, y_n\right) = \mathbf{R}^{mn} \cdot \vec{E}_{\text{inc}}\left(x_m, y_n\right), \tag{9.1}$$

where $\vec{E}_{\text{inc/ref}}\left(x_m, y_n\right) = E_{\text{inc/ref},x}\left(x_m, y_n\right)\hat{x} + E_{\text{inc/ref},y}\left(x_m, y_n\right)\hat{y}$, being $E_{\text{inc/ref},x}$ and $E_{\text{inc/ref},y}$ the incident or the reflected electric field components over the $\hat{x}$ and $\hat{y}$ directions, respectively. Moreover, the reflection coefficient matrix for the (m, n)th RA element is expressed as follows:

$$\mathbf{R}^{mn} = \begin{pmatrix} \rho_{xx}^{mn} & \rho_{xy}^{mn} \\ \rho_{yx}^{mn} & \rho_{yy}^{mn} \end{pmatrix}, \tag{9.2}$$

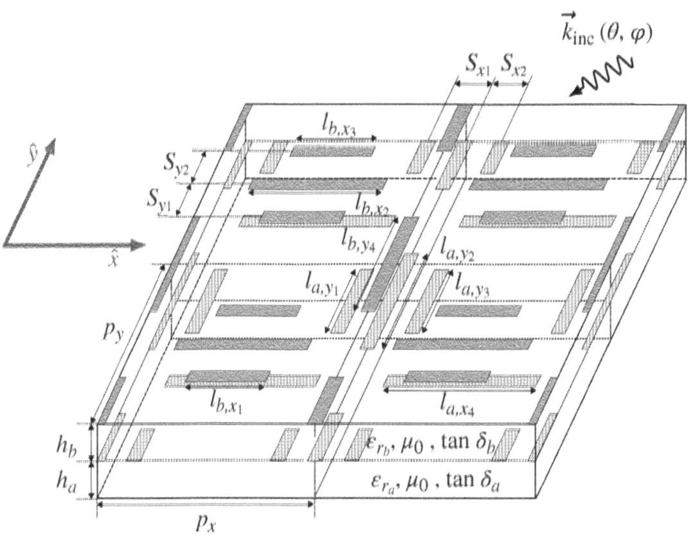

Figure 9.4 Example of an RA unit cell based on parallel and coplanar dipoles in two layers of metallization.

being ρ_{xx}^{mn} and ρ_{yy}^{mn} the copolar reflection coefficients that relate the collinear reflected and incident electric field components over $\hat{x}$ and $\hat{y}$ directions, respectively; meanwhile, ρ_{xy}^{mn} and ρ_{yx}^{mn} are the cross-polar reflection coefficients that relate the orthogonal reflected and incident electric field components. For the sake of clarity, we will omit the dependence on (m, n) from this point forward. Therefore, we will note $\mathbf{R}^{mn}$ just as $\mathbf{R}$. In addition, we will refer to the reflection coefficients, in general, as ρ.

Matrix $\mathbf{R}$ depends on several variables, such as: frequency (f), angle of the impinging plane wave (incident angle (θ, ϕ)), geometric features (cell periodicity $-p_x$ and p_y-, dipoles dimensions –length and width, noted as $\vec{D}$–, inter-dipole separation $-\vec{S}-$, see Figure 9.4), as well as substrate properties. In principle, all those variables might be used as input variables of the SVR-based surrogate models (noted as $\vec{x}$). The reflection coefficients ρ are, in general, complex numbers, meanwhile, SVR outputs a real-valued function, $f(\vec{x})$. Consequently, at least eight models must be produced to estimate $\mathbf{R}$ (real and imaginary parts, or amplitude and phase, of each matrix element ρ). We may use a black-box system modeling of SVR, as illustrated in Figure 9.5. Nonetheless, to avoid the *curse of dimensionality* [57, 58] it is convenient to carefully select which of those variables are included in $\vec{x}$ and which are used as fixed-value parameters to produce different SVR models. For the sake of clarity, we will consider the input and the output of the SVR, respectively, as $\vec{x}$ and $\tilde{y}$ (or $\tilde{\rho}$), ignoring, unless for exemplifying purposes, the precise physical variables involved.

For the sake of exemplification, we will use in this section the RA unit cell in Figure 9.4, the same as in [50], that is comprised of eight parallel and coplanar dipoles in two layers of metallization. Each polarization is controlled by a set of four dipoles and both sets are shifted half a unit-cell period from each other. The samples used to train the SVMs are computed using an inhouse electromagnetic code based on the spectral domain Method of Moments with local periodicity assumption, presented in [59], in compact notation MoM–LP. These samples are calculated for a set of different angles of incidence (θ, φ) of the impinging wave. A different SVR is yielded for each (θ, φ) combination. To also alleviate the complexity of the SVR modeling, we keep constant most of the degrees of freedom (DoFs) of the unit cell. First, we consider for both layers the use of the commercially available substrate CuClad 233, with a thickness of 0.787 mm

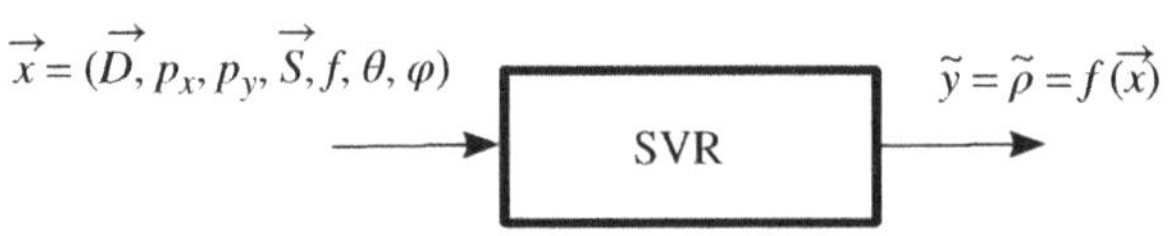

Figure 9.5 Black-box system model of SVR applied on an RA unit cell.

and a complex relative permittivity $\varepsilon_r = 2.33 - j3.029 \cdot 10^{-3}$. The unit cell is of approximately half of a wavelength at the frequency of analysis, $f = 25.5$ GHz, which corresponds to $p_x = p_y = 5.84$ mm. In addition, we also reduce the number of geometrical DoFs of the unit cell. In particular, we fix the dipole widths to 0.5 mm and the separation between dipoles to 4 mm, leaving as DoFs just two geometrical variables, T_x and T_y, each one controlling the dipole lengths for each polarization that act as input variables of the SVR ($\vec{x} = \left(T_x, T_y\right)$). The variables T_x and T_y are allowed to vary in the interval [0.1, 5] mm and their relation with dipole lengths are as follows (see Figure 9.4):

$$\begin{aligned} &\ell_{a,x_4} = T_x;\ \ \ell_{b,x_1} = \ell_{b,x_3} = 0.63\,T_x;\ \ \ell_{b,x_2} = 0.93\,T_x, \\ &\ell_{b,y_4} = 0.95\,T_y;\ \ \ell_{a,y_1} = \ell_{a,y_3} = 0.58\,T_y;\ \ \ell_{a,y_2} = T_y. \end{aligned} \tag{9.3}$$

Regarding the output of the SVM models used for exemplification, $\tilde{y}$, we model the real and imaginary parts of the reflection coefficients that form the matrix in (9.2) for each incidence angle. Note that the results given in this section have been obtained using, in sequential mode, a workstation with 2 Intel Xeon E5-2650v3 CPU at 2.3 GHz and 256 GB of RAM.

This section is organized as follows. In Section 9.2.2.1, we describe the theoretical foundations of SVMs applied to general regression problems. We next illustrate in Section 9.2.2.2 what considerations have to be accounted for to get the optimal model. Section 9.2.2.3 is devoted to the description of an efficient grid search for obtaining the optimal model.

9.2.2.1 SVM Theoretical Background

SVMs are supervised learning structures well-suited to handle both classification and regression problems [60]. The regression performed by an SVM is usually referred to as SVR. In a general regression problem, the SVR uses a set of inputs $\left(\vec{x}_i\right)$ and outputs $\left(y_i\right)$ to find a function f, or model, that best relates inputs and outputs not originally used during the training process of the SVM. In the next paragraphs, we briefly summarize the foundations of SVMs applied to SVR problems.

We start by noting the L-D real input space as $\chi \subseteq \mathbb{R}^L$ and the output space as the set of real numbers $\mathbb{R}$. In addition, any element of χ is noted in general as $\vec{x} \in \chi$ and any element of the output space as $y \in \mathbb{R}$. Formally speaking, SVMs use a training set $T = \left\{\left(\vec{x}_i, y_i\right)\right\}_{i=1,2,\ldots,N_r}$ to find a mathematical function $f : \chi \to \mathbb{R}$, such that the output for a new element $\vec{x} \in \chi$ is estimated as follows:

$$\tilde{y} = f\left(\vec{x}\right), \tag{9.4}$$

where f is the SVR function and $\tilde{y}$ is its estimation of the actual output y corresponding to $\vec{x}$.

For the sake of clarity, we start by explaining the case where the regression function f is linear. In such case, it takes the form:

$$f\left(\vec{x}\right) = \langle \vec{w}, \vec{x} \rangle + b, \tag{9.5}$$

where $\langle \cdot, \cdot \rangle$ notes the dot product in χ and $b \in \mathbb{R}$. In addition, $\vec{w} \in \chi$ may be interpreted in terms of a vector decomposition of the normal to the regression function, $\vec{n}$, that appears in Figure 9.6. In that plot, vector $\vec{n}$, which is not necessary for unit norm, is decomposed into a unit vector in the $-\hat{y}$ direction (orthogonal to χ space) and $\vec{w} \in \chi$. As it is illustrated in Figure 9.6, seeking f by considering the minimization of the Euclidean norm to the square of $\vec{w}$, i.e., $||\vec{w}||^2 = \langle \vec{w}, \vec{w} \rangle$, assures obtaining a *flat* regression function. Assuming that the training data are properly fitted to a linear function, the statistical learning theory states that the optimal regression function is maximally flat [60]. In this regard, the optimality of f is measured in terms of its generalization properties, i.e., how well it will fit data of the same nature but not initially included in the training set. Figure 9.6 graphically suggests this idea since a flatter f is expected to find more new samples with a regression error below a given threshold (ε).

In the above paragraph, it is suggested the idea that the regression function to find should properly fit the data and, at the same time, be maximally flat. This motivated [61] to pose the problem in terms of a minimization of a regularized risk functional, $R_{\text{reg}}[f]$, which considers the empirical errors and the flatness of f:

$$R_{\text{reg}}[f] \propto \frac{||\vec{w}||^2}{2} + C\sum_{i=1}^{N_r} c\left(\vec{x}_i, y_i, f\left(\vec{x}_i\right)\right), \tag{9.6}$$

where $||\vec{w}||^2/2$ acts as a regularization term whose minimization assures to provide *flatness* of f; $c\left(\vec{x}, y, f\left(\vec{x}\right)\right)$ is a cost (or loss) function associated with the regression error; and C is a positive constant that determines a trade-off

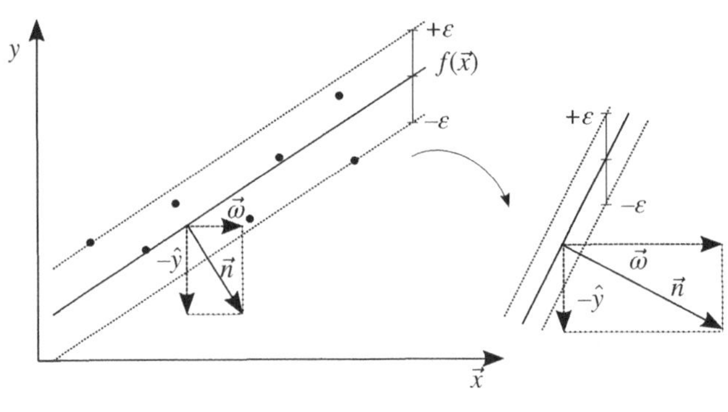

Figure 9.6 1D geometrical interpretation of SVR parameter $\vec{\omega}$ and its relation with *flatness* of f.

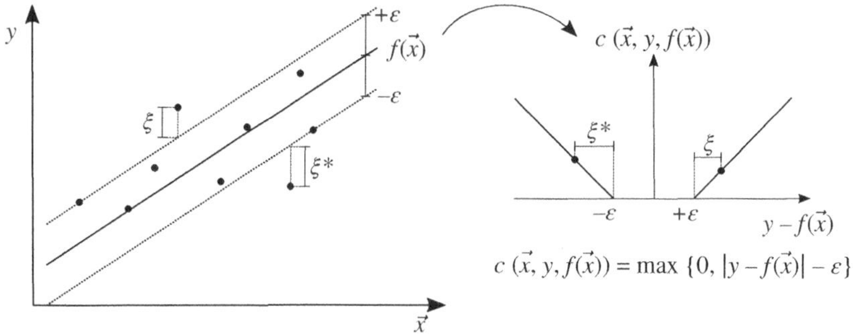

Figure 9.7 Illustration of the ε-insensitive loss function and the corresponding slack variables, ξ and ξ^*.

between minimizing the empirical risk (regression error) and providing good generalization properties [60].

The most popular cost function is the ε-insensitive loss function, which is illustrated in Figure 9.7 and defined as follows:

$$c\left(\vec{x}, y, f\left(\vec{x}\right)\right) = \max\left\{0, \left|y - f\left(\vec{x}\right)\right| - \varepsilon\right\}. \tag{9.7}$$

Note that the cost function defined in (9.7) does not penalize regression errors below a given $\varepsilon \geq 0$. Therefore, the parameter ε, which must be chosen beforehand, is related to the maximum accuracy achievable by the SVR model. This is used in Section 9.2.2.2 to estimate the variance of the SVR model error.

Using the ε-insensitive loss function, we may arrive at the constrained minimization problem stated in [61]:

$$\begin{aligned} &\text{minimize } \frac{\|\vec{w}\|^2}{2} + C\sum_{i=1}^{N_r}\left(\xi_i + \xi_i^*\right), \\ &\text{subject to } \begin{cases} y_i - \langle\vec{w}, \vec{x}_i\rangle - b \leq \varepsilon + \xi_i \\ \langle\vec{w}, \vec{x}_i\rangle - b - y_i \leq \varepsilon + \xi_i^* \\ \xi_i, \xi_i^* \geq 0, \end{cases} \end{aligned} \tag{9.8}$$

where the slack variables ξ_i and ξ_i^* account for the regression errors that are above ε. Figure 9.7 illustrates the definition of the slack variables and its relation with the ε-insensitive loss function.

The solution of the optimization problem in (9.8) is a function f of the form [60]:

$$f(\vec{x}) = \langle\vec{w}, \vec{x}\rangle + b = \sum_{i=1}^{N_s}\left(\alpha_i^- - \alpha_i^+\right)\langle\vec{x}_i, \vec{x}\rangle + b, \tag{9.9}$$

where $\vec{x}_i$ are the support vectors (SVs); N_s is the number of SVs; α_i^- and α_i^+ are the optimal Lagrange multipliers; and b is the offset.

Oftentimes, the input and output data may be better related by means of a nonlinear function instead of a linear one. In such cases, the above insights are valid, though the input data should be mapped to a high-dimensional space, called feature space ($\mathcal{H}$), where the regression may be accurately performed using a linear function. Noting the mapping from input to feature space as $\vec{\phi} : \chi \rightarrow \mathcal{H}$, we may write the nonlinear regression function as follows:

$$f(\vec{x}) = \langle \vec{w}, \vec{\phi}\,(\vec{x}) \rangle + b = \sum_{i=1}^{N_s} \left[\left(\alpha_i^- - \alpha_i^+\right) \langle \vec{\phi}\,(\vec{x}_i)\,, \vec{\phi}\,(\vec{x}) \rangle \right] + b, \tag{9.10}$$

where $\vec{w}, \vec{\phi}\,(\vec{x})\,, \vec{\phi}\,(\vec{x}_i) \in \mathcal{H}$. In general, this approach is not feasible from a computational viewpoint, since the dimensionality of $\mathcal{H}$ may be extremely high. Nonetheless, some maps $\vec{\phi}$ allow to express the dot product between elements in $\mathcal{H}$ by means of a kernel function, $K(\cdot, \cdot)$, evaluated on elements in χ as follows:

$$K\left(\vec{x}, \vec{x}'\right) = \langle \vec{\phi}\,(\vec{x})\,, \vec{\phi}\left(\vec{x}'\right) \rangle. \tag{9.11}$$

Equation (9.11) is often referred to as the *kernel trick*, and it was initially used to extend the generalized portrait hyperplane classifier to nonlinear SVMs. The kernel trick is very useful for the efficient evaluation of (9.10) because all the feature vectors appear there only as arguments of dot products. Using the kernel trick, the nonlinear regression function f takes the form:

$$f(\vec{x}) = \langle \vec{w}, \vec{\phi}\,(\vec{x}) \rangle + b = \sum_{i=1}^{N_s} \left[\left(\alpha_i^- - \alpha_i^+\right) K(\vec{x}_i, \vec{x}) \right] + b, \tag{9.12}$$

where $\vec{w}$ is no longer given explicitly and the optimization problem consists of finding the flattest function in the feature space, not in the input space.

In this chapter, we use the ε-insensitive loss function since it provides a sparse representation of function f in terms of the SVs, i.e., it requires a smaller value of N_s than other loss functions [60]. This sparsity comes from the fact that the samples that fall within the ε-*tube* ($|y - f\,(\vec{x})\,| \leq \varepsilon$) do not contribute to regularized risk in (9.6) and, thus, they will not be part of the SVs. In addition, we use a Gaussian kernel (also known as radial basis function, RBF), which follows the equation

$$K(\vec{x}, \vec{x}') = \exp\left(-\gamma\, \|\vec{x} - \vec{x}'\|^2\right), \tag{9.13}$$

where γ is a tunable parameter that is related to the inverse of the width of the bell function. Equation (9.12) expresses function f as a linear combination of versions of the kernel function plus an offset. Therefore, low values of γ yield a mixture of wide RBFs and, very likely, a soft function f. On the contrary, high values of γ produce a combination of narrow RBFs and, very likely, a very rich function f.

9.2.2.2 Model Selection, Expected Accuracy, and Training

From Section 9.2.2.1, it comes that the SVR models used in this chapter directly depend on the selection of parameters C, γ, and ε. The model accuracy and its generalization properties are going to depend on the adequate choice of these parameters. For instance, high values of C may yield to an SVR that tries to fit very well the training data, though it might yield considerable errors for new samples (bad generalization properties). This behavior is commonly known as *overfitting*. On the other hand, low values of C may produce an SVR that does not properly fit the data, neither training nor new data.

The main goal of the SVR model selection is to find the optimal parameters (C, γ), noted as (C^*, γ^*), that are supposed to minimize the relative error (RE) of the estimation over a data set that was not initially included as part of the training set. We may define the error, $e(\cdot)$, of the regression function f at point $\vec{x}$ as follows:

$$e(\vec{x}) = y - f(\vec{x}), \tag{9.14}$$

where $f(\vec{x})$ is the output of the SVR and y is the actual output.

Lets assume that the SVR is computed over a set of M input samples $\vec{x}_i, i = 1, 2, \ldots, M$. This computation yields a set of output samples, y_i, a set of samples of $f, f(\vec{x}_i)$, and the corresponding set of samples of error function, $e_i = e(\vec{x}_i)$. In this manner, we may define the vectors of samples: $\vec{y} = (y_1, y_2, \ldots, y_M)$ and $\vec{e} = (e_1, e_2, \ldots, e_M)$. To quantify the accuracy of the SVR over the set of M samples, we use a RE defined by

$$\text{RE (dB)} = 10\log\left(\frac{\|\vec{e}\|^2}{\|\vec{y}\|^2}\right) = 10\log\left(\frac{\sum_{i=1}^{M}|e_i|^2}{\sum_{i=1}^{M}|y_i|^2}\right), \tag{9.15}$$

that expresses the relation between the Euclidean norms of the error vector, $\|\vec{e}\|$, and of the estimated output vector, $\|\vec{y}\|$. In this manner, small errors associate accurate approximations of y by regression function f independently of the level of the function to be estimated. Ideally, the selected optimal parameters (C^*, γ^*) should produce a model that minimizes (9.15) over samples not used during the training process.

As well as the selection of (C^*, γ^*), it is necessary to choose the size of the insensitive zone, ε, that relates to the maximum achievable accuracy of the SVR and to the sparsity of the representation in terms of SVs. Regarding the latter, higher values of ε will produce that many data fall inside the insensitive zone, or ε-tube, thus not becoming part of the set of SVs, which translates to a sparse representation of function f. With respect to accuracy, all samples of the function that fall inside the ε-tube do not contribute to the loss function of (9.7), though they actually increase the regression error of (9.15). Assuming that, asymptotically, all the training samples, except the SVs, fall within the ε-tube and that the samples of $e(\vec{x})$ are uniformly distributed in the interval $[-\varepsilon, +\varepsilon]$, it is possible to estimate the variance of samples of the error function by $\tilde{\sigma}_e^2 = E\left\{|e|^2\right\} = \varepsilon^2/3$, where $E\{\cdot\}$ stands for the expected value. Then, using (9.15), the mean squared value of RE over a set of N_r training samples may also be estimated as

$$E\left\{\mathrm{RE}^2\right\} = E\left\{\frac{\|\vec{e}\,\|^2}{\|\vec{y}\,\|^2}\right\} = \frac{N_r\tilde{\sigma}_e^2}{\|\vec{y}\|^2}. \tag{9.16}$$

Therefore, given a set of N_r samples and a desired precision, in terms of $E\left\{\mathrm{RE}^2\right\}$, the value of ε ends up to be approximated by

$$\varepsilon \approx \sqrt{\frac{3}{N_r}}\|\vec{y}\|\sqrt{E\left\{\mathrm{RE}^2\right\}}. \tag{9.17}$$

In practice, Eq. (9.16) becomes a lower threshold for the quadratic value of the RE that is only achieved for a sufficiently large N_r (in principle, $N_r \to \infty$). Consequently, the approximation of ε given by (9.17) is more accurate the larger N_r is. In addition, a proper value of N_r is also necessary to achieve the desired SVR accuracy. The selection of N_r for modeling a RA cell is discussed in Section 9.4.2.1.

The efficient selection of the optimal parameters (C^*, γ^*) is a crucial question that depends on the problem under study and usually needs a trial and error procedure. To select (C^*, γ^*), we use the cross-validation procedure where the whole data set, composed of N samples, is divided into three nonoverlapping subsets: training (e.g., $\leq$70%, i.e., $N_r \leq 0.7N$), validation (e.g., 15%, i.e., $N_v = 0.15N$) and test (e.g., 15%, i.e., $N_t = 0.15N$). During the search of (C^*, γ^*), each evaluated pair (C, γ) is used along with the training set to train the SVM, i.e., to obtain the SVR by solving the optimization problem posed in (9.8). We use the library described in [62] for training the SVM, which is an efficient and widely used tool. At the end of the training associated with each (C, γ), we compute the RE yielded by the surrogate model over the validation set. This is performed over a limited size grid on the (C, γ) plane, usually about a few tenths (in logarithm to base 2) above and below 0 for both C and γ. (C^*, γ^*) are selected as those which minimize the RE over the validation set. Finally, the error of the selected optimum model is evaluated with the test set to estimate the error produced on new samples. This process ensures that the sets employed to compute the error during the grid search, validation set, and to test the final model, test set, remain hidden during the training process and are not actual training samples, which enhances the generalization properties of the regression. Other more sophisticated cross-validation schemes can be used to provide more accurate estimations on the test error.

Figure 9.8 shows the RE (in dB) of SVM models trained with different values of (C_2, γ_2) for the estimation of $\mathrm{Re}\left\{\rho_{yx}\right\}$ and incidence $\left(\theta = 20°, \varphi = 20°\right)$ related to the unit cell described at the beginning of this section (see Figure 9.4).

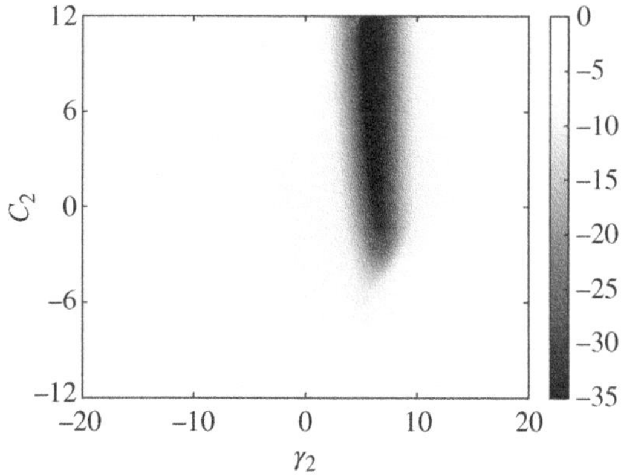

Figure 9.8 Relative error (dB) in estimating $\text{Re}\{\rho_{yx}\}$ for an incident angle $(\theta = 20°, \varphi = 20°)$ over the validation set when performing an exhaustive grid search in the (C_2, γ_2) plane with grid spacing $\Delta C_2 = \Delta\gamma_2 = 0.2$.

These results are taken from [50]. Note that we note $C_2 = \log_2 C$ and $\gamma_2 = \log_2 \gamma$. In this case, the number of samples is $N_r = 1750$, $N_v = N_t = 375$ and the grid spacing considered for both C_2 and γ_2 is $\Delta C_2 = \Delta\gamma_2 = 0.2$. The grid search produces $C_2^* = 10.4$, $\gamma_2^* = 5.4$, and RE $= -36.05$ dB, over the validation set. This search, or model selection, took 50,006 s (about 13 h and 54 min). Considering that we estimate 10 output variables per incident angle and that the number of incident angles may be substantial, this brings to light the necessity of using an efficient grid search.

9.2.2.3 Efficient Grid Search

In this section, we detail a strategy to speed up the search of the optimal model (C_2^*, γ_2^*). To this end, we may first note that the shape of the RE in the (C_2, γ_2) plane is characterized by a narrow valley along C_2, as shown in Figure 9.8, and that this shape is similar for different output parameters and incident angles. Furthermore, the training time cost for large values of C_2 in the area of interest can be three orders of magnitude higher than for small values of C_2. This is illustrated for the RA cell used as example in Figure 9.9, which plots the time cost of the training for each point on the (C_2, γ_2) grid. All this suggests that it is very convenient to limit the grid search to certain domain that, in general, can be expressed as $\gamma_2 \in [\gamma_{\min}, \gamma_{\max}]$ and $C_2 \in [C_{\min}, C_{\max}]$. In the case of analyzing the RA cell used as an example, the suggested search domain by Prado et al. [50] is $\gamma_2 \in [-20, +20]$ and $C_2 \in [-10, +6]$. Nonetheless, the optimal search domain may vary depending on the problem at hand.

Another way of reducing the time cost of the grid search consists in decreasing N_r. For instance, Figure 9.10 shows, for the analyzed RA cell, the RE on the (C_2, γ_2) grid for the estimation of $\text{Im}\{\rho_{xy}\}$ with incidence $(\theta = 10°, \varphi = 20°)$ and using three different training set sizes ($N_r = 87$, 350, 1750). The search domain is the suggested above, $\gamma_2 \in [-20, +20]$ and $C_2 \in [-10, +6]$, with grid spacing 1 for

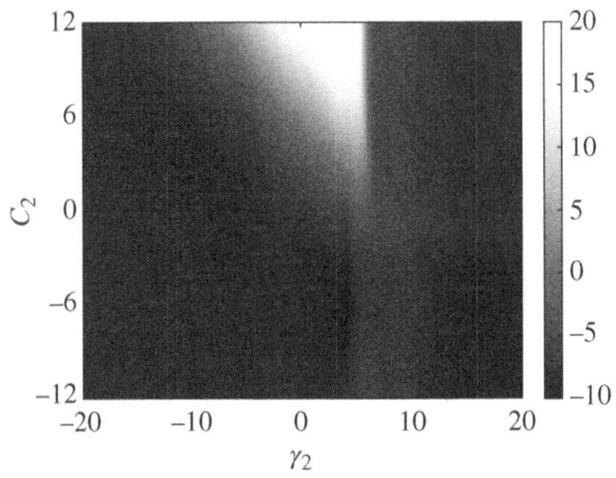

Figure 9.9 Training time in logarithmic units $(10 \log (t_{\text{train}}(\text{s})))$ for each point of the grid search of Figure 9.8.

both C_2 and γ_2, i.e., $\Delta C_2 = \Delta \gamma_2 = 1$. The seeking took 1.2 s for $N_r = 87$, 10.0 s for $N_r = 350$, and 196.4 s for $N_r = 1750$. Figure 9.10 displays that the valley where the optimum is located shifts a bit its position and becomes deeper for increasing N_r, though the region where it is placed remains approximately the same. Note that the scale is intentionally different for each case to clearly place the valley. Therefore, a grid search with a reduced number of samples (note the great resemblance between Figures 9.10b and 9.10c) may yield a good starting point to get (C_2^*, γ_2^*) at a feasible time cost.

The use of a local optimizer starting close to (C_2^*, γ_2^*) can give a precise estimation of that point. In this regard, we suggest to employ the Nelder–Mead simplex method [63] due to its computational efficiency. This algorithm is a direct search method since it tries to minimize a scalar-valued nonlinear function using only its values and not its derivative. At each iteration, it stores a simplex of $L + 1$ vertices, where L is the dimension of the input space. In the case at hand, the simplex is a triangle because $L = 2$. At each iteration, the function is evaluated 1, 2, or $L + 2$ times, depending on the steps taken by the method [64]. In reality, the steps performing $L + 2$ evaluations are very rare and, thus, most frequent steps perform just 1 or 2 evaluations [64]. Given the RE over (C_2, γ_2) plane and an adequate starting point, the Nelder–Mead method would need few function evaluations to come to the minimum. Nevertheless, this method may sometimes stagnate on a local minimum whose associated RE is higher than expected. In those cases, a better starting point should be used to restart the algorithm. Furthermore, the starting point given by a grid search performed with a reduced number of samples (see Figure 9.10) may drive the Nelder–Mead method to one of those undesired local minimums. This downside may be overtaken by applying Nelder–Mead to look for the minimum using all the training samples in the vicinity of the minimum yielded by the reduced-sample set search. It is also of interest that the Nelder–Mead method avoids searching on large values of C_2 since those values considerably slow down

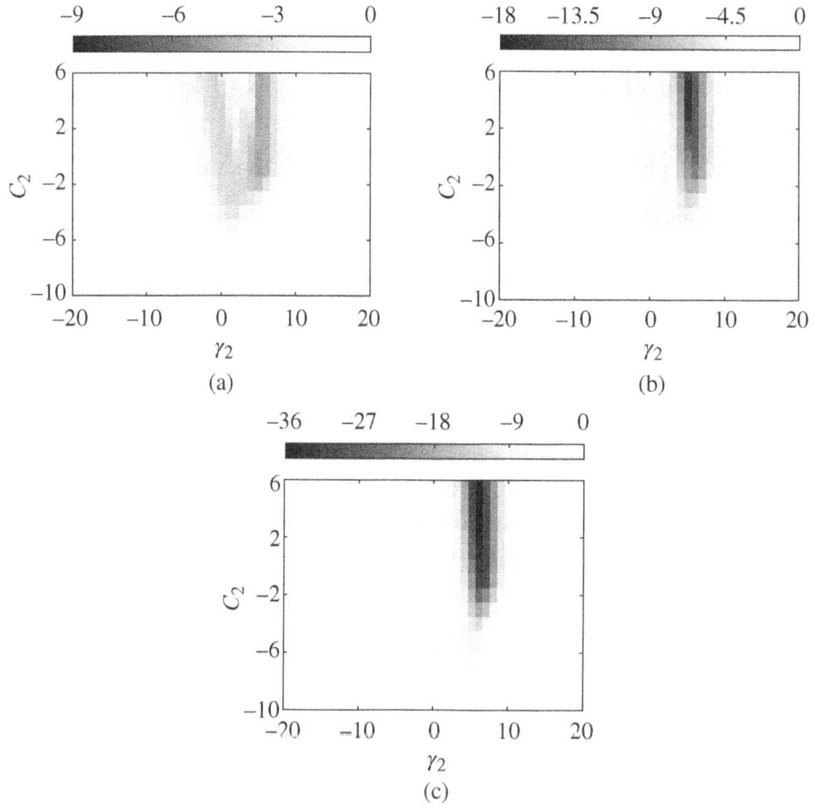

Figure 9.10 Relative error (dB) in estimating $\mathrm{Im}\{\rho_{xy}\}$ for oblique incidence ($\theta = 10^\circ, \varphi = 20^\circ$) over the validation set ($N_v = 375$) when performing an exhaustive grid search in the (C_2, γ_2) plane with grid spacing $\Delta C_2 = \Delta\gamma_2 = 1$, and using (a) 5%, (b) 20%, and (c) 100% samples of the training pattern set ($N_r = 1750$).

the error function evaluation. In principle, the Nelder–Mead method does not allow to perform a bounded search, though it has a variant[1] that inserts a wrapper function around the RE function, allowing a bounded search.

The suggested strategy to find $\left(C_2^*, \gamma_2^*\right)$ may be summarized as follows:

1) Perform a wide grid search, using 20% of the training samples, over $\gamma_2 \in \left[\gamma_{\min}, \gamma_{\max}\right]$ and $C_2 \in \left[C_{\min}, C_{\max}\right]$ with given grid spacings ΔC_2 and $\Delta\gamma_2$. The value of $\left(C_2, \gamma_2\right)$ where the minimum RE on the validation set is attained at this point is noted as $\left(C_2^{1)}, \gamma_2^{1)}\right)$.

1 Matlab function `fminsearchbnd.m` by John D'Errico.

2) Carry out a localized grid search, using all the training samples and the same resolution than in 1, over the vicinity of $\left(C_2^{1)}, \gamma_2^{1)}\right)$. The value of (C_2, γ_2) where the minimum RE on the validation set is achieved at this step is noted as $\left(C_2^{2)}, \gamma_2^{2)}\right)$.
3) If $\left(C_2^{2)}, \gamma_2^{2)}\right)$ provides the requested precision on the validation data, then approximates the optimal parameters as $\left(C_2^* = C_2^{2)}, \gamma_2^* = \gamma_2^{2)}\right)$. Otherwise, use the training data to run the Nelder–Mead algorithm starting at $\left(C_2^{2)}, \gamma_2^{2)}\right)$ and enforcing $C_2 \leq C_{\max}$ to find (C_2^*, γ_2^*) that minimizes the validation error.

Finally, given the selected model (C_2^*, γ_2^*), the error is estimated by means of the error over the test set. Using this algorithm, the model selection time for the example at hand (shown in Figures 9.8 and 9.9) has been reduced from 50,006 to 65 s. The presented efficient grid search strategy yields $C_2^* = 3.8096$ and $\gamma_2^* = 5.7188$ and RE$= -35.32$ dB, over the validation set. Thus, this technique reduces the time cost of the exhaustive grid search about three orders of magnitude while it slightly increases the RE (less than 1 dB in the present example).

9.3 RIS-1: Analysis and Optimization

This section contains a detailed explanation on how to analyze the field radiated by an RIS, either in NF or FF, and then, how to optimize this radiated field to meet specific design/operation requirements.

The starting point of the analysis is the tangential field, which is defined as the tangential component of the electric field onto the RIS aperture, $\vec{E}_{\text{ap}}$, that accounts for the reflected field, $\vec{E}_{\text{ap},r}$. To compute $\vec{E}_{\text{ap}}$, it is assumed that both the incident field, $\vec{E}_{\text{inc}}$, and the response of the unit element of the RIS are known beforehand. Several methods can be applied to compute $\vec{E}_{\text{inc}}$, such as the use of a $\cos^q(\theta)$ function or direct use of full-wave simulations of the feed. Note that the BS is far from the RIS, so the illumination will be almost uniform and any of those approximations is usually valid. Then, the relation between $\vec{E}_{\text{inc}}$ and $\vec{E}_{\text{ap}}$ at any element of the RIS is given by the electromagnetic response of the element, i.e., by its reflection coefficient matrix **R** (defined in Eq. (9.2) and used in Eq. (9.1), where $\vec{E}_{\text{ap}}$ is noted as $\vec{E}_{\text{ref}}$).

There are two methods to compute the **R** matrix: phase-only approximation or complete characterization of the element. The first approximation involves considering the element as an ideal phase shifter, meaning that it only adds a phase shift without losses. This approximation is usually valid when working with the copolar component of the radiated field and a single polarization, as it does not allow for the cross-polar component of the field. For improved accuracy, it is necessary to use more complex methods to fully characterize the response of the element, matrix **R**. In this regard, the element is modeled as a four-port network that becomes a two-port network in reflection (two ports are nullified by the ground plane). This type of analysis is often carried out under local periodicity considerations. When analyzing the whole RIS, this approximation yields accurate results when the variation in the dimensions of the elements changes smoothly across the surface. This ensures that close elements have similar geometries. These approaches are commonly based on implementations of the MoM in the Spectral Domain and Floquet Theorem or electromagnetic software, which allows for the full characterization of the element response in a periodic structure.

Once the tangential field at the RIS aperture is computed, the field radiated by the RIS can be calculated in any of its field regions (either NF or FF) using various methods. Therefore, the response of the element relates $\vec{E}_{\mathrm{inc}}$ to $\vec{E}_{\mathrm{ap}}$ and consequently to the radiated field, $\vec{E}_{\mathrm{rad}}$. Assuming that the incident field is constant along the element surface, and determined by the BS, the radiated field will depend solely on the response of each element (just its phase for ideal phase shifters). Hence, it is possible to obtain collimated or shaped patterns through the appropriate selection of the element phase shift. This approach is usually valid when focused beams (in the NF) or high-gain beams (in the FF) are required, as these solutions can be reached analytically and the beams are typically robust against design errors. However, if more complex beam patterns are required, such as shaped or contoured patterns, or the design must fulfill bandwidth and/or cross-polarization requirements, optimization techniques must be employed.

Within the optimization techniques, two main approaches can be distinguished. On the one hand, there is a direct optimization of the element geometry, which aims to adjust one or more geometric parameters of the element topology to achieve the desired radiation pattern. In this method, a complete characterization of the element is necessary because the optimization variables are geometric parameters that directly affect its electromagnetic response. The main drawback of this method is the high computational cost required to analyze each RIS element during the optimization stage.

On the other hand, a simplification of the above approach is to consider the element, once again, as an ideal phase shifter. Therefore, it is only necessary to obtain the phase shift introduced by each element to achieve the desired radiation pattern. This technique is known as POS and it is widely used in the synthesis of SFA antennas (such as RA, transmitarray, or metasurfaces). It is worth noting that the POS technique is independent of the element geometry since it only synthesizes phase responses. Hence, it requires a subsequent design process where the obtained phase response is physically implemented by adjusting the geometry of the elements. Although POS is widely used and provides successful results, it has some limitations. Unlike direct optimization techniques, its output is not directly the antenna layout. It does not allow simultaneous synthesis of two polarizations or take into account further requirements, such as bandwidth or polarization purity, among others.

9.3.1 Radiated Field by a RIS

Before delving into the analysis of the field radiated by a RIS, either NF or FF, let us define a scenario geometry to better understand the analysis. Figure 9.11a depicts the classical use of RIS antennas to enhance coverage in outdoor scenarios, while Figure 9.11b shows the antenna optics used to support the following analysis. We set the origin of the coordinate system $(\hat{x}, \hat{y}, \hat{z})$ at the center of the RIS aperture. The formulation is generic and independent of the feed placement, which is located at (x_f, y_f, z_f), and the radiated field is computed in the hemisphere $z > 0$ regarding the system shown in Figure 9.11b. The RIS elements are distributed in a 2D regular grid of periodicity p_x and p_y in the $\hat{x}$ and $\hat{y}$ directions, respectively. Then, the pointing direction of the outgoing beam is given by (θ_0, ϕ_0), which is defined by the radiation system $(\hat{x}_r, \hat{y}_r, \hat{z}_r)$, and it is related to $(\hat{x}, \hat{y}, \hat{z})$ by

$$\begin{pmatrix} \hat{x}_r \\ \hat{y}_r \\ \hat{z}_r \end{pmatrix} = \begin{pmatrix} \cos\theta_0 \cos^2\phi_0 + \sin^2\phi_0 & \cos\phi_0 \sin\phi_0 (\cos\theta_0 - 1) & -\sin\theta_0 \cos\phi_0 \\ \cos\phi_0 \sin\phi_0 (\cos\theta_0 - 1) & \cos\theta_0 \sin^2\phi_0 + \cos^2\phi_0 & -\sin\theta_0 \sin\phi_0 \\ \sin\theta_0 \cos\phi_0 & \sin\theta_0 \sin\phi_0 & \cos\theta_0 \end{pmatrix} \cdot \begin{pmatrix} \hat{x} \\ \hat{y} \\ \hat{z} \end{pmatrix}. \tag{9.18}$$

9.3.1.1 Electric Field on the RIS Aperture

According to Huygens' Principle or the Classical Theory of Aperture Antennas [65], the only fields contributing to the radiation of an aperture antenna, like a

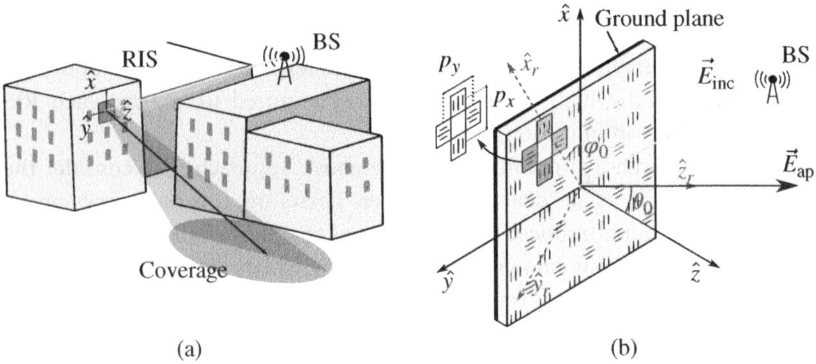

Figure 9.11 (a) Sketch of a RIS used in an outdoor scenario to enhance coverage and (b) its antenna optics definition.

RIS, are the tangential fields, $\vec{E}_{ap}$. To calculate the radiated field, $\vec{E}_{rad}$, we start with the incident field emitted by the BS, which can be expressed for a point (x, y) on the antenna surface as shown in Eq. (9.19a). This expression is valid for both linear polarizations radiated by the BS (either V-polarization, V-pol or H-polarization, H-pol, which are normally associated with the $\hat{x}$ and $\hat{y}$ directions, respectively), and allows us to decompose $\vec{E}_{inc}(x, y)$ into the two Cartesian components projected onto the RIS surface. Similarly, the tangential field at a point (x, y) on the aperture surface can be expressed in terms of its Cartesian components (see Figure 9.11b) [22].

$$\vec{E}_{inc}^{X/Y}(x, y) = E_{inc,x}^{X/Y}(x, y)\hat{x} + E_{inc,y}^{X/Y}(x, y)\hat{y}, \tag{9.19a}$$

$$\vec{E}_{ap}^{X/Y}(x, y) = E_{ap,x}^{X/Y}(x, y)\hat{x} + E_{ap,y}^{X/Y}(x, y)\hat{y}. \tag{9.19b}$$

As mentioned earlier, the relationship between $\vec{E}_{inc}$ and $\vec{E}_{ap}$ is defined by the response of the element. In this case, we will model the element as a four-port network (as shown in Figure 9.12), and its response is given by **A**. The two inputs are $\vec{a}_1$ and $\vec{a}_2$, which are defined by Eqs. (9.20a) and (9.20b), and the two outputs $\vec{b}_1$ and $\vec{b}_2$ defined by Eqs. (9.20c) and (9.20d).

$$\vec{a}_1 = a_{x,1}\hat{x} + a_{y,1}\hat{y}, \tag{9.20a}$$

$$\vec{a}_2 = a_{x,2}\hat{x} + a_{y,2}\hat{y}, \tag{9.20b}$$

$$\vec{b}_1 = b_{x,1}\hat{x} + b_{y,1}\hat{y}, \tag{9.20c}$$

$$\vec{b}_2 = b_{x,2}\hat{x} + b_{y,2}\hat{y}. \tag{9.20d}$$

The relationship between the inputs ($\vec{a}_1$ and $\vec{a}_2$) and outputs ($\vec{b}_1$ and $\vec{b}_2$) is given by the matrix **A**, which represents the response of the network (or RIS element). This relationship can be expressed in matrix form as follows

$$\begin{pmatrix} \vec{b}_1 \\ \vec{b}_2 \end{pmatrix} = \mathbf{A} \cdot \begin{pmatrix} \vec{a}_1 \\ \vec{a}_2 \end{pmatrix}. \tag{9.21}$$

In principle, matrix **A** is a generic matrix that allows defining the behavior in terms of transmission and/or reflection, resulting in a combination of

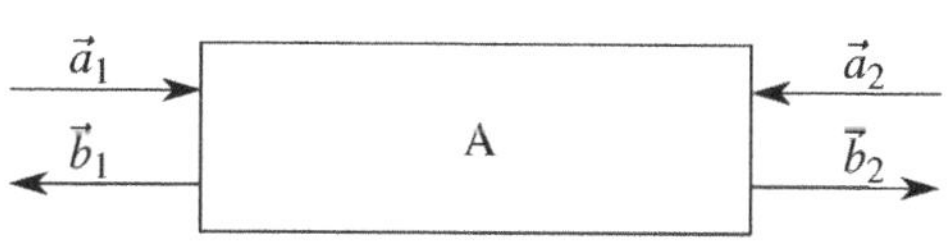

Figure 9.12 Element of the RIS modeled as a four-port network.

transmission matrix (**T**) and reflection matrix (**R**). Therefore, **A** can be expressed as

$$\mathbf{A} = \begin{pmatrix} \mathbf{R} & \mathbf{T} \\ \mathbf{T} & \mathbf{R} \end{pmatrix}, \tag{9.22}$$

where **T** and **R** matrices characterize the complete element in transmission or reflection, since they are composed of direct coefficients (ρ_{xx}, ρ_{yy}, τ_{xx}, and τ_{yy}) and cross-coefficients (ρ_{xy},ρ_{yx}, τ_{xy}, and τ_{yx}).

In this case, let us consider that the antenna is transmitting, $\vec{a}_2 = 0$, thus Eq. (9.22) can be expressed in terms of the incident and tangential fields expressed in Eqs. (9.19a) and (9.19b), just by considering that $\vec{a}_1 = \vec{E}_{\text{inc}}$, $\vec{b}_1 = \vec{E}_{\text{ap},r}$, and $\vec{b}_2 = \vec{E}_{\text{ap},t}$, yielding:

$$\begin{pmatrix} E_{\text{ap},r,x} \\ E_{\text{ap},r,y} \\ E_{\text{ap},t,x} \\ E_{\text{ap},t,y} \end{pmatrix} = \begin{pmatrix} \rho_{xx} & \rho_{xy} \\ \rho_{yx} & \rho_{yy} \\ \tau_{xx} & \tau_{xy} \\ \tau_{yx} & \tau_{yy} \end{pmatrix} \cdot \begin{pmatrix} E_{\text{inc},x} \\ E_{\text{inc},y} \end{pmatrix}. \tag{9.23}$$

It is important to note that Eq. (9.23) is valid for transmission and reflection. In the case of transmission (for example, in a transmissive RIS), it allows characterizing the response in transmission of the element while also considering the reflection produced within the element, such as that due to an impedance mismatch. In the case of a conventional RIS, which operates only in reflection and has a ground plane, the transmission terms (τ) would be null. Similarly, the field $E_{\text{ap},r,x/y}$ is associated with the reflected tangential field, while the field $E_{\text{ap},t,x/y}$ is associated with the transmitted tangential field.

Henceforth and for the shake of simplicity, we will only focus on the analysis of reflective RIS, which allows us to eliminate the terms τ and the field $\vec{E}_{\text{ap},t}$. In the same way, the matrix **A** is reduced to a single reflection matrix **R**. Since there are different methods to obtain the reflection coefficients, we will define the matrix **R** based on the calculation method. More specifically, in this chapter, we will focus on three cases: the $\mathbf{R}^{\text{MoM}}$ calculated using MoM, $\mathbf{R}^{\text{SVR}}$ computed using SVR or SVM models, and $\mathbf{R}^{\text{PO}}$, which considers the elements as ideal phase shifters and only includes direct coefficients ($\rho_{xy} = \rho_{yx} = 0$). In addition, ρ_{xx} and ρ_{yy}, as they do not introduce losses, will be simplified to a phase term $e^{j\phi_{xx}}$ or $e^{j\phi_{yy}}$, while $|\rho_{xx}| = |\rho_{yy}| = 1$, as follows

$$\mathbf{R}^{\text{MoM/SVR}} = \begin{pmatrix} \rho_{xx} & \rho_{xy} \\ \rho_{yx} & \rho_{yy} \end{pmatrix} \rightarrow \mathbf{R}^{\text{PO}} = \begin{pmatrix} e^{j\,\phi_{xx}} & 0 \\ 0 & e^{j\,\phi_{yy}} \end{pmatrix} \tag{9.24}$$

It is worthy noting that either $\mathbf{R}^{\text{MoM}}$ or $\mathbf{R}^{\text{SVR}}$ provide a more accurate response. Both matrices are obtained under local periodicity assumption so that they consider the losses due to the substrate and the mutual coupling between elements. Moreover, both matrices factor in the field re-radiated by the element's metallization, as well as the field reflected by the substrate and ground plane [50, 66].

9.3.1.2 Radiated Field of an RIS

Here are different methods to calculate the radiated field distribution of a RIS. Primarily, being an aperture antenna, any of the existing methods can be used. For the analysis, we take the generic aperture shown in Figure 9.13 as reference, which has uniform polarization with the electric field vector in the $\hat{x}$ direction, and its tangential field for a point (x, y) is $\vec{E}_{\text{ap}}$. If the aperture is much larger than the wavelength at the operating frequency, the radiated field can be computed as follows [67]:

$$\vec{E}_{\text{rad}}(\vec{r}) = \frac{1}{4\pi}\int_S \vec{E}_{\text{ap}}\frac{e^{-jkr}}{r}\left[\left(jk + \frac{1}{r}\right)\vec{u}_z \cdot \vec{r} + jk\vec{u}_z \cdot \vec{s}\right] \mathrm{d}_x \mathrm{d}_y. \tag{9.25}$$

This expression allows to calculate the radiated field at an observation point $\vec{r}$, regardless of whether it is in the NF or FF region. $\hat{u}_z$ is the unit vector normal to the aperture, in this case, it is defined along the $\hat{z}$-direction, while $\vec{s}$ is the unit vector normal to the wavefront. Depending on the distance from the observation point and the size of the aperture (D), three radiated field regions are defined [68]. If $r < 0.62\sqrt{D^3/\lambda}$, the observation point is in the reactive NF, whereas if the distance falls between $0.62\sqrt{D^3/\lambda} < r < 2D^2/\lambda$, the region is defined as radiated NF or Fresnel region. Finally, if $r > 2D^2/\lambda$, the observation point is in the FF or Fraunhofer region. If the field is computed in the reactive NF region, Eq. (9.25) might be solved exactly. However, when calculating the field in the Fresnel or Fraunhofer regions, Eq. (9.25) can be simplified. Particularly, for the Fresnel region, the term $1/r$ inside the parentheses tends to cancel out when compared to k, and the same term outside the parentheses can be approximated by $1/R$. Moreover, the dot product $\vec{u}_z \cdot \vec{r}$ becomes $\vec{u}_z$, and R can be approximated by $\cos\theta$ [67]. Similarly, the variation of r across the aperture in terms of phase e^{jkr} can be expressed by a linear and quadratic term. Those considerations simplify

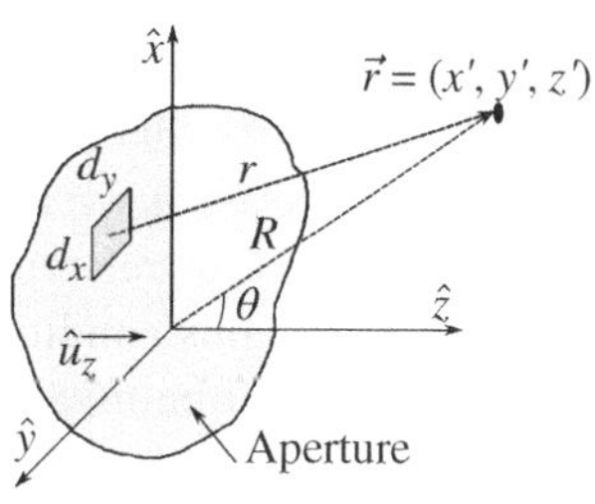

Figure 9.13 Scheme of an arbitrary aperture.

Eqs. (9.25) and (9.26). Then, the phase term (9.27) can be expressed as a Taylor expansion, where the terms above the first order can be neglected [67].

$$\vec{E}_{\text{rad}}(\vec{r}) \approx \frac{j}{2\lambda}\frac{e^{-jkz}}{R}\int_S \vec{E}_{\text{ap}} e^{-jk\left(\frac{(x-x')^2+(y-y')^2}{2z}\right)}\left(\cos\theta + \vec{u}_z \cdot \vec{s}\right) d_x d_y, \tag{9.26}$$

$$r \approx z + \left(\frac{(x-x')^2+(y-y')^2}{2z}\right). \tag{9.27}$$

If we consider that $R \to \infty$ and limit the observation area to around $\hat{z}$ such that $z \approx R$ and $\cos\theta = 1$, (9.26) simplifies to (9.28), which is the well-known expression for the FF.

$$\vec{E}_{\text{FF}}(\vec{r}) \approx \frac{j}{\lambda}\frac{e^{-jkR}}{R}\int_S \vec{E}_{\text{ap}} e^{-jk\sin\theta(x\cos\phi + y\sin\phi)} d_x d_y. \tag{9.28}$$

Both (9.26) and (9.28), whether for the NF or FF, can be applied to calculate the field radiated by the RIS. However, (9.26) requires high computational cost due to the integration of the quadratic phase term (9.27). Therefore, an alternative method is proposed for calculating the NF of the RIS. In this case, it is suggested to calculate the NF using a combination of Huygens' Principle along with the Principle of Superposition. To do so, the aperture of the RIS is divided into subdomains. Typically, each subdomain is associated with an element of the RIS, so the aperture is subdivided into $N_s = N_x \times N_y$ subdomains. Moreover, each of these subdomains is treated as a small radiating aperture, henceforth referred to as a subaperture. Now, the NF radiation from the RIS at $\vec{r}$ is the contribution of each of the subapertures at $\vec{r}$. Hence, the NF of the RIS ($\vec{E}_{\text{NF}}$) can be defined according to (9.29) [69].

$$\vec{E}_{\text{NF}}(\vec{r}) = \sum_{i=1}^{N_s} \vec{E}_{\text{NF},i}(\vec{r}). \tag{9.29}$$

The calculation of the contribution of the subaperture ith ($\vec{E}_{\text{NF},i}$) can be obtained by applying the Classical Theory of Planar Apertures for NF calculations [65]. However, as is the case with (9.26), this formulation is relatively complex and requires high computational cost. It should be noted that this computation needs to be applied for each element of the RIS (N_s times), which is typically composed of hundreds or even thousands of elements. To alleviate the computational cost, we consider that the observation point $\vec{r}$ is within the Fresnel region of the RIS. However, if we take a single subaperture (or subdomain) as the only radiating source of reference, $\vec{r}$ will be in the FF of the subaperture. As will be seen below, this condition allows to simplify the formulation at the cost of establishing a minimum distance at which the model is valid (9.30). Nevertheless, this should not impose a limitation on applicability since the RIS is intended for communications and coverage enhancement applications, where it is typical for the user to be at a much greater distance than $|\vec{r}_{\min}|$ ($|\vec{r}| \gg |\vec{r}_{\min}|$).

$$|\vec{r}_{\min}| = \frac{2\left(p_x^2 + p_y^2\right)}{\lambda_0} + \lambda_0, \tag{9.30}$$

$$\vec{E}_{\mathrm{NF}}(\vec{r}) = \sum_{l=1}^{N_s} \vec{E}_{\mathrm{FF},l}(\vec{r}). \tag{9.31}$$

With this consideration, (9.29) and (9.30) lead to (9.31), where the contribution of the ith subaperture can be calculated using the Equivalence Principles as $\vec{r}$ pertains to the FF [65]. To compute the radiation of the ith subaperture, we set Figure 9.14 as reference. The observation point referred to the center of the ith subaperture is given by $\vec{r}$, which is defined as (9.32). Therefore, the contribution $\vec{E}_{\mathrm{FF},i}$ can be expressed in terms of the propagation from the center of the ith subaperture to the observation point $\vec{r}_i$, and the active element pattern $\vec{F}(\theta_i, \varphi_i)$.

$$\vec{r}_i = \vec{r} - \vec{r}_{\mathrm{sap},i} = r_i \hat{r}_i, \tag{9.32}$$

$$\vec{E}_{\mathrm{FF},i} = \frac{jk}{4\pi}\frac{e^{-jkr_i}}{r_i}\vec{F}(\theta_i, \varphi_i). \tag{9.33}$$

Now, the calculation of the contribution of the ith subaperture can be computed using Love's Equivalence Principle [65] (9.34), allowing us to express $\vec{E}_{\mathrm{FF},i}$ in terms of its θ_i- and φ_i- components (9.35a)–(9.35b).

$$\vec{E}_{\mathrm{FF},i}(\vec{r}) = \frac{jk}{4\pi}\frac{e^{-jkr_i}}{ri}\left(E_{\theta_i}\hat{\theta}_i + E_{\varphi_i}\hat{\varphi}_i\right), \tag{9.34}$$

$$E_{\theta_i} = P_{x_i}\cos\varphi_i + P_{y_i}\sin\varphi_i - \eta\cos\theta\left(Q_{x_i}\sin\varphi_i - Q_{y_i}\cos\varphi_i\right), \tag{9.35a}$$

$$E_{\varphi_i} = \cos\theta_i\left(P_{y_i}\cos\varphi_i - P_{x_i}\sin\varphi_i\right) - \eta\left(Q_{y_i}\sin\varphi_i + Q_{x_i}\cos\varphi_i\right). \tag{9.35b}$$

Each of these components requires computing the spectrum functions P_x, P_y, Q_x, and Q_y using (9.36a)–(9.36d), which are directly related to the electric and magnetic tangential fields ($\vec{E}_{\mathrm{ap}}$ and $\vec{H}_{\mathrm{ap}}$) in the integration domain S (the subaperture

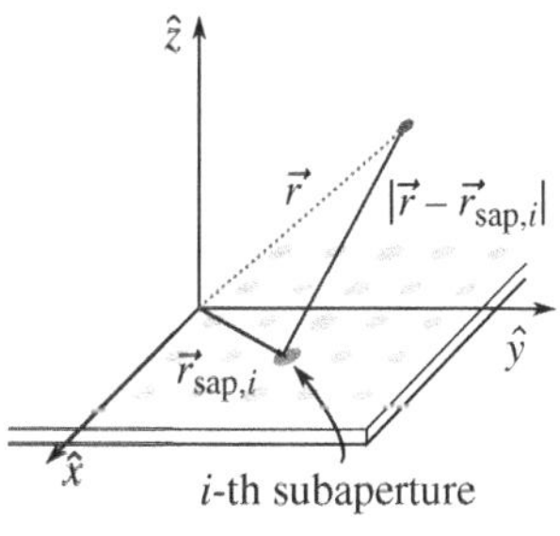

Figure 9.14 Reference system to compute the contribution of a single subaperture of the RIS.

surface in the (x', y') plane).

$$P_x(x, y) = \iint_S E_{ap,x}(x', y') e^{-jk_0(x' \sin\theta\cos\varphi + y' \sin\theta\sin\varphi)} dx' dy', \tag{9.36a}$$

$$P_y(x, y) = \iint_S E_{ap,y}(x', y') e^{-jk_0(x' \sin\theta\cos\varphi + y' \sin\theta\sin\varphi)} dx' dy', \tag{9.36b}$$

$$Q_x(x, y) = \iint_S H_{ap,x}(x', y') e^{-jk_0(x' \sin\theta\cos\varphi + y' \sin\theta\sin\varphi)} dx' dy', \tag{9.36c}$$

$$Q_y(x, y) = \iint_S H_{ap,y}(x', y') e^{-jk_0(x' \sin\theta\cos\varphi + y' \sin\theta\sin\varphi)} dx' dy'. \tag{9.36d}$$

It is worth noting that it is common to use the Second Equivalence Principle as it only requires the calculation of P_x and P_y (which depend on the electric tangential field), eliminating the dependence on the magnetic tangential field $\vec{H}_{ap}$.

In the case of the FF calculation, (9.28) provides a computationally efficient solution. This expression can be also written in terms of the Love's First Equivalence Principle [24], similar to NF (9.34)–(9.35b). However, the spectrum functions for the θ- and φ- components no longer refer to a subaperture (subdomain) of the RIS but to the entire aperture. Therefore, the spectrum functions can be calculated using (9.36a)–(9.36d) while considering that the integration domain S is the whole RIS [24].

Considering that $u = k_0 \sin\theta\cos\varphi$ and $v = k_0 \sin\theta\sin\varphi$, the spectrum functions can be rewritten as (9.37a)–(9.37b). Now, it is readily seen that the spectrum functions are as the Fourier transform of the electric and magnetic tangential field in the RIS aperture.

$$P_{x/y}(u, v) = \iint_S E_{ap,x/y}(x', y') e^{-jk_0(ux' + vy')} dx' dy', \tag{9.37a}$$

$$Q_{x/y}(u, v) = \iint_S H_{ap,x/y}(x', y') e^{-jk_0(ux' + vy')} dx' dy'. \tag{9.37b}$$

The integration domain S is divided into N_s subdomains (note that $N_s = N_x \times N_y$ elements with periodicity $p_x \times p_y$), it is possible to discretize the integration of S, so that (9.37) can be rewritten as (9.38a) and (9.38b). Note that F is a constant associated with the active element pattern [24].

$$P_{x/y}(u, v) = F \cdot \sum_{m=1}^{N_x} \sum_{n=1}^{N_y} E_{ap,x/y}(x_m, y_n) e^{-jk_0(ux_m + vy_n)}, \tag{9.38a}$$

$$Q_{x/y}(u, v) = F \cdot \sum_{m=1}^{N_x} \sum_{n=1}^{N_y} H_{ap,x/y}(x_m, y_n) e^{-jk_0(ux_m + vy_n)}. \tag{9.38b}$$

The double summation in (9.38a) and (9.38b) can be expressed as a 2D inverse discrete Fourier transform (IDFT2) (9.39a)–(9.39b), making it possible to apply the 2D inverse fast Fourier transform and achieve a very efficient radiation pattern calculation [24].

$$P_{x/y}(u,v) = N_s \cdot F \cdot \text{IDFT2}\left[E_{\text{ap},x/y}(x_m, y_n)\right], \tag{9.39a}$$

$$Q_{x/y}(u,v) = N_s \cdot F \cdot \text{IDFT2}\left[H_{\text{ap},x/y}(x_m, y_n)\right]. \tag{9.39b}$$

Finally, from the FF in spherical coordinates, it is possible to calculate the copolar and cross-polar components by applying Ludwig's Third Definition.

9.3.2 Intersection Approach Framework

Direct optimization methods are based on defining a function, which serves as the error cost function, and seeking its optimization. However, the intersection approach (IA) is an alternative based on finding the intersection of two sets [70]. In particular, one set is composed of all radiated fields (NF or FF) that meet the desired specifications ($\mathcal{M}$), and the second set is made up of all fields that can be radiated by the antenna's optics ($\mathcal{R}$). The IA aims to find the intersection of these two sets using an iterative process based on alternate projections. For this purpose, the following operations are defined in (9.40) for each iteration of the IA.

$$\vec{E}_{\text{rad}}^{\,i+1} = \mathcal{B}\left[\mathcal{F}\left(\vec{E}_{\text{rad}}^{\,i}\right)\right]. \tag{9.40}$$

The forward projector $\mathcal{F}$ is defined, which allows projecting a point from $\mathcal{R}$ onto the set $\mathcal{M}$, while the backward projector $\mathcal{B}$ retrieves a point from $\mathcal{M}$ onto $\mathcal{R}$. Both projectors are applied to the radiated field $\vec{E}_{\text{rad}}$ calculated in iteration ith. Using (9.40), the IA seeks the intersection between $\mathcal{M}$ and $\mathcal{R}$, meaning a point (radiated field) that belongs simultaneously to both sets (can be radiated by the antenna's optics and meets the specifications). If an intersection is not possible, the IA will aim for at least one radiated field whose distance to set $\mathcal{M}$ is minimized.

The IA can be easily explained in terms of the electric tangential field $\vec{E}_{\text{ap}}$ and the radiated field ($\vec{E}_{\text{rad}}$), as shown in Figure 9.15. As in most local optimization algorithms, an initial starting point is required, which is associated with the tangential field $\vec{E}_{\text{ap}}^{\,0}$ calculated using (9.23) but considering only the reflection matrix $\mathbf{R}$, either $\mathbf{R}^{\text{MoM/SVR}}$ or $\mathbf{R}^{\text{PO}}$ (9.24).

In both cases, the optimization variables are given or related to the matrix $\mathbf{R}$. If the POS technique is applied, where the elements are modeled as ideal phase shifters, the optimization variables or DoFs of the synthesis for linear polarization X are $\overline{f_X} = \{\phi_{xx}^1, \dots, \phi_{xx}^l, \dots, \phi_{xx}^{N_s}\}$, while the variables for polarization Y would be $\overline{f_Y} = \{\phi_{yy}^1, \dots, \phi_{yy}^l, \dots, \phi_{yy}^{N_s}\}$. Since ideal phase shifters do not establish any relationship with the element topology, control of the response for each

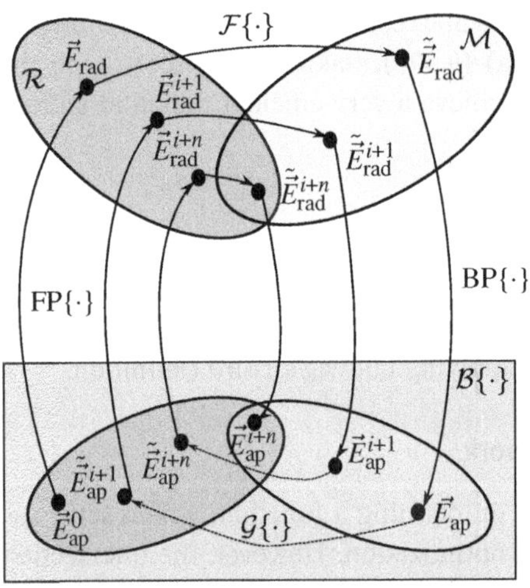

Figure 9.15 Flowchart of the classical intersection approach.

polarization must be done independently. However, if $\mathbf{R}^{\mathrm{MoM}}$ or $\mathbf{R}^{\mathrm{SVR}}$ matrices are used, the phase of the reflection coefficients is not synthesized; instead, direct optimization of the geometry of the element is performed. In this case, it is important to consider that a topology or geometry of the element (such as patches, dipoles, split rings, etc.) must be established. The optimization variables (DoFs) are determined by the geometric parameters of the element. This means that the optimization variables will now be $\overline{f} = \{T_1, \dots, T_L, \dots, T_M\}$, each of them representing a physical parameter of the cell. For example, for an element based on a single patch (as shown in Figure 9.16), each side of the patch can be set as a DoF, so $\overline{f} = \{T_1(W_x), T_2(W_y)\}$, where T_1 is associated with W_x and T_2 to W_y.

As the optimization variables modify the geometry of the element, the electromagnetic response is directly affected, and thus, $\mathbf{R}$ (with all its elements) is also modified. Therefore, it is necessary to include an analysis within the optimization process that allows us to get the $\mathbf{R}$ matrix associated with the geometry of the nth

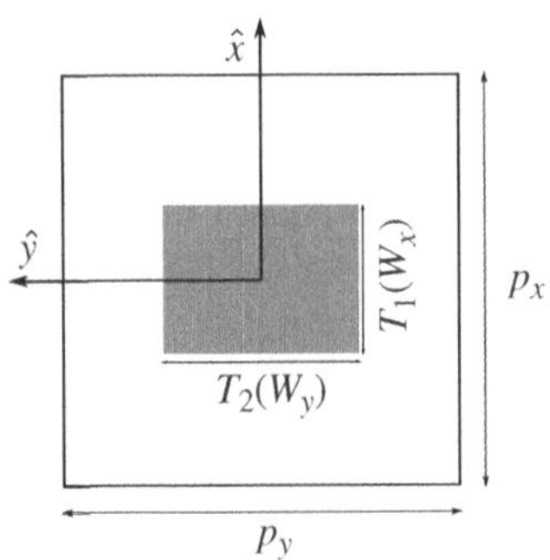

Figure 9.16 Example of a RIS element based on a patch topology considering the optimization variables T_1 and T_2.

element. This analysis is typically based on MoM considering local periodicity (MoM–LP), which is much more precise than the phase-only consideration because it also takes into account other effects, such as the angle of incidence. In this way, for the nth element of the RIS, a matrix $\mathbf{R}^n(\overline{f}, \theta, \phi)$ will be calculated, which is a function of the element geometry and angle of incidence (θ, ϕ). As expected, the analysis of $\mathbf{R}$ must be performed element by element at each iteration of the optimization process, significantly increasing the computational cost of the IA.

An alternative that figures on all the considerations mentioned but improves the efficiency of calculating $\mathbf{R}^n$, is to use surrogate modes of the cell response. In particular, SVR allows reducing the calculation of all matrices $\mathbf{R}^{\text{SVR}}$ by more than three orders of magnitude, exhibiting a high degree of concordance with MoM results [50].

It is worth noting that using $\overline{f}$ as optimization variables, allows to control the two linear polarizations simultaneously, as well as to reduce cross-polarization or to perform in-band enhancements.

Returning to the flowchart in Figure 9.15, the radiated field $\vec{E}_{\text{rad}}$ is calculated from the tangential field $\vec{E}_{\text{ap}}^{0}$ using a forward propagation (FP) model. This FP is defined as a propagation model that relates the tangential fields in the aperture to the radiated fields, i.e., (9.31) in the case of imposing conditions in the NF or (9.28) in the case of FF. It is common for the radiated field E_{rad} to belong to the set $\mathcal{M}$ but not meet the desired specifications. Therefore, the forward projector $\mathcal{F}$ is applied to obtain the projection of $\vec{E}_{\text{rad}}$ onto the set $\mathcal{R}$. This projection $\tilde{\vec{E}}_{\text{rad}}$ is a radiated field that meets the specifications. It is essential to note that having a radiated field that meets the specifications does not necessarily mean that the field can be radiated by the antenna. The specifications are imposed in terms of the radiated field but not on the antenna's optics. Starting from the field $\tilde{\vec{E}}_{\text{rad}}$, the backward projector $\mathcal{B}$ is applied, which is divided into different steps. First, a backward propagation (BP) must be applied to retrieve the source that generates $\tilde{\vec{E}}_{\text{rad}}$, i.e., the tangential field $\vec{E}_{\text{ap}}$. When working in the FF and, in some NF approaches, the FP–BP implementation is done using inverse operators, such as the fast Fourier transform (FFT)/inverse Fourier transform (IFFT) [42, 71]. This efficient implementation is typical in the IA since FP and BP are the most computationally costly operations. The tangential field $\vec{E}_{\text{ap}}$ retrieved from $\tilde{\vec{E}}_{\text{rad}}$ might not be achievable by the RIS optics. The tangential field depends on $\vec{E}_{\text{inc}}$ (which is fixed by the BS) and the response of the element given by $\mathbf{R}$. The $\mathbf{R}^n$ to reach a given $\vec{E}_{\text{ap}}$ might yield a not feasible geometry. Therefore, constraints are added through the process $\mathcal{G}$ on the antenna geometry, so that a feasible field, $\tilde{\vec{E}}_{\text{ap}}$, similar to the required $\tilde{\vec{E}}_{\text{ap}}$, will radiate a new $\vec{E}_{\text{rad}}$ that is closer to meeting the specifications than in the last guess. This process is done iteratively until $\vec{E}_{\text{rad}}$ meets both conditions.

As mentioned earlier, the backward projector is typically implemented using direct/inverse operators, such as the FFT/IFFT. This implementation is commonly referred to as the classic version of the IA. However, this implementation has some disadvantages. First, in the case of the NF, the classic version based on FFT/IFFT can only impose field constraints on parallel planes to the aperture, which can be a significant limitation, especially for the RIS design. Typically, the shaped coverage of the RIS does not meet this specification, so alternatives must be sought. Second, the IA is a local optimization algorithm, which means it is not exempt from the presence of local minima, also known as traps, which can endanger the convergence to a valid solution. There are mainly two sources of traps in the IA: the number of DoFs and the non-convexity of the used sets. The first of these problems is easily solvable by reducing the number of DoFs in the early stages of the synthesis process. In the case of a POS, $\overline{f}_{X/Y}$ is usually reduced using the fictitious reduction technique [22], which is based on fictitiously reducing $\overline{f}_{X/Y}$ by setting a low edge taper illumination on the RIS. Since the elements with low illumination barely contribute to the radiated field, they might be ignored during the optimization process without compromising the accuracy of the results. While this technique provides good results in the synthesis of SFA antennas or conventional arrays, it is more limited in RIS synthesis. The RIS is illuminated by a BS, which is typically located at several tens or hundreds of meters, therefore, the illumination taper is usually much higher than in SFA antennas limiting the fictitious reduction achievable. Conversely, in the direct layout optimization, this technique is not necessary since $\overline{f}$ (geometric variables of the element) are directly selected.

Regarding the nonconvexity of the sets, the solution is somewhat more complex. The problem can be addressed by working with radiated intensity ($|\vec{E}_{\text{rad}}|^2$) instead of radiated field ($|\vec{E}_{\text{rad}}|$), or in the case of FF, directly with the antenna gain (G) (9.41) [42]. The proportionality factor between gain and radiated intensity remains constant throughout the synthesis process, so the degree of nonlinearity of the optimization problem does not change. Using the radiated intensity (or the gain) helps to improve the convergence of the IA by convexifying set $\mathcal{M}$ so that traps are alleviated [42]. This solution, although effective, directly leads to the need for redefining the backward projector since, if its original form is used, then it is not possible to retrieve the field at the aperture from the corresponding radiated intensity (or gain) by an IFFT, which results in a significant efficiency loss. The alternative to the classic version of the IA is to implement the recovery of the field at the aperture through an optimization algorithm, giving rise to the generalized intersection approach (GIA) [42, 72].

$$G \propto |\vec{E}_{\text{rad}}|^2. \tag{9.41}$$

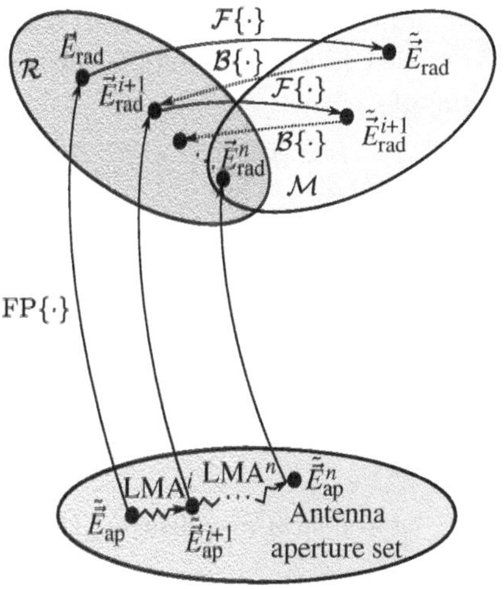

Figure 9.17 Sketch of the generalized intersection approach integrating the LM algorithm and its iterative process.

9.3.3 Generalized Intersection Approach

In this approach, the Levenberg–Marquardt (LM) algorithm is used to implement the backward projector $\mathcal{B}$ [72], modifying the flowchart from Figure 9.15 to 9.17. First, we focus on the projector $\mathcal{F}$, which is common to both the classic and generalized versions. This operator projects $\vec{E}_{\text{rad}}$ (belonging to $\mathcal{R}$) onto a point in the set $\mathcal{M}$ (field that meets the specifications, $\tilde{\vec{E}}_{\text{rad}}$). Normally, specifications are defined based on two templates, which define the lower (T_{lwr}) and upper (T_{upr}) limits of the constraints. The operator $\mathcal{F}$, to obtain $\tilde{\vec{E}}_{\text{rad}}$, performs a field trimm on $\vec{E}_{\text{rad}}$ according to T_{lwr} and T_{upr}, following the condition given in (9.42). This operation is performed point by point and can be implemented according to (9.43). Figure 9.18 graphically shows the trimming operation performed by $\mathcal{F}$ over $|\vec{E}_{\text{rad}}(\vec{r})|^2$. Note that the operator $\mathcal{F}$ is independent of whether it is the generalized or classic version. It can be applied to either depending on whether working with the radiated

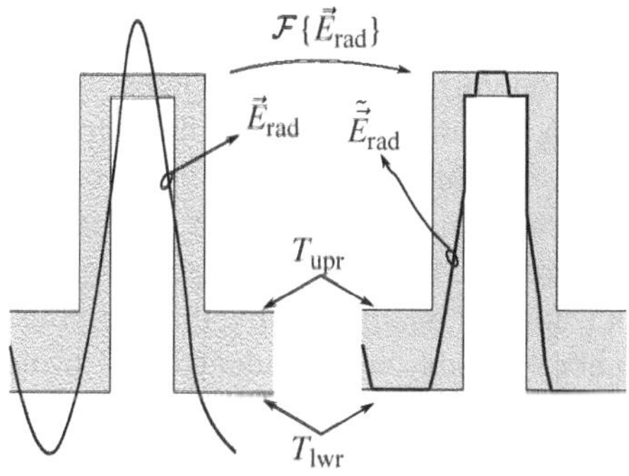

Figure 9.18 Projection of the field $|\vec{E}_{rad}(\vec{r})|^2$ (set $\mathcal{R}$) onto the set of valid radiated fields $\mathcal{M}$, resulting on $|\tilde{\vec{E}}_{rad}(\vec{r})|^2$.

field or the radiated intensity, just replacing $T^2_{\text{upr/lwr}}$ by $T_{\text{upr/lwr}}$. Similarly, it might be applied directly to gain (G) instead of radiated intensity, defining the masks in terms of G.

$$T^2_{\text{lwr}}(\vec{r}) \leq |\tilde{\vec{E}}_{\text{rad}}(\vec{r})|^2 \leq T^2_{\text{upr}}(\vec{r}), \tag{9.42}$$

$$\mathcal{F}\{|\vec{E}_{\text{rad}}(\vec{r})|^2\} = \begin{cases} T^2_{\text{upr}}(\vec{r}), & \text{if} \quad |\vec{E}_{\text{rad}}(\vec{r})|^2 \geq T^2_{\text{upr}}(\vec{r}) \\ T^2_{\text{lwr}}(\vec{r}), & \text{if} \quad |\vec{E}_{\text{rad}}(\vec{r})|^2 \leq T^2_{\text{lwr}}(\vec{r}) \\ |\vec{E}_{\text{rad}}(\vec{r})|^2, & \text{otherwise} \end{cases}. \tag{9.43}$$

To apply the LM algorithm as the backward projector $\mathcal{B}$, the GIA defines a measurement of the distance, d, from a point in the set $\mathcal{R}\,(r)$ to another in $\mathcal{M}\,(m)$ noted as follows:

$$d \rightarrow \text{dist}(m, r) = \text{dist}(|\tilde{\vec{E}}_{\text{rad}}|^2, |\vec{E}_{\text{rad}}|^2). \tag{9.44}$$

The Euclidean weighted norm for square-integrable functions serves as a distance definition, and it can be easily implemented through the use of the weighted Euclidean metric (9.45). A widely used definition for functional d is based on a weighted Euclidean metric where a weighting function w is used to leverage the square of the difference between $|\tilde{\vec{E}}_{\text{rad}}|^2$ and $|\vec{E}_{\text{rad}}|^2$ along the integration domain.

$$d(\vec{r}) = \text{dist}^2\left(|\tilde{\vec{E}}_{\text{rad}}(\vec{r})|^2, |\vec{E}_{\text{rad}}(\vec{r})|^2\right) = \int_{\Omega} w(\vec{r})\left(|\tilde{\vec{E}}_{\text{rad}}(\vec{r})|^2 - |\vec{E}_{\text{rad}}(\vec{r})|^2\right)^2 d\Omega. \tag{9.45}$$

The functional d is defined for the entire space of the radiated field calculation, with $\vec{r}$ being an observation point within Ω (observation domain wherein the radiated field is computed). The evaluation of d is performed point by point, so Ω can be discretized into N_Ω. N_Ω will depend on whether $\vec{E}_{\text{rad}}$ is in the NF region, $\vec{r} \in (x, y, z)$, or in the FF, where $\vec{r}$ is typically represented as $\vec{r} \in (\theta, \varphi)$ or $\vec{r} \in (u, v)$. In either case, (9.45) can be solved numerically, allowing d to be calculated as (9.46) (where $\Delta_{\Omega,i}$ is the sampling interval used). (9.46) is similar to (9.47), which means the square of a residual $z(x)$ discretized over M points. The relationship between (9.46) and (9.47) is important since (9.46) can be defined as a cost function for a local optimizer, such as LM. Observing the flowchart of Figure 9.17, the LM algorithm operates on the set of fields in the aperture. Note that constraints on the recovered tangential field $\vec{E}_{\text{ap}}$ have been omitted since they are not necessary in this case. The LM starts with a tangential field and seeks to reduce d using the DoFs defined by $\bar{f}_{X/Y}$ or $\bar{f}$, depending on whether it is a POS or direct layout optimization. It is worth noting that with this implementation, the tangential field is not retrieved from the radiated field, avoiding the need for implementing the BP. Instead, the LM algorithm directly operates on the tangential field by the relation given in (9.23) with $\mathbf{R}$ (either $\mathbf{R}^{\text{POS}}$ or $\mathbf{R}^{\text{MoM/SVR}}$).

The LM goal is to progressively reduced d in each iteration. Therefore, in each IA iteration, only a few LM iterations are needed (typically, three iterations of LM are sufficient). Broadly speaking, the starting point of the LM is the radiated field in the ith iteration of the IA ($\vec{E}_{\text{rad}}^{i}$), and the LM outputs the radiated field in the $(i+1)$th iteration ($\vec{E}_{\text{rad}}^{i+1}$) (by using $\overline{f}_{X/Y}$ or $\overline{f}$ as optimization variables), which should be closer to $\tilde{\vec{E}}_{\text{rad}}$. After several iterations of the IA, the intersection between sets $\mathcal{M}$ and $\mathcal{R}$ might be achieved, resulting in a radiated field by the RIS that meets the specifications. Nonetheless, if the above intersection between $\mathcal{M}$ and $\mathcal{R}$ is not finally attained, at least the distance between $\vec{E}_{\text{rad}}$ and $\tilde{\vec{E}}_{\text{rad}}$ will have been minimized.

$$d(\vec{r}) = \sum_{i=1}^{N_\Omega} w(\mathbf{r_i})\left(|\tilde{\vec{E}}_{\text{rad}}(\vec{r})|^2 - |\vec{E}_{\text{rad}}(\vec{r})|^2 \right)^2 \Delta_{\Omega,i}, \tag{9.46}$$

$$\chi(x) = \sum_{i=1}^{M} \left(z_i(x)\right)^2. \tag{9.47}$$

9.4 SVR-Based Design of RIS-1 to Enhance 5G mmWave NF Coverage

9.4.1 Definition of Scenario and Single-Layer Unit Cell

A conventional indoor/outdoor scenario with a blind zone has been considered to deploy an RA-based RIS, see Figure 9.19. The RIS will be installed in the wall. The RIS panel has a rectangular shape, and its dimensions are 40 cm × 40 cm. The central operating frequency of the RIS is set to 27.6 GHz (in the 5G FR2 band)

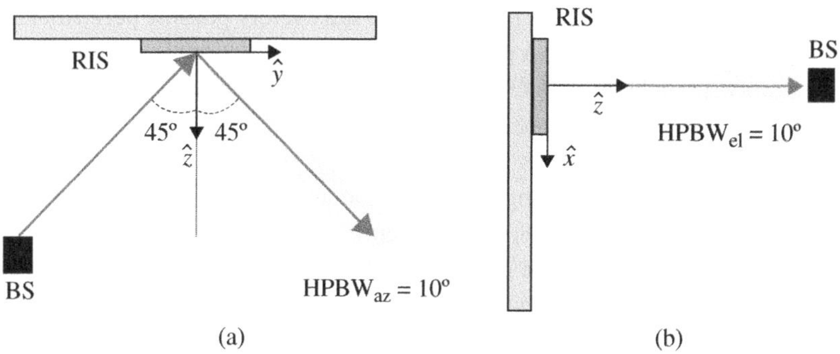

Figure 9.19 Sketch of the proposed deployment scenario: (a) azimuth and (b) elevation views, with the RIS fixed on the wall (the z-axis is normal to its surface), showing the beam requirements of the RIS.

and a bandwidth requirement of 400 MHz has been considered to ensure that the RIS can operate over a standard 5G channel. According to the proposed scenario, the BS and the RIS are at the same height, separated by a distance of 15 m and forming an angle of 45° in the azimuth plane. Therefore, the feed (BS antenna) is located far away from the RIS, which is a typical situation in this kind of scenario. Considering the reference coordinate system shown in Figure 9.19, where the origin of coordinates coincides with the geometrical center of the RIS panel, the feeder antenna is located at $(x_f, y_f, z_f) = (0, -10.61, 10.61)$ m.

The electric requirements of the RIS to ensure proper coverage of the blind zone in the proposed scenario are as follows: the reflected beam in dual-LP must the point at 45° in azimuth and 0° in elevation, and the half-power beamwidth (HPBW) must be 10° in both azimuth and elevation, see Figure 9.19a and 9.19b. Note that, if a 40 cm × 40 cm RA panel is designed to produce a pencil beam, it will provide only around 2° HPBW coverage in both elevation and azimuth planes. Therefore, the RIS must be designed to broaden the reflected beam in order to comply with the scenario specifications in Figure 9.19a. For this purpose, a pattern synthesis process will be carried out to obtain a shaped beam with the required pointing and HPBW.

The key point of the proposed design is that the blind zone is located at a distance of 10 m from the RIS. According to the operating frequency and panel size, the 5G users in the blind zone are located in the NF region of the RIS. Therefore, FF radiation patterns cannot be used to properly characterize the RIS performance in the deployment scenario. The beam requirements must be accomplished in the NF radiation patterns of the RIS, so the pattern synthesis must be conducted in the NF.

The average incidence angles on the cells of the RIS, in spherical coordinates, are $\theta_{\text{inc}} = 45°$ and $\phi_{\text{inc}} = 90°$ (note that the x-axis is used as the reference for $\phi_{\text{inc}} = 0$). Consequently, the reflective cells must exhibit a satisfactory phasing performance under strong oblique incidence. A positive aspect is that variations in the incidence angles are very small: between 44.5° and 45.5° in θ_{inc}, and between 88.9° and 91.1° in ϕ_{inc}. This feature allows a more straightforward design process of the RA-based RIS, as will be explained later.

The RIS has been designed using dual-LP single-layer RA cells based on orthogonal groups of parallel dipoles [38, 39], see Figure 9.20, which enable independent control of the phase response in each LP (Horizontal, H, and Vertical, V). The RIS periodicity is $p_x \times p_y$. The phase shifts introduced in V-pol are controlled by the groups of three rectangular dipoles that are oriented in the x-axis direction, while the phase shifts in H-pol are achieved by a proper design of the groups of dipoles parallel to the y-axis. The orthogonal sets of dipoles are arranged as shown in Figure 9.20a, with half-a-period shifting along x and y axes, in order to place

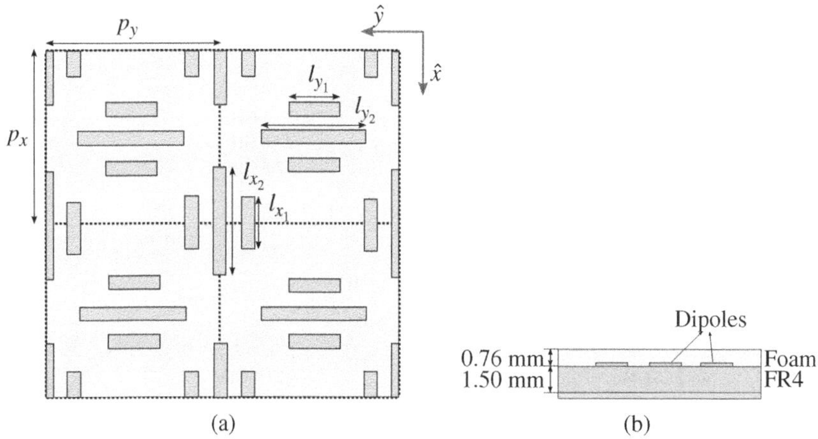

Figure 9.20 (a) Upper view showing the periodic structure of the RA-based RIS and (b) lateral view showing the stack-up configuration.

both sets of dipoles on a single substrate layer. In each group of three dipoles, the lateral ones have equal length, to reduce the levels of cross-polarization.

The RIS stack-up consists of two dielectric layers, see Figure 9.20b. The lower substrate is FR4 with 1.5 mm thickness, $\epsilon_r = 3.628$ and $\tan\delta = 0.0098$. This substrate has printed dipoles on its upper side, and it is backed by the ground plane. The upper substrate (radome) is a foam with 0.762 mm thickness, $\epsilon_r = 1.532$ and $\tan\delta = 0.012$, which is used to protect the dipoles printed on the FR4.

The geometric parameters of the RIS cell have been optimized to provide a linear phase response of the reflection coefficients associated with H and V polarizations at the operating frequencies of the RIS. The unit-cell dimensions are set to $p_x = 5$ mm and $p_y = 4.1$ mm, both values smaller than $\lambda/2$ at the 27.4–27.8 GHz band. The reason for choosing $p_y < p_x$ instead of $p_y = p_x$ is to improve the phasing performance of the cells under strong oblique incidence along the y-axis, as it is the case of the scenario shown in Figure 9.19. The width of the rectangular dipoles is fixed to 0.25 mm and the separation between neighbor parallel dipoles is 0.6 mm (measured from center to center). The lengths ℓ_{x_1} and ℓ_{x_2} are adjusted cell by cell to control the phase introduced in V-pol, while the lengths ℓ_{y_1} and ℓ_{y_2} are used to implement the required phase shifts in H-pol. To have a single design variable for each linear polarization (ℓ_{x_2} and ℓ_{y_2}), a fixed scale factor is considered between the lengths of the central and lateral dipoles in each group, so that $\ell_{x_1} = 0.56\ell_{x_2}$ and $\ell_{y_1} = 0.74\ell_{y_2}$.

The analysis of the unit cell has been performed using an in-house electromagnetic code based on MoM–LP [66]. The results of the amplitude and phase of the copolar reflection coefficients, ρ_{xx} and ρ_{yy}, associated with H and V polarizations at

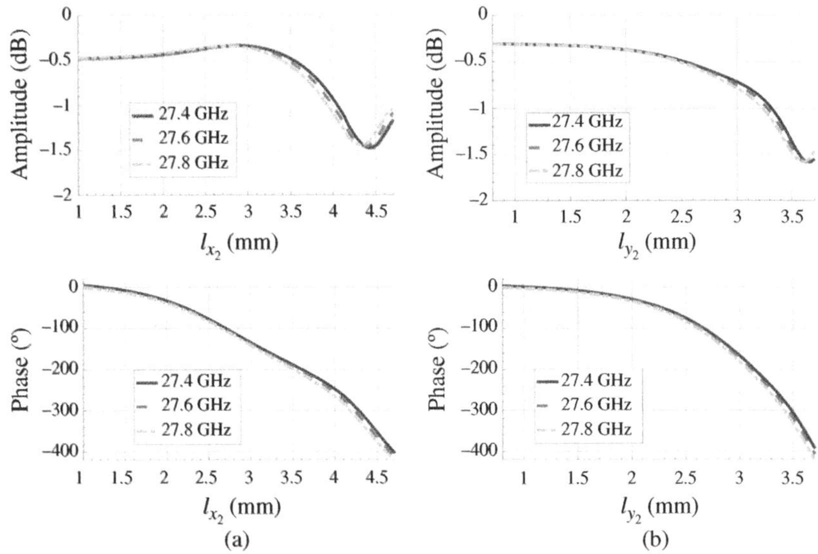

Figure 9.21 Simulated phase and amplitude of the reflection coefficient at several frequencies under $\theta_{inc} = 45°$ and $\phi_{inc} = 90°$ incidence: (a) V-pol and (b) H-pol.

the central and extreme operating frequencies of the RIS (27.4, 27.6, and 27.8 GHz) are shown in Figure 9.21 as a function of the central dipole length that controls each polarization (ℓ_{x_2} and ℓ_{y_2}). The simulations have been carried out considering that the incidence angles on the cell are $\theta_{inc} = 45°$ and $\phi_{inc} = 90°$, which are associated with the center of the RIS panel in the scenario shown in Figure 9.19. Similar performance is achieved in both polarizations, with a phase variation range of 400° and a smooth slope of the phase curve.

As abovementioned, the effect of the incidence angles in the reflection coefficients of the RIS cells can be considered negligible since variations in θ_{inc} and ϕ_{inc} along the RIS panel are of the order of $\pm 1°$. As shown in Figure 9.21, the cell exhibits robust behavior under the strong oblique incidence conditions of the selected scenario, keeping more than 360° of phase variation. To obtain the layout of the 40 cm × 40 cm RIS panel, the length of the printed dipoles must be designed cell by cell according to the objective phase distributions of the reflection coefficient achieved by the pattern synthesis.

9.4.2 Unit-Cell Modeling Based on SVR

In this section, we are going to obtain surrogate models to estimate the values of the reflection coefficients of matrix **R** for the unit cell represented in Figure 9.20 using an incidence angle ($\theta = 45°, \varphi = 90°$) and operating frequency of 27.6 GHz.

The input variables of the SVR are two geometrical variables: ℓ_{x_2} and ℓ_{y_2} (see Figure 9.20). Moreover, ℓ_{x_2} varies freely in the interval [0.9625, 4.7] mm, meanwhile, ℓ_{y_2} takes values in the interval [0.75768, 3.69984] mm. Therefore, the input space is $\chi \in [0.9625, 4.7]$ mm$\times$ [0.75768, 3.69984] mm. In addition, we have generated different models for the real and imaginary parts of the reflection coefficients (eight in total), from which it is straightforward to obtain the magnitude and phase. Nonetheless, the shape of the magnitude for copolar coefficients is smoother than their real and imaginary parts. Consequently, the direct estimation of $|\rho_{xx}|$ and $|\rho_{yy}|$ from their correspondent models is much more precise than from their real and imaginary parts, which justifies considering two extra models for the magnitude of those coefficients. In contrast, the shape of the magnitude of the cross-polar coefficients exhibits a deep resonance, more striking than on the real and imaginary parts, that prevents them from using that strategy with their modeling. In summary, we consider 10 output variables: $|\rho_{xx}|$, $|\rho_{yy}|$, $\text{Re}\{\rho_{xx}\}$, $\text{Im}\{\rho_{xx}\}$, $\text{Re}\{\rho_{xy}\}$, $\text{Im}\{\rho_{xy}\}$, $\text{Re}\{\rho_{yx}\}$, $\text{Im}\{\rho_{yx}\}$, $\text{Re}\{\rho_{yy}\}$, and $\text{Im}\{\rho_{yy}\}$.

Our developments are based on the theory summarized in Section 9.2.2.1. We will initially discuss the relation between the number of training samples, N_r, and the achieved accuracy versus computational time cost. We will show that the adequate selection of N_r is of main importance for getting accurate results in a feasible time. Finally, we note that the results given in this section have been obtained using, in sequential mode, a laptop with Intel i7-5600U CPU at 2.6 GHz and 8 GB of RAM.

This section is divided into two subsections. In Section 9.4.2.1, we get into details of the aforementioned accuracy-time discussion. In Section 9.4.2.2, we compare the results of the surrogate models based on SVRs with those of MoM, after using a number of training patterns that is the one selected from the previous discussion.

9.4.2.1 Discussion on the Number of Training Patterns, Time Cost and Achieved Precision

The grid search strategy described in Section 9.2.2.3 to obtain (C^*, γ^*) yields values of C and γ close to the optimal ones with a dramatic reduction on the time cost compared to an exhaustive search. Nonetheless, as was pointed out in that subsection, the value of N_r determines the time cost of the SVM training and, thus, the time cost of the grid search. In addition, the accuracy of the regression provided by the models is also dependent on N_r. Consequently, it is of interest to study the relation between the number of training patterns, the achieved accuracy and the time cost. For the study covered here, we have used a set of $N = 2000$ samples that have been split into validation and test sets with 300 samples each, i.e., $N_v = N_t = 0.15N = 300$, and training sets with different number of samples. Cardinalities of the training sets are: $N_r = 20, 100, 200, 300, 400, 600, 800, 1000, 1200,$

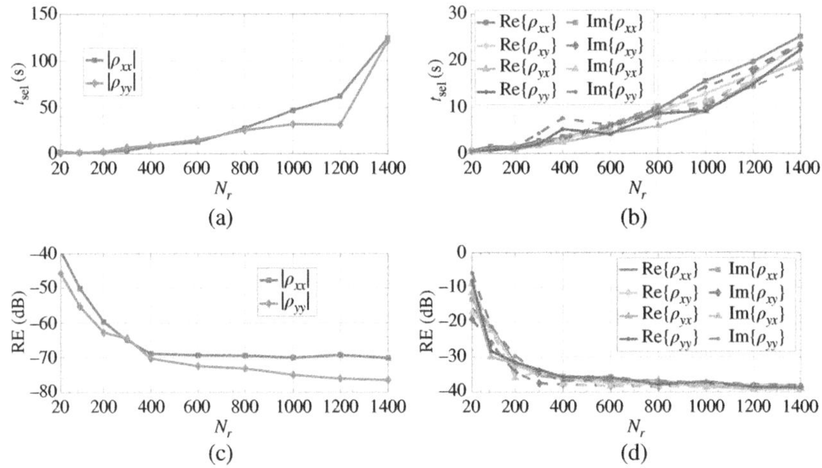

Figure 9.22 Evolution of the model selection time, t_{sel} (s), and of the relative error, RE (dB), over the test set, using $N_v = 300$, with the number of training patterns, N_r. (a) t_{sel} versus N_r for the magnitude of copolar coefficients. (b) t_{sel} versus N_r for the real and imaginary parts of the reflection coefficients. (c) RE versus N_r for the magnitude of copolar coefficients. (d) RE versus N_r for the real and imaginary parts of the reflection coefficients.

and 1400 that represent, respectively, 1%, 5%, 10%, 15%, 20%, 30%, 40%, 50%, 60%, and 70% of the whole data set.

On the one hand, the model selection time cost (t_{sel}) increases with N_r. This is shown in Figure 9.22a,b, where the time cost of the model selection for every estimated output variable is plotted versus N_r. Due to the different time scales achieved for the magnitude of the copolar coefficients and the real and imaginary components of both copolar and cross-polar coefficients, we show both results in different plots. On the other hand, the RE decreases with N_r very fast for moderate values of N_r and slower for higher values of N_r. This result is illustrated in Figure 9.22c,d where the RE over the test set for every estimated output variable is plotted vs. N_r, showing that the error rapidly decreases with N_r until approximately $N_r = 400$, where the RE almost stagnates, i.e., its slope approaches to zero.

As a consequence, from Figure 9.22 it seems that a good trade-off between training time and RE may be achieved for $N_r \approx 400$, which is the value that we have used in this work to finally model the cell. In addition, the value of N_r to attain a nice trade-off between time cost and accuracy is very dependent upon the problem at hand. For instance, in [50] the authors found optimal $N_r \approx 1750$ for the RA cell used there (the same used in Section 9.2.2.1), which contrasts with the optimal value of $N_r \approx 400$ found in the present study. Note that the RE for the real and imaginary parts of all the reflection coefficients is very similar and it asymptotically tends to –40 dB. Meanwhile, RE for the magnitude of both direct coefficients

Table 9.1 Relative error over the test set and model selection time of the best SVM model for each output cross-polar variable when employing $N_r = 400$, $N_v = 80$, and $N_t = 80$.

Variable	$\|\rho_{xx}\|$	$\|\rho_{yy}\|$	$\text{Re}\{\rho_{xx}\}$	$\text{Im}\{\rho_{xx}\}$	$\text{Re}\{\rho_{yy}\}$	$\text{Im}\{\rho_{yy}\}$
RE (dB)	−68.3	−69.1	−36.3	−35.5	−34.7	−35.4
t_{sel} (s)	5.30	8.75	2.77	2.83	1.74	6.73

Table 9.2 Relative error over the test set and model selection time of the best SVM model for each output crosspolar variable when employing $N_r = 400$, $N_v = 80$, and $N_t = 80$.

Variable	$\text{Re}\{\rho_{xy}\}$	$\text{Im}\{\rho_{xy}\}$	$\text{Re}\{\rho_{yx}\}$	$\text{Im}\{\rho_{yx}\}$
RE (dB)	−36.6	−38.4	−35.9	−37.1
t_{sel} (s)	2.52	2.53	1.67	2.26

has bigger differences, though considering extremely low values of the error, tending the error of $|\rho_{xx}|$ to −70 dB and to −80 dB for $|\rho_{yy}|$. Note that $E\left\{\text{RE}^2\right\}$ was set to −40 dB for the real and imaginary parts and to −80 dB for the magnitude of direct coefficients, and that using (9.17), the value of ε is adaptively computed as a function of the Euclidean norm of the output training data vector ($\|\vec{y}\|$), for a given value of N_r (see Section 9.2.2.2).

Table 9.1 summarizes the results for both model selection time cost and RE over the test set for all the estimated output variables associated to the copolar coefficients when using $N_r = 400$, $N_v = 80$, and $N_t = 80$. Table 9.2 summarizes the same results as Table 9.1, though for cross-polar coefficients. Slight differences between the results plotted in Figure 9.22 and summarized in Tables 9.1 and 9.2 may appear because the validation and the test sets used in both simulations are different.

9.4.2.2 Reflection Coefficients

In this section, we compare the results of the reflection coefficients estimated by the SVRs with those computed using MoM. For the sake of comparison between SVR and MoM results, we represent 1D plots of the magnitude and phase for different coefficients along straight lines in the considered input space $\chi \in [0.9625, 4.7]\,\text{mm} \times [0.75768, 3.69984]\,\text{mm}$ composed by 300 samples. We selected those lines to get striking fluctuations of the coefficients. For instance, Figure 9.23 compares the results of SVR with MoM for ρ_{xx} and ρ_{yx} along the straight line given by $\ell_{x_2} = 1.27\ell_{y_2}$ (a diagonal crossing the whole input domain).

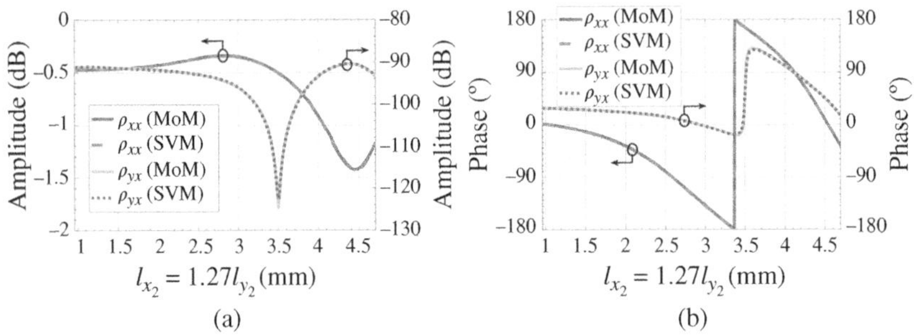

Figure 9.23 Comparison of the magnitude (dB) and phase (°) of ρ_{xx} and ρ_{yx} over 300 samples along the line given by $\ell_{x_2} = 1.27\ell_{y_2}$. (a) ρ_{xx} (dB) and ρ_{yx} (dB). (b) $\angle\rho_{xx}$ (°) and $\angle\rho_{yx}$ (°).

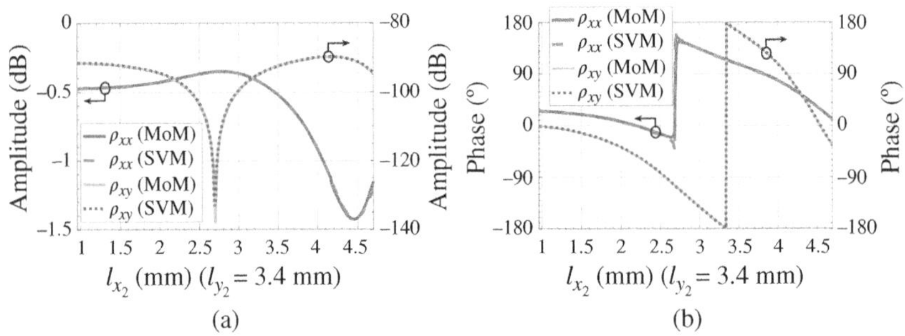

Figure 9.24 Comparison of the magnitude (dB) and phase (°) of ρ_{xx} and ρ_{xy} over 300 samples along the line given by ℓ_{x_2} for constant ℓ_{y_2} =3.4 (mm). (a) ρ_{xx} (dB) and ρ_{xy} (dB). (b) $\angle\rho_{xx}$ (°) and $\angle\rho_{xy}$ (°).

Particularly, Figure 9.23a displays the magnitude in dB of the reflection coefficients, i.e., $20\log(|\rho|)$, and Figure 9.23b shows the phase in degrees (°). As expected from the results in Tables 9.1 and 9.2, both figures exhibit a high degree of agreement, and it is worth mentioning the ability of the SVR to predict and follow the resonance of ρ_{yx} at ℓ_{x_2} =3.5 mm.

Figure 9.24 contrasts the SVR and MoM outcomes for ρ_{xx} and ρ_{xy} along a straight line on ℓ_{x_2} for constant ℓ_{y_2} =3.4 (mm). Figure 9.24a displays the magnitude in dB of the coefficients and Figure 9.24b shows the phase in degrees. Similar conclusions as those yielded from Figure 9.23 are drafted about the accuracy and ability of the SVR to estimate resonance. Nonetheless, in this case, the phase predicted by the SVR is a bit more spiky than that yielded by MoM.

Finally, Figure 9.25 puts alongside the SVR and MoM calculation of ρ_{yy} and ρ_{yx} along a straight line on ℓ_{y_2} for constant ℓ_{x_2} =4.6 (mm). Figure 9.25a displays the magnitude in dB of the coefficients and Figure 9.25b shows the phase in degrees.

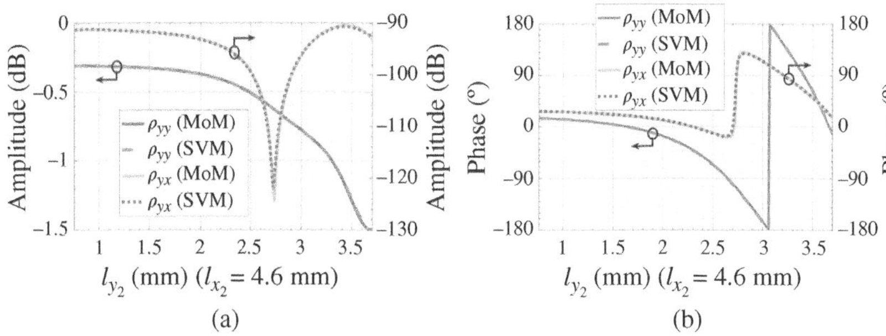

Figure 9.25 Comparison of the magnitude (dB) and phase (°) of ρ_{yy} and ρ_{yx} over 300 samples along the line given by ℓ_{y_2} for constant ℓ_{x_2} =4.6 (mm). (a) ρ_{yy} (dB) and ρ_{yx} (dB). (b) $\angle\rho_{yy}$ (°) and $\angle\rho_{yx}$ (°).

The precision of the SVR shown in Figure 9.25 is similar to the one obtained in Figures 9.23 and 9.24.

9.4.3 RIS-1 Designed Based on Intersection Approach Framework

Considering the defined geometry of RIS-1, a preliminary design of a beam that focuses the radiated power at the center of the coverage area is made. In this initial approximation, the analytical solution for a high-gain beam in the direction (θ_b, φ_b) defined by (9.48) is used [22]. This expression allows to calculate the phase shift that the elements must introduce, using geometric optics. It takes into account only the position of the nth element inside the RIS (referred to the system in Figure 9.11b as (x_l, y_l)), the distance from the BS (feed) to the nth element (d_n), the vacuum wave number k_0, and the outgoing direction of the beam (θ_b, φ_b). Equation (9.48) is applied independently for each of the two linear polarizations to either compute ϕ_V (phase of the coefficient ρ_{xx}) or ϕ_H (phase of the coefficient ρ_{yy}). However, considering that the coverage has to be identical for both polarizations, the solution for the phase-shift distribution across the RIS surface is the same in both polarizations, as shown in Figure 9.26.

$$\phi^n_{V/H} = k_0 \left(d_n - \left(x_n \cos\varphi_b + y_n \sin\varphi_b\right) \sin\theta_b\right). \tag{9.48}$$

The radiated field produced by the phase-shift distributions in Figure 9.26 redirects the incident signal in the desired direction, obtaining the beam with the highest gain achievable for the given aperture. Although this solution improves coverage by increasing the signal level, it is not sufficient to meet the specifications of the scenario. The HPBW of the beam is approximately 2° in both elevation and azimuth, while the coverage requirements state 10° × 10° in both directions to cover the entire blind zone, see Figure 9.27. This field

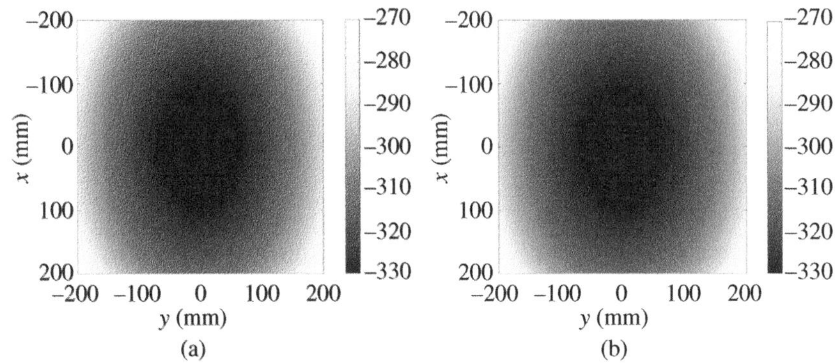

Figure 9.26 Phase-shift distribution on the RIS surface obtained with the analytical approach: (a) V-pol and (b) H-pol.

pattern has been calculated at a distance of $R = 10$ m. As this analytical solution does not meet the specifications in terms of radiated pattern, and no alternative is known to achieve the desired pattern, it is necessary to use more complex optimization techniques. In this case, POS has been employed, which simplifies the optimization complexity by modeling the elements as ideal, lossless phase shifters. Thus, each element of the RIS will have its $\mathbf{R}_n^{\mathrm{PO}}$, and for each polarization, it is only necessary to synthesize ϕ_V^n or ϕ_H^n independently. The goal of the V-pol and H-pol synthesis, respectively, the V and H synthesis, is to obtain a phase-shift distribution that achieves an NF pattern at 10 m generating a uniform field distribution within a 10° × 10° angular margin. A field distribution is considered uniform if the peak-to-peak ripple over the entire angular range is less than 3 dB. Although the optimization complexity is reduced by using POS instead of direct layout optimization, it is still a complex synthesis in terms of radiated pattern requirements. The starting point is a pencil beam with just a 2° HPBW, which needs to be widened to the required 10° in both directions. In addition, as shown in Figure 9.27, it is not just about widening the main lobe; the nulls that fall within the desired angular margin also need to be shifted.

The strategy to approach the synthesis process is based on a multistage optimization [69]. This strategy involves dividing the synthesis into different sub-syntheses, where intermediate objectives are sought. These intermediate objectives are usually radiation patterns that are laxer than the desired specifications. In this way, in the early syntheses, objectives close to the starting point will be sought, and in the later stages, objectives closer to the desired specifications will be pursued. A multistage process is particularly useful when the desired field pattern is far from the starting point, as in this RIS design. The subdivision of the synthesis helps strengthen the algorithm convergence and even reach a valid solution, since the desired goal in a single-stage synthesis would be impossible

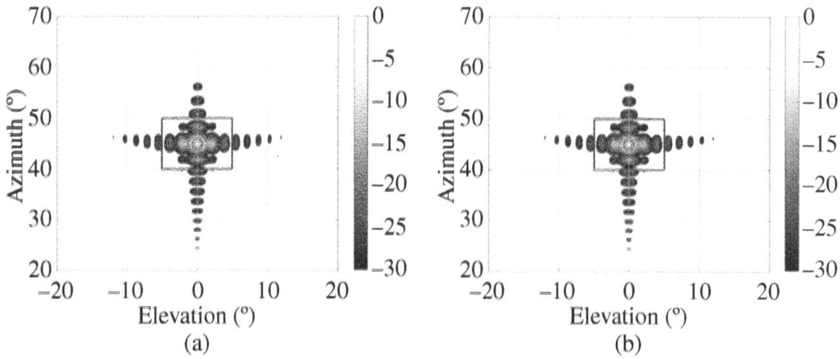

Figure 9.27 Near-field pattern (dB) (normalized to its maximum) radiated by the analytical approach at 10 m: (a) V-pol and (b) H-pol. The dark gray-line rectangle shows the limits of the coverage area.

to reach. In the case of the V-pol, 8 syntheses are required, while 10 stages are needed for the H-pol. In each of these syntheses, an average of 300 iterations of the GIA are needed. In addition, 3 iterations of the LM algorithm are used in each GIA iteration. In the multistage process, the goal is to gradually widen the main beam in azimuth while keeping the width in elevation. Later, as it finishes azimuth widening, it also starts elevation widening, to finally adjust both angular margins to the desired range. Finally, the secondary lobe level is adjusted.

The output of V-pol is the phase-shift distribution across the surface that the elements must introduce to obtain an NF pattern that meets the specifications for V-pol. In the case of H-pol, the output is the phase-shift distribution for the H-pol. Figure 9.28 shows the phase-shift distribution obtained for each case, while

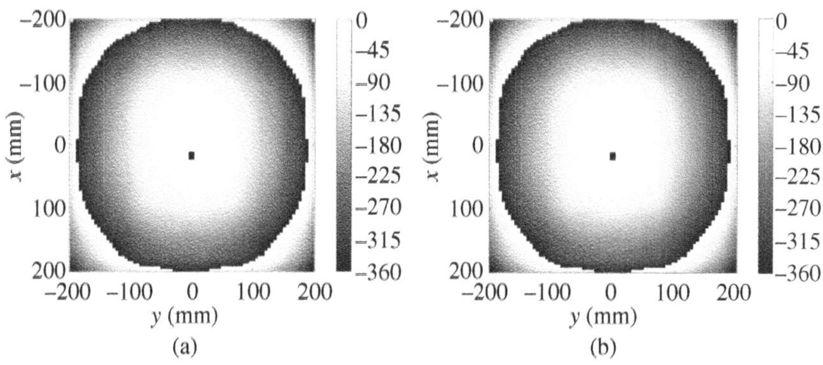

Figure 9.28 Phase-shift distribution on the RIS surface obtained after the multistage synthesis process carried out through the GIA: (a) V-pol and (b) H-pol.

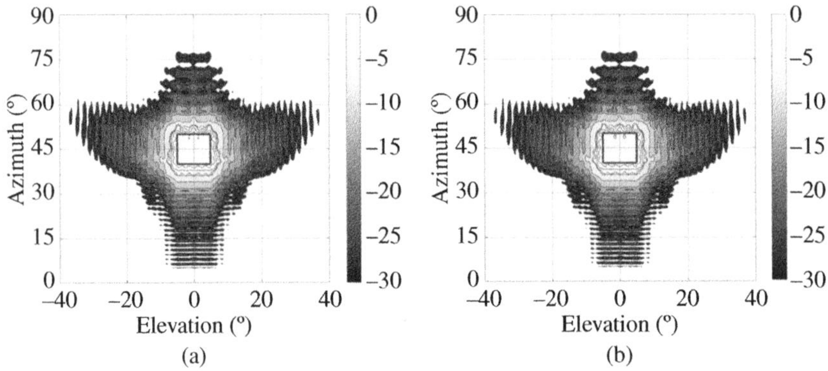

Figure 9.29 Near-field pattern (dB) (normalized to its maximum) radiated by the phase-shift distribution obtained with the GIA at 10 m: (a) V-pol and (b) H-pol. The dark gray-line rectangle shows the limits of the coverage area.

Figure 9.29 shows the NF pattern at 10 m reached after the synthesis. Since the synthesis is based on POS, it is only carried out at a single frequency, in this case, at the central one (27.6 GHz).

First, the phase-shift distribution of the elements along the aperture follows a smooth transition, thus, most elements will fulfill the local periodicity assumption, which yields physically feasible results. Second, in terms of the NF pattern, a significant increase in the coverage area has been achieved compared to the analytical case (see Figure 9.30). In particular, the high-gain beam, in both

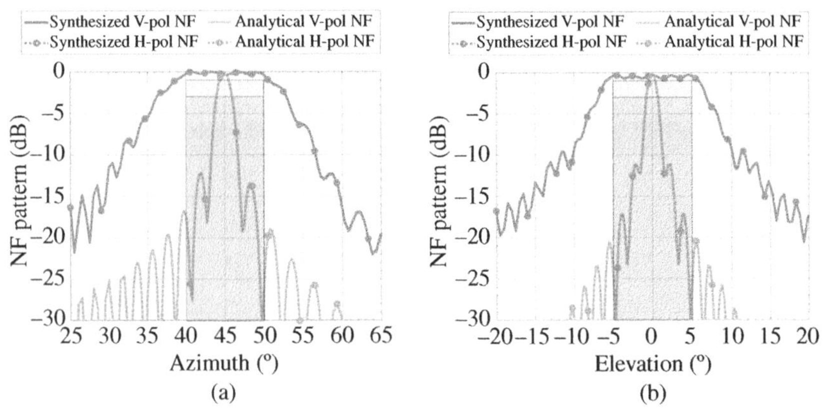

Figure 9.30 Comparison of main cuts of the NF pattern at 10 m in (a) azimuth (b) elevation for the high-gain and synthesized beam for both linear polarizations at 27.6 GHz. The shadowed area represents the angular coverage area and the admitted ripple for 3 dB (dark) and 1 dB (light).

Table 9.3 Comparison of the compliance for different ripple requirements for the high-gain and synthesized beam at 27.6 GHz.

RIS-1	1 dB (%)	1.5 dB (%)	3 dB (%)	4.5 dB (%)
V-pol High-gain beam	0.85	1.11	2.23	3.61
V-pol Synthesized beam	99.20	100	100	100
H-pol High-gain beam	0.85	1.11	2.23	3.61
H-pol Synthesized beam	97.89	100	100	100

polarizations, barely attained a 3.61% compliance within the coverage area when evaluating a 3 dB peak-to-peak ripple. Logically, this compliance dropped dramatically to 0.85% when evaluating 1 dB ripple. After synthesis, the compliance of 1 dB ripple is practically 100% in both polarizations, being 99.2% and 97.9% for V-pol and H-pol, respectively. The coverage angular range is increased for just 2° to 10° in both elevation and azimuth, and also the 3 dB specification is perfectly met, achieving a solution that practically reduces the ripple by 2 dB. Figure 9.30 shows the main cuts (elevation 0° and azimuth 45°) for both the high-gain beam and the synthesized beam. The improvement in the ripple and angular margin reached through the synthesis can be better appreciated in this representation. Table 9.3 provides a comparison before and after the synthesis for the central frequency and different ripple requirements compliance.

9.4.4 RIS-1 Design Process

To obtain the layout of the 40 cm × 40 cm RIS panel, the dimensions of the printed elements have been adjusted cell by cell in order to obtain the required phase-shift distributions on the RIS surface computed by the multistage POS synthesis at 27.6 GHz, which a shown in Figure 1.16. According to the RA cell configuration and its operating principle, both described in Section 9.4.1, the phase shift introduced by each cell can be controlled by a proper adjustment of the lengths of the three dipoles associated to each linear polarization. To simplify this process, only two design variables are considered: ℓ_{x_2} and ℓ_{y_2}, which represent the lengths of the central dipoles associated with V-pol and H-pol, respectively (as explained in Section 9.4.1, a fixed scale factor is considered between the lengths of the central and lateral dipoles for each polarization, so that $\ell_{x_1} = 0.56\ell_{x_2}$ and $\ell_{y_1} = 0.74\ell_{y_2}$).

The adjustment of the dipole lengths to obtain the final layout of the RIS has been performed by applying the SVR models of the unit cell described in Section 9.4.2. This represents a significant difference with respect to previous RIS implementations based on RAs [38, 39], where the design of the unit-cell elements is performed by applying an electromagnetic analysis routine based on MoM [59]. Note that the SVR models of the unit cell account for the real incidence angles from the BS on the RIS panel. In this case, due to the large distance between the RIS and the BS, the average incidence angles on the RIS ($\theta_{inc} = 45°$ and $\varphi_{inc} = 90°$) were considered to simplify the design process. Since variations in θ_{inc} and φ_{inc} along the RIS panel are of the order of $\pm1°$, the effect of the incidence angles in the reflection coefficients of the RIS cells can be considered as negligible. However, note that the unit-cell characterization by the SVR models can be extended to consider more complex scenarios with larger variations in the incidence angles on the RIS.

Figure 9.31 shows the final lengths of the printed dipoles in each cell obtained after the design process based on the SVR models. The lengths of the lateral and central dipoles associated to V-pol are represented in Figure 9.31a–9.31b, respectively, while Figure 9.31c–9.31d depict the lengths of the lateral and central dipoles associated with H-pol.

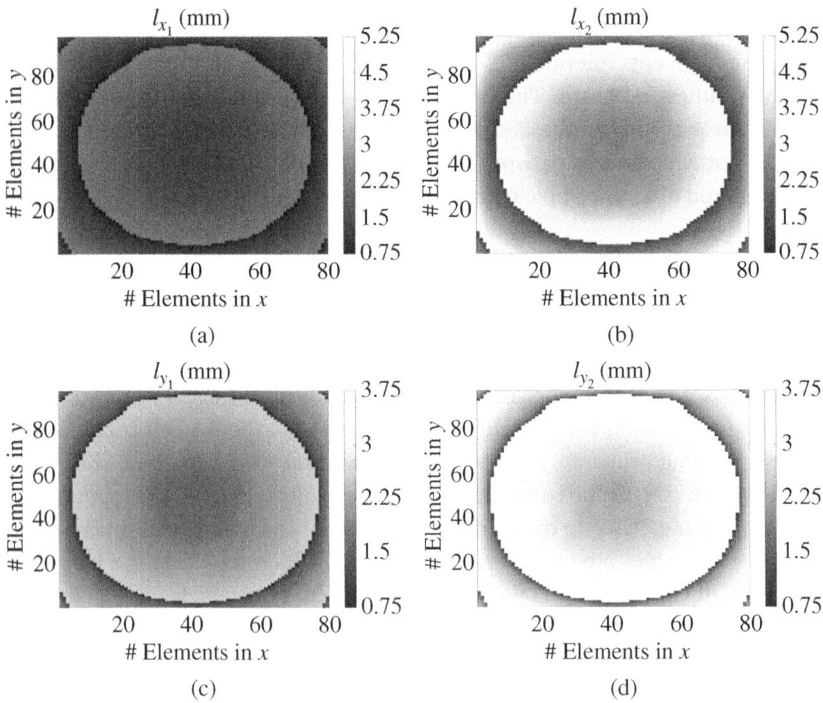

Figure 9.31 Length distributions (in mm) obtained after the design process of the RIS for the: (a) lateral dipoles for V-pol, (b) central dipoles for V-pol, (c) lateral dipoles for H-pol, and (d) central dipoles for H-pol.

Once the designed layout of dipoles is obtained, the in-band behavior of the RIS can be evaluated. To do so, the tangential field at the aperture is first obtained by analyzing, element by element, the real response of the unit cell. This means taking into account the angle of incidence and the incident field. This involves computing the matrix $\mathbf{R}^{\text{MoM}}$ or $\mathbf{R}^{\text{SVR}}$. The propagation models developed in Section 9.3.1.2 are then applied to the tangential field. Figure 9.32 shows the main cuts of the coverages radiated by the RIS at the designed frequency (27.6 GHz), and the extreme frequencies 27.4 and 27.8 GHz, which correspond to a standard 400 MHz channel in 5G. At the lower extreme frequency, the behavior is quite similar to that of the central frequency, while at the upper extreme frequency, the ripple increases. Nevertheless, in all cases, the coverage within the 400 MHz channel fulfills the specifications of a maximum ripple of 3 dB (see Table 9.4).

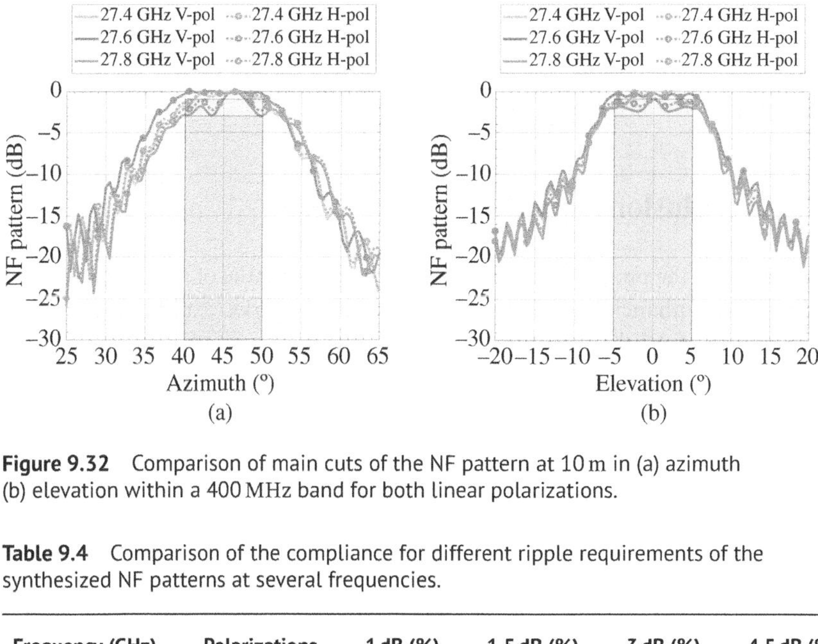

Figure 9.32 Comparison of main cuts of the NF pattern at 10 m in (a) azimuth (b) elevation within a 400 MHz band for both linear polarizations.

Table 9.4 Comparison of the compliance for different ripple requirements of the synthesized NF patterns at several frequencies.

Frequency (GHz)	Polarizations	1 dB (%)	1.5 dB (%)	3 dB (%)	4.5 dB (%)
27.4	V	33.56	65.41	100	100
27.6	V	99.20	100	100	100
27.8	V	3.18	6.18	96.12	100
27.4	H	48.38	88.16	96.12	100
27.6	H	97.89	100	100	100
27.8	H	5.12	36.94	100	100

It is important to note that the synthesis is performed at a single frequency. Moreover, the NF pattern is a shaped beam, which is much less robust to the in-band response than pencil beams (high-gain beams) or focused NF beams. Therefore, the obtainment of this in-band response is due to the linearity of the unit-cell frequency response. Hence, the design stage is especially crucial since it allows to enhance the features of the RIS. Furthermore, the phase-shift distribution for each polarization is reached independently, so the design is also responsible for ensuring that both phase-shift distributions are implemented with minimal error, thereby enhancing the in-band response in both polarizations. In most of the RA-based RIS-1 designed by the authors, bandwidths between 5 and 10% have been achieved [38, 39]. In this case, the bandwidth is reduced by a second effect: the coverage is generated in the specular direction, which involves having a second reflection in the same direction that outside the design frequency, deteriorates (by increasing the ripple) the obtained coverage. If the coverage were to be generated in another direction, it is expected that the behavior over the 400 MHz bandwidth would be practically invariant and the RIS could even operate in a larger bandwidth.

9.5 Conclusions and Road Map

The results of the proposed RIS-1 design show the potential of the RA technology for coverage-enhancing applications in 5G FR2 networks, exploiting the use of SVR models of the RA cells to carry out the design process of the RIS elements. The designed RIS-1 provides a reflected-shaped beam in the NF that complies with the specifications of pointing (0° in elevation and 45° in azimuth) and beamwidth (HPBW of 10° in both azimuth and elevation) of the proposed deployment scenario. Also, the 3-dB ripple criterion within the coverage region is fulfilled in a bandwidth of 400 MHz, associated with a standard 5G channel, and the RIS presents the same radiation pattern in both horizontal and vertical polarizations (associated with y- and x- RIS axes, respectively), which can be used to exploit polarization diversity techniques. The radiation patterns of the RIS are quite stable within the 400-MHz operating bandwidth, and the small variations, which are more appreciable at the upper frequencies, are produced by the effect of the specular reflection of the incident signal that falls within the coverage region of the shaped beam.

In conclusion, both the design techniques and the results provided in this chapter demonstrates the feasibility and advantages of using RIS-1 to improve wireless coverage in mmWave 5G networks. RA-based RISs represent an energetically efficient and low-cost solution to the problem of blind zones, as they are able to provide a smart reflection of the incident signal to concentrate the power received from the BS in a specific coverage area. The latter can be achieved by the application of proper POS techniques, as those based in the gIA framework, to obtain the required phase distribution on the RIS that enforces specific requirements in the RIS radiation patterns (in terms of beam pointing, beamwidth, coverage ripple, etc.) in the NF or the FF regions depending on the RIS deployment scenario. The selection of an appropriate unit-cell configuration is another key aspect of the RIS design. The unit cell should provide a satisfactory phasing performance within a sufficiently large bandwidth, with low losses and high angular stability. In this chapter, we have selected an RIS formed by unitary cells composed of orthogonal sets of coplanar parallel dipoles. We have designed the RIS by using SVR models whose accuracy to predict the amplitude and phase of the reflection coefficients has been verified by comparison with a conventional electromagnetic analysis tool based on MoM–LP. The low errors obtained after this process confirm the validity of this approach and the large potential of applying the SVR models for the design of RA-based RIS with advanced capabilities.

The work presented in this chapter should be understood as the first step toward integrating ML techniques into RIS design. In this case, SVR models have been used for the design of passive RIS with shaped beams, enabling a much faster modeling of the unit cell than any conventional technique. The time saving obtained in RIS analysis opens up different lines of action, where ML (or SVM in this case) can greatly contribute. Figure 9.33 contextualizes the work developed in this chapter within the road map envisioned for the integration of RIS plus SVM in mmWave communications. In addition, it depicts the potential scenarios to address within this line.

First, characterizing the unit cells using these models does not have to be limited solely to the element design stage; it can be directly integrated into the RIS design

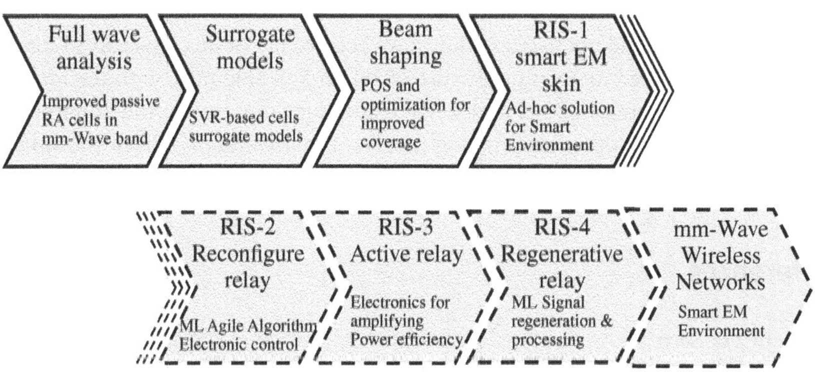

Figure 9.33 ML-based RIS development road map.

and optimization stage. That is, the low computational cost of the unit cell with SVR models allows them to be combined with direct optimization techniques of the elements' geometry. In this case, instead of synthesizing the phase of the reflection coefficient, geometric variables of the unit cell itself would be used as optimization variables, considering the whole reflection matrix. This allows a direct design and optimization of the RIS layout. These optimization techniques require including an electromagnetic analysis of each element in each iteration, making the computational cost significantly high. Considering that RISs are typically electrically large apertures to capture sufficient power from the BS, this makes the problem very costly from a computational viewpoint. However, thanks to the time reduction provided by the use of SVR models, computational costs are notably decreased, making the problem manageable in terms of computational time. A second advantage of these optimization techniques are not only that shaped diagrams can be obtained for the desired coverage but also by directly optimizing the layout (element geometry), it is possible to impose other figures of merit, such as bandwidth requirements, polarization requirements, or working with two linear or circular polarizations simultaneously, addressing more complex designs.

However, ML techniques are not only limited to the design stages of passive RIS but also have high potential when combined with active RIS (from RIS-2 to RIS-4). In active technology-based RIS, performance such as beam steering or beam adaptation to channel conditions is sought. In these types of RISs, reaching a real-time response that allows modifying the antenna radiation characteristics is essential. A first step in this direction would be to include SVR models that allow relating the phase (and the amplitude) response of the active element to some control variables, e.g., voltages or binary codes. Hence, depending on the channel needs, the antenna could map the control variables agilely and efficiently, quickly modifying its radiation pattern. This first step could be considered the seed of the new paradigm of smart environments, where antennas must gain the ability to adapt the radiation to communication needs. Therefore, antennas must be real-time adaptive and not fixed, as they have been so far. In this sense, SVM models can be considered to open up the use of ML in RIS antennas, with the aim of gradually using increasingly complex ML techniques that enable the antenna to make more complex decisions making possible the integration within the smart environment paradigm.

References

1 Buffi, A., Serra, A.A., Nepa, P. et al. (2010). A focused planar microstrip array for 2.4 GHz RFID readers. *IEEE Transactions on Antennas and Propagation* 58 (5): 1536–1544.

2 García-Fernández, M., Álvarez-Narciandi, G., Álvarez-López, Y., and Las-Heras, F. (2023). Array-based ground penetrating synthetic aperture radar on board an unmanned aerial vehicle for enhanced buried threats detection. *IEEE Transactions on Geoscience and Remote Sensing* 61: 1–18. https://doi.org/10.1109/TGRS.2023.3272982.

3 Zhang, P., Li, L., Zhang, X. et al. (2019). Design, measurement and analysis of near-field focusing reflective metasurface for dual-polarization and multi-focus wireless power transfer. *IEEE Access* 7: 110387–110399.

4 Busari, A., Mumtaz, S., Al-Rubaye, S., and Rodriguez, J. (2018). 5G millimeter-wave mobile broadband: performance and challenges. *IEEE Communications Magazine* 56 (6): 137–143.

5 Hong, W., Jiang, Z.H., Yu, C. et al. (2021). The role of millimeter-wave technologies in 5G/6G wireless communications. *IEEE Journal of Microwaves* 1 (1): 101–102.

6 Rappaport, T.S., Xing, Y., MacCartney, G.R. et al. (2017). Overview of millimeter wave communications for fifth-generation (5G) wireless networks–with a focus on propagation models. *IEEE Transactions on Antennas and Propagation* 65 (12): 6213–6230.

7 Rappaport, T.S., Xing, Y., Kanhere, O. et al. (2019). Wireless communications and applications above 100 GHz: opportunities and challenges for 6G and beyond. *IEEE Access* 7: 78729–78757. https://doi.org/10.1109/ACCESS.2019.2921522.

8 Salous, S., Degli Esposti, V., Fuschini, F. et al. (2016). Millimeter-wave propagation: characterization and modeling toward fifth-generation systems. *IEEE Antennas and Propagation Magazine* 58 (6): 115–127.

9 Shafi, M., Zhang, J., and Tataria, H. (2018). Microwave vs. millimeter-wave propagation channels: key differences and impact on 5G cellular systems. *IEEE Communications Magazine* 56 (12): 14–20.

10 Zheng, K., Wang, D., Han, Y. et al. (2020). Performance and measurement analysis of a commercial 5G millimeter-wave network. *IEEE Access* 8: 163996–164011.

11 Andrews, J.G., Buzzi, S., Choi, W. et al. (2014). What will 5G be? *IEEE Journal on Selected Areas in Communications* 32 (6): 1065–1082. https://doi.org/10.1109/JSAC.2014.2328098.

12 Rappaport, T.S., Sun, S., Mayzus, R. et al. (2013). Millimeter wave mobile communications for 5G cellular: it will work! *IEEE Access* 1: 335–349. https://doi.org/10.1109/ACCESS.2013.2260813.

13 Wang, P., Fang, J., Yuan, X. et al. (2020). Intelligent reflecting surface-assisted millimeter wave communications: joint active and passive precoding design. *IEEE Transactions on Vehicular Technology* 69 (12): 14960–14973.

14 Wu, Q. and Zhang, R. (2020). Towards smart and reconfigurable environment: Intelligent reflecting surface aided wireless network. *IEEE Communications Magazine* 58 (1): 106–112.

15 Di Renzo, M., Zappone, A., Debbah, M. et al. (2020). Smart radio environments empowered by reconfigurable intelligent surfaces: how it works, state of research, and the road ahead. *IEEE Journal on Selected Areas in Communications* 38 (11): 2450–2525. https://doi.org/10.1109/JSAC.2020.3007211.

16 Nemati, M., Park, J., and Choi, J. (2020). RIS-assisted coverage enhancement in millimeter-wave cellular networks. *IEEE Access* 8: 188171–188185. https://doi.org/10.1109/ACCESS.2020.3031392.

17 Shafique, K. and Alhassoun, M. (2024). Going beyond a simple RIS: trends and techniques paving the path of future RIS. *IEEE Open Journal of Antennas and Propagation* 5 (2): 256–276. https://doi.org/10.1109/OJAP.2024.3360900.

18 Maci, S. (2023). Reflective intelligent surfaces and metasurface antennas. *2023 IEEE Radio and Antenna Days of the Indian Ocean (RADIO)*, 1–1. Balaclava, Mauritius, IEEE.

19 Oliveri, G., Zardi, F., Rocca, P. et al. (2023). Constrained design of passive static EM skins. *IEEE Transactions on Antennas and Propagation* 71 (2): 1528–1538.

20 Amri, M.M., Tran, N.M., and Choi, K.W. (2021). Reconfigurable intelligent surface-aided wireless communications: adaptive beamforming and experimental validations. *IEEE Access* 9: 147442–147457. https://doi.org/10.1109/ACCESS.2021.3124319.

21 Bagheri, A., Safaei, M., Araghi, A. et al. (2023). Mathematical model and real-world demonstration of multi-beam and wide-beam reconfigurable intelligent surface. *IEEE Access* 11: 19613–19621. https://doi.org/10.1109/ACCESS.2023.3248501.

22 Huang, J. and Encinar, J.A. (2007). *Reflectarray Antennas*. Hoboken, NJ: Wiley. ISBN: 978-0-471-74275-4.

23 Pozar, D.M., Targonski, S.D., and Syrigos, H.D. (1997). Design of millimeter wave microstrip reflectarrays. *IEEE Transactions on Antennas and Propagation* 45 (2): 287–296.

24 Arrebola, M., Encinar, J.A., and Barba, M. (2008). Multifed printed reflectarray with three simultaneous shaped beams for LMDS central station antenna. *IEEE Transactions on Antennas and Propagation* 56 (6): 1518–1527. https://doi.org/10.1109/TAP.2008.923360.

25 Zhou, M., Sτrensen, S.B., Kim, O.S. et al. (2013). Direct optimization of printed reflectarrays for contoured beam satellite antenna applications. *IEEE Transactions on Antennas and Propagation* 61 (4): 1995–2004.

26 Chaharmir, M.R., Shaker, J., and Legay, H. (2010). Dual-band Ka/X reflectarray with broadband loop elements. *IET Microwaves, Antennas and Propagation* 4 (2): 225–231.

27 Deng, R., Mao, Y., Xu, S., and Yang, F. (2015). A single-layer dual-band circularly polarized reflectarray with high aperture efficiency. *IEEE Transactions on Antennas and Propagation* 63 (7): 3317–3320.

28 Martinez-de-Rioja, D., Florencio, R., Martinez-de-Rioja, E. et al. (2020). Dual-band reflectarray to generate two spaced beams in orthogonal circular polarization by variable rotation technique. *IEEE Transactions on Antennas and Propagation* 68 (6): 4617–4626.

29 Rengarajan, S.R. (2013). Reflectarrays of rectangular microstrip patches for dual-polarization dual-beam radar interferometers. *Progress In Electromagnetics Research* 133: 1–15.

30 Carrasco, E., Barba, M., and Encinar, J.A. (2012). X-band reflectarray antenna with switching-beam using pin diodes and gathered elements. *IEEE Transactions on Antennas and Propagation* 60 (12): 5700–5708.

31 Mei, P., Zhang, S., and Pedersen, G.F. (2020). A low-cost, high-efficiency and full-metal reflectarray antenna with mechanically 2-D beam-steerable capabilities for 5G applications. *IEEE Transactions on Antennas and Propagation* 68 (10): 6997–7006.

32 Yu, Z.-Y., Zhang, Y.-H., He, S.-Y. et al. (2022). A wide-angle coverage and low scan loss beam steering circularly polarized folded reflectarray antenna for millimeter-wave applications. *IEEE Transactions on Antennas and Propagation* 70 (4): 2656–2667.

33 Costanzo, S., Venneri, F., Borgia, A., and Massa, G.D. (2020). Dual-band dual-linear polarization reflectarray for mmWaves/5G applications. *IEEE Access* 8: 78183–78192.

34 Cuong, H.D., Le, M.-T., and Dinh, N.Q. (2020). A reflectarray antenna using crosses and square rings for 5G millimeter-wave application. *2020 International Conference on Advanced Technologies for Communications (ATC)*, Vietnam.

35 Zhang, L., Zhang, J., He, Y. et al. (2022). A single-layer 10—30 GHz reflectarray antenna for the internet of vehicles. *IEEE Transactions on Vehicular Technology* 71 (2): 1480–1490.

36 Kitayama, D., Kurita, D., Miyachi, K. et al. (2019). 5G radio access experiments on coverage expansion using metasurface reflector at 28 GHz. *2019 IEEE Asia-Pacific Microwave Conference (APMC)*, 435–437. Singapore. IEEE.

37 Lu, Z., Fang, Y., Yi, H., and Li, L. (2019). Broadband reflectarray for millimeter wave coverage enhancement in indoor NLOS scenario. *2019 Cross Strait Quad Regional Radio Science and Wireless Technology Conference (CSQRWC)*, 1–3. IEEE.

38 Martinez-de-Rioja, E., Vaquero, Á.F., Arrebola, M. et al. (2022). Enhancement of 5G millimeter-wave coverage in indoor scenarios by passive shaped-beam reflectarray panels. *2022 16th European Conference on Antennas and Propagation (EuCAP)*, 1–5. Madrid, Spain. IEEE.

39 Martinez-de-Rioja, E., Vaquero, Á.F., Arrebola, M. et al. (2023). Passive intelligent reflecting surfaces based on reflectarray panels to enhance 5G millimeter-wave coverage. *International Journal of Microwave and Wireless Technologies* 15 (1): 3–14.

40 Rocca, P., Da Rù, P., Anselmi, N. et al. (2022). On the design of modular reflecting EM skins for enhanced urban wireless coverage. *IEEE Transactions on Antennas and Propagation* 70 (10): 8771–8784.

41 Prado, D.R., Arrebola, M., Pino, M.R., and Las-Heras, F. (2017). Improved reflectarray phase-only synthesis using the generalized intersection approach with dielectric frame and first principle of equivalence. *International Journal of Antennas and Propagation* 2017 (1): 3829390.

42 Zornoza, J.A. and Encinar, J.A. (2004). Efficient phase-only synthesis of contoured-beam patterns for very large reflectarrays. *International Journal of RF and Microwave Computer-Aided Engineering* 14 (5): 415–423. https://doi.org/10.1002/mmce.20039.

43 Martinez-de-Rioja, D., Encinar, J.A., Martinez-de-Rioja, E. et al. (2022). A simple beamforming technique for intelligent reflecting surfaces in 5G scenarios. *2022 International Workshop Antenna Technology (iWAT)*, 249–252. Dublin, Ireland. IEEE.

44 3GPP (2020). 3GPP Release 16 Description. *Tech. Rep. 3GPP TR 21.916 v16.0.0.* 3GPP. https://www.3gpp.org/release-16 (accessed 3 October 2024).

45 Encinar, J.A., Arrebola, M., de la Fuente, L.F., and Toso, G. (2011). A transmit-receive reflectarray antenna for direct broadcast satellite applications. *IEEE Transactions on Antennas and Propagation* 59 (9): 3255–3264. https://doi.org/10.1109/TAP.2011.2161449.

46 Caputo, D., Pirisi, A., Mussetta, M. et al. (2009). Neural network characterization of microstrip patches for reflectarray optimization. 3rd *European Conference on Antennas and Propagation (EuCAP)*, 2520–2522 (23—27 March 2009). Berlin, Germany.

47 Robustillo, P., Zapata, J., Encinar, J.A., and Arrebola, M. (2012). Design of a contoured-beam reflectarray for a EuTELSAT European coverage using a stacked-patch element characterized by an artificial neural network. *IEEE Antennas and Wireless Propagation Letters* 11: 977–980.

48 Robustillo, P., Zapata, J., Encinar, J.A., and Rubio, J. (2012). ANN characterization of multi-layer reflectarray elements for contoured-beam space antennas in the Ku-band. *IEEE Transactions on Antennas and Propagation* 60 (7): 3205–3214. https://doi.org/10.1109/TAP.2012.2196941.

49 Robustillo, P., Zapata, J., Encinar, J.A. et al. (2014). Accurate characterization of multi-resonant reflectarray cells by artificial neural networks. *The 8th European Conference on Antennas and Propagation (EUCAP)*, 2297–2299 (6—11 April 2014). The Hague, The Netherlands.

50 Prado, D.R., López-Fernández, J.A., Barquero, G. et al. (2018). Fast and accurate modeling of dual-polarized reflectarray unit cells using support vector machines. *IEEE Transactions on Antennas and Propagation* 66 (3): 1258–1270. https://doi.org/10.1109/TAP.2018.2790044.

51 Prado, D.R., López-Fernández, J.A., Arrebola, M., and Goussetis, G. (2019). Support vector regression to accelerate design and crosspolar optimization of shaped-beam reflectarray antennas for space applications. *IEEE Transactions on Antennas and Propagation* 67: 1659–1668. https://doi.org/10.1109/TAP.2018.2889029.

52 Florencio, R., Boix, R.R., Losada, V. et al. (2012). Comparative study of reflectarrays based on cells with three coplanar dipoles and reflectarrays based on cells with three stacked patches. *2012 6th European Conference on Antennas and Propagation (EUCAP)*, 3707–3710. Prague, Czech Republic. IEEE.

53 Florencio, R., Encinar, J.A., Boix, R.R. et al. (2015). Reflectarray antennas for dual polarization and broadband telecom satellite applications. *IEEE Transactions on Antennas and Propagation* 63 (4): 1234–1246.

54 Chaharmir, M.R., Shaker, J., and Legay, H. (2009). Broadband design of a single layer large reflectarray using multi cross loop elements. *IEEE Transactions on Antennas and Propagation* 57 (8): 3363–3366.

55 Cordente, F., Martinez-de-Rioja, E., Arboleya, A., and Encinar, J.A. (2023). Dual-band and dual-polarized reflectarray for intelligent reflecting surface applications in millimeter-wave 5G. *2023 17th European Conference on Antennas and Propagation (EuCAP*, 1–4. Florence, Italy. IEEE.

56 Soroka, R., Martinez-de-Rioja, E., Arboleya, A., and Encinar, J.A. (2022). Design of FSS-backed reflectarray cells for coverage-enhancing panels with suppressed out-of-band reflections in millimeter-wave 5G. *2022 16th European Conference on Antennas and Propagation (EuCAP)*, 1–5. Madrid, Spain. IEEE.

57 Massa, A., Oliveri, G., Salucci, M. et al. (2018). Learning-by-examples techniques as applied to electromagnetics. *Journal of Electromagnetic Waves and Applications* 32 (4): 516–541.

58 Prado, D.R., López-Fernández, J.A., Arrebola, M., and Goussetis, G. (2021). On the use of the angle of incidence in support vector regression surrogate models for practical reflectarray design. *IEEE Transactions on Antennas and Propagation* 69 (3): 1787–1792. https://doi.org/10.1109/TAP.2020.3015707.

59 Florencio, R., Boix, R.R., Carrasco, E. et al. (2013). Efficient numerical tool for the analysis and design of reflectarrays based on cells with three parallel dipoles. *Microwave and Optical Technology Letters* 55 (6): 1212–1216.

60 Schölkopf, B. and Smola, A.J. (2001). *Learning With KERNELS*, 1e. Cambridge, MA: The MIT Press.

61 Vapnik, V. (1999). *The Nature of Statistical Learning Theory*, 2e. New York, NY: Springer.

62 Chang, C.-C. and Lin, C.-J. (2011). LIBSVM: A library for support vector machines. *ACM Transactions on Intelligent Systems and Technology (TIST)* 2 (3): 27:1–27:27. https://www.csie.ntu.edu.tw/cjlin/libsvm.

63 Nelder, J.A. and Mead, R. (1965). A simplex method for function minimization. *The Computer Journal* 7 (4): 308–313.

64 Lagarias, J.C., Reeds, J.A., Wright, M.H., and Wright, P.E. (1998). Convergence properties of the Nelder—Mead simplex method in low dimensions. *SIAM Journal on Optimization* 9 (1): 112–147.

65 Balanis, C.A. (2016). *Antenna Theory: Analysis and Design*. Wiley.

66 Florencio, R., Boix, R.R., and Encinar, J.A. (2013). Enhanced MoM analysis of the scattering by periodic strip gratings in multilayered substrates. *IEEE Transactions on Antennas and Propagation* 61 (10): 5088–5099. https://doi.org/10.1109/TAP.2013.2273213.

67 Sherman, J. (1962). Properties of focused apertures in the Fresnel region. *IRE Transactions on Antennas and Propagation* 10 (4): 399–408. https://doi.org/10.1109/TAP.1962.1137900.

68 Hansen, R. and Bailin, L. (1959). A new method of near field analysis. *IRE Transactions on Antennas and Propagation* 7 (5): 458–467. https://doi.org/10.1109/TAP.1959.1144713.

69 Vaquero, Á.F., Pino, M.R., and Arrebola, M. (2023). Adaptive field-to-mask procedure for the synthesis of metalens antennas with complex near-field coverage patterns in 5G scenarios. *IEEE Transactions on Antennas and Propagation* 71 (9): 7158–7171. https://doi.org/10.1109/TAP.2023.3291431.

70 Bucci, O.M., Franceschetti, G., Mazzarella, G., and Panariello, G. (1990). Intersection approach to array pattern synthesis. *IEE Proceedings H (Microwaves, Antennas and Propagation)* 137 (6): 349–357. https://doi.org/10.1049/ip-h-2.1990.0064.

71 Feito, S.M., Vaquero, Á.F., Manzillo, F.F. et al. (2023). Comparison of classic and generalized intersection approach for the synthesis of transmitarrays with near-field constraints. *2023 International Conference on Electromagnetics in Advanced Applications (ICEAA)*, 566–570. https://doi.org/10.1109/ICEAA57318.2023.10297882.

72 Vaquero, Á.F., Arrebola, M., Pino, M.R. et al. (2021). Demonstration of a reflectarray with near-field amplitude and phase constraints as compact antenna test range probe for 5G new radio devices. *IEEE Transactions on Antennas and Propagation* 69 (5): 2715–2726. https://doi.org/10.1109/TAP.2020.3030969.

10

AI at the Physical Layer for Wireless Network Security and Privacy

Aly S. Abdalla[1*], *Bo Tang*[2], *and Vuk Marojevic*[1]

[1]*Department of Electrical and Computer Engineering, Mississippi State University, MS, USA*
[2]*Department of Electrical and Computer Engineering, Worcester Polytechnic Institute, Worcester, MA, USA*

CHAPTER MENU

10.1 Introduction

Amidst an increasingly interconnected world, network security and privacy have become paramount [1, 2]. The incessant menace of cyberattacks, data breaches, and privacy violations looms over organizations and individuals alike. In an ever-evolving world where network security and privacy stand as formidable challenges, the integration of artificial intelligence (AI) technologies holds great potential in fortifying defenses and safeguarding the confidentiality and integrity of vital information [3]. Within the realm of AI, an extensive spectrum of groundbreaking technologies awaits exploration, encompassing statistical learning theory, natural language processing, and deep learning (DL). These

*The work of A.S. Abdalla and V. Marojevic was in part supported by the NSF and Office of the Under Secretary of Defense (OUSD) – Research and Engineering, under Grant ITE2326898, as part of the NSF Convergence Accelerator Track G: Securely Operating Through 5G Infrastructure Program.

Artificial Intelligence for Future Networks, First Edition.
Edited by Mohammad A. Matin, Sotirios K. Goudos, and George K. Karagiannidis.

cutting-edge advancements pave the way for innovative approaches to address network security and privacy concerns [4]. This amalgamation empowers systems to learn, analyze, and make intelligent decisions devoid of explicit programming [5]. AI's capability for managing vast data volumes, identifying patterns, and executing real-time decisions renders it an indispensable asset in the ongoing battle against cyber threats [6]. By analyzing large sets of data, identifying potential threats in real-time, and detecting anomalies, AI empowers organizations to proactively address security concerns. AI can identify malicious actors in a network, thwarting their attempts before they have the chance to do severe damage [7]. It can automate the detection of suspicious activity across networks, helping organizations keep an eye on activities that may be harmful or illegal and respond to threats faster [8]. Since AI is completely automated, it does not require human input to operate effectively; this means that mistakes are less likely to happen, and data remains confidential. In addition, AI-based solutions can adapt to new threats, making them a moving target for hackers. These advancements empower systems to learn, analyze, and make intelligent decisions autonomously without explicit programming.

Within the realm of wireless networks, AI can wield transformative influence by enhancing threat detection and prevention, automating security operations, and facilitating adaptive defense mechanisms [9]. Moreover, AI-driven privacy solutions offer promising avenues to address mounting concerns surrounding data privacy [10]. It can be effectively leveraged for secure data analytics, for implementing robust access controls, and for prioritizing user-centric privacy protection in wireless networks [11, 12].

10.2 Network Security and Privacy Threats and Vulnerabilities

Numerous threats and vulnerabilities lurk within the dynamic landscape of communication networks, including beyond 5G (B5G) and 6G, capable of compromising the integrity and confidentiality of sensitive data, and the availability of these networks [13]. By investigating this field, we discover a multitude of risks that demand attention from both academia and industry.

10.2.1 Security Threats

- *Eavesdropping*: Unauthorized individuals can secretly intercept wireless communications, accessing sensitive information without the knowledge of the parties involved. This can lead to the exposure of personal data, financial details, or confidential business information [14, 15].

- *Jamming*: Deliberate interference with wireless signals can disrupt network connectivity and communication. Attackers aim to disrupt legitimate transmissions, causing service outages and hindering critical operations [16].
- *Reply attack*: Attackers capture and replay previously recorded data transmissions to deceive the network or gain unauthorized access. By exploiting vulnerabilities in security protocols, they can bypass authentication mechanisms and compromise network resources [17].
- *Network spoofing*: This crafty attack involves the creation of malicious wireless access points that cunningly imitate legitimate networks. The attackers behind this scheme set up these deceptive access points to tempt unsuspecting users into connecting to them, unknowingly falling into the trap. Once connected to these rogue access points, users become vulnerable to various malicious activities orchestrated by the attackers. The attacker can request subscriber identities, locations, or other private information or intercept and manipulate data transmissions between devices [18].
- *Man-in-the-middle (MITM) attack*: MITM attackers position themselves between two communicating parties, intercepting and potentially altering the exchanged data. It also allows them to impersonate legitimate users and gain unauthorized access to networks and services [19].
- *Sidelink attack*: Sidelink attacks target direct communications among nearby devices in wireless networks. Attackers exploit vulnerabilities in the communication protocols, aiming to disrupt or manipulate the exchanged data. This can disrupt vehicular safety or other mission critical applications enabled by device-to-device communications and IoT networks [20].
- *Malware*: Malware, a persistent and ever-evolving menace, encompasses various forms of malicious software designed to infiltrate wireless networks and devices. It sneaks into networks through infected email attachments, deceptive downloads, or compromised websites. Once inside the network, malware wreaks havoc by compromising security measures and pilfering sensitive data. Financial loss, data breaches, and unauthorized access are some of the potential consequences [21].

10.2.2 Identifying and Assessing Network Security and Privacy Threats

The identification and assessment of communication network security and privacy threats play a vital role in safeguarding the integrity and confidentiality of sensitive information. By understanding the potential risks and vulnerabilities, organizations can develop effective strategies to mitigate these threats.

For identifying and assessing network security and privacy threats, a comprehensive analysis of the communication network and its components is essential.

This entails examining the various communication protocols, devices, and systems in use. By evaluating the network architecture and topology, potential entry points for threats can be identified. Threat modeling is used to identify and categorize threats and anticipate potential attack scenarios [22].

The assessment phase encompasses vulnerability scans and penetration tests [23]. Vulnerability scans aim to identify known vulnerabilities, misconfigurations, or outdated software versions in the network infrastructure. Penetration tests simulate real-world attacks to uncover potential weaknesses in the network's defenses. These assessments aid in identifying vulnerabilities that could be exploited by attackers. Risk assessment then evaluates the potential implications of identified threats and attacks. This involves assessing the likelihood of an attack occurring and the potential consequences such as data breaches, unintended system access, and network failure. Risk assessment is employed to prioritize new security measures.

The identification and assessment process for active eavesdropping revolves around monitoring communication traffic for any suspicious activity. This includes analyzing data packets and network logs to identify unauthorized interception attempts [24]. For jamming attacks, the identification and assessment process involves RF sensing and monitoring of signal strength and signal quality in the communication networks. Analyzing system logs for any signal disruptions or anomalies helps identify such attacks [25]. In the case of spoofing attacks, the identification and assessment process revolves around monitoring unauthorized access points and unfamiliar devices within the network. Authorization and authentication logs of the system and device fingerprints are examined for anomalies that may indicate the presence of spoofed networks or devices [18]. The identification and assessment process of MITM attacks focuses on analyzing network traffic patterns for any unauthorized relays. Denial of service (DoS) attacks may be identified by monitoring the network performance and processing the feedback from end devices. The network's response to common DoS attacks can be evaluated by performing stress tests that simulate heavy traffic [26]. Monitoring repeated or duplicated data packets within the network can be used for identification and assessment of replay attacks. Communication networks that identify potential replay attacks can take early actions to maintain the integrity and authenticity of data transmissions [27]. In the case of sidelink attacks, the identification and assessment procedures focus on analyzing sidelink network traffic and behavior. By evaluating the effectiveness of sidelink security mechanisms and detecting any anomalies or vulnerabilities, networks can proactively defend against sidelink attacks and protect critical wireless communication channels [28].

10.2.3 Exploiting Vulnerabilities: Techniques and Attack Vectors

Exploiting vulnerabilities and attack vectors refers to the process of taking advantage of weaknesses, entry points, or flaws in a system or network to gain unauthorized access, manipulate data, inject malicious inputs, disrupt services, take advantage of insecure network connections, introduce rogue devices, or adhoc networks, or other malicious activities [29]. In the context of wireless communication systems, it entails identifying and capitalizing on security weaknesses within the network, devices, or protocols to compromise the system's integrity and confidentiality. The vulnerabilities for common wireless attacks are summarized in continuation.

- *Eavesdropping*: The eavesdropping attack can take advantage of weak or no encryption or transmission patterns by capturing and analyzing wireless transmissions, monitoring network traffic, and intercepting data packets [24].
- *Jamming*: The jamming attack exploits the lack of jamming detection mechanisms, such as frequency diversity, and vulnerable communication channels,

Table 10.1 Identifying and assessing communication network security and privacy threats.

Attack type	Description	Vulnerability	Identification process	Assessment process
Eavesdropping	Unauthorized interception of wireless communication	Weak encryption protocols	Monitoring wireless network traffic for suspicious activity	Analyzing captured network packets for unauthorized data access
Spoofing	Impersonation of a legitimate network or device	Lack of authentication mechanisms	Monitoring for unauthorized access points and unfamiliar devices	Examining network logs and device fingerprints for anomalies
Man-in-the-Middle	Interception and alteration of communication	Lack of end-to-end encryption	Analyzing network traffic patterns for unauthorized relays	Inspecting network logs and analyzing packet headers
Denial of Service	Disruption or degradation of network or service	Network congestion or resource exhaustion	Monitoring network performance and abnormal traffic patterns	Conducting stress tests to simulate heavy traffic load
Jamming	Intentional interference with wireless signals	Transmission interference	Monitoring signal strength and quality in the network	Analyzing network logs for signal disruptions and abnormalities
Replay Attack	Capture and replay of legitimate data packets	Weak or nonexistent message integrity checks	Monitoring for repeated or duplicated data packets	Examining network logs for duplicated or out-of-sequence packets

e.g., control channels and signals, by sending high-power interference signals on specific frequency bands, targeting critical communication links, disrupting wireless communication transmissions [25].

- *Spoofing*: The spoofing attack is enabled by weak mutual authentication mechanisms and the implicit trust that end devices have in the network. Rogue access points can exploit this by impersonating authorized access points to lure user equipment to connect to them [18].
- *Replay attack*: The replay attack benefits from a lack of message integrity checks and timestamp verification as well as insufficient session validation. The attacker intercepts and records legitimate messages, then replays them to deceive the system of gaining unauthorized access to services and data [27].
- *Man-in-the-middle attack*: The MITM attack leverages weak authentication and encryption mechanisms and insecure key exchange protocols. The attacker positions itself strategically between two or more communicating parties for intercepting, altering, and relaying or rerouting communication data [26].

Table. 10.1 summarizes the vulnerabilities, identification, and assessment processes for the abovediscussed attacks on wireless communication networks.

10.3 Fundamentals of AI for Network Security and Privacy

AI-based security and privacy solutions take advantage of the unique abilities of AI to process vast amounts of data, learn from patterns, and make intelligent decisions without explicit programming. By leveraging AI algorithms, networks can proactively detect, mitigate, and respond to cyber threats, safeguarding sensitive data and ensuring privacy [3, 4, 6].

AI-enabled secure communication systems are capable of continuously analyzing network data and detecting signs of suspicious behavior or potential security breaches. This proactive stance allows for timely mitigation of threats before they cause major harm. AI solutions rely on their adaptability and constant learning capacity; specifically, as the security and privacy threats evolve, AI algorithms can update their knowledge base to recognize and respond to emerging attack patterns. This innate intelligence fosters a robust defense against ever-changing security and privacy attacks.

AI solutions can be classified into these learning paradigms: supervised, unsupervised, reinforcement learning, generative adversarial networks (GANs), federated learning (FL), and ensemble learning. Figure 10.1 presents different AI algorithms with their different applications that can be applied to empowering network security and privacy.

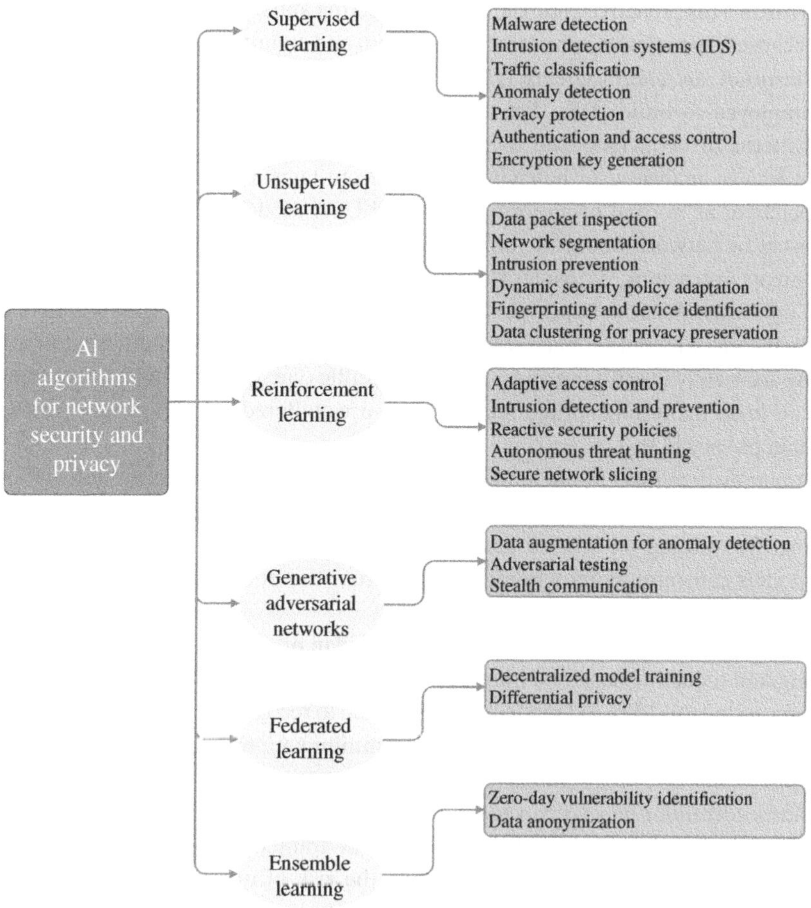

Figure 10.1 The AI algorithms classifications with their network security and privacy applications.

10.3.1 Supervised Learning

Supervised learning can be applied to many applications, including enhancing the security and privacy of communication networks. This category of AI algorithms relies on labeled training data to learn patterns and make predictions [30]. In the context of communication system security and privacy, supervised learning can be leveraged in various ways:

- *Malware detection*: Supervised learning models can be trained on large data sets containing labeled examples of known malware and benign files. These models can then classify incoming files or network traffic to identify potential malware threats. This proactive approach aids in detecting and mitigating malware before it causes harm to the communication system and compromise user privacy [31].

- *Intrusion detection systems (IDSs)*: Supervised learning algorithms can be employed to build IDSs that analyze network traffic and identify abnormal patterns or behaviors. The IDSs can also used be for user behavior analytics (UBA) to analyze user behavior patterns from tagged data and classify users' behavior as normal or abnormal. IDS and UBA models can detect suspicious users or network behavior indicative of unauthorized access, potential cyber threats, intrusions, or insider threats [32].
- *Traffic classification*: Supervised learning can be used to analyze the content and metadata of incoming network traffic and classify the data into different types. By accurately identifying and filtering out traffic types, communication systems can implement appropriate security measures tailored to each type, ensuring data privacy [33].
- *Anomaly detection*: Many anomaly detection systems that identify unusual or suspicious activities within communication networks leverage supervised learning. By training on labeled data sets representing normal behavior, these AI models can raise alerts when deviations from the norm occur, helping in the early detection of security breaches [34].
- *Privacy protection*: Supervised learning can aid in privacy protection by identifying and redacting sensitive information in data. For example, natural language processing (NLP)-based models can be trained to recognize and mask personally identifiable information in text-based communications to ensure compliance with data privacy regulations [35].
- *Authentication and access control*: User authentication mechanisms can benefit from supervised learning. Models can be trained to recognize patterns of legitimate user login behavior, reducing the risk of unauthorized access and enhancing overall system security [36].
- *Encryption key generation*: Supervised learning algorithms can be employed for generating strong encryption keys for secure communication between users and systems. By learning from annotated data of secure key pairs, these models can produce encryption keys that are resistant to attacks [37].

Overall, supervised learning solutions offer powerful tools to enhance the security and privacy of communication systems. By learning from labeled data, these models can accurately identify and respond to threats, making communication networks more resilient and secure. However, it is essential to ensure the quality and diversity of designated data sets to develop effective and robust supervised learning models. In addition, regular updates and retraining of these models are necessary to adapt to evolving cyber threats and maintain the highest level of security and privacy.

10.3.2 Unsupervised Learning

Unsupervised learning can efficiently identify patterns, anomalies, and underlying structures within the data without the need for labeled examples [38]. Some key applications of unsupervised learning in wireless network security domain includes the following.

- *Data packet inspection*: Unsupervised learning models can analyze data packets in real time to identify unusual or malicious patterns that may indicate an ongoing attack. Unsupervised feature extraction methods, such as principal component analysis and t-distributed stochastic neighbor embedding, reduce the dimensionality of the data while retaining its essential information [39]. This is particularly useful for visualizing high-dimensional network data and identifying anomalies. By autonomously detecting deviations from established baselines, such as unusual spikes in data transfer or unexpected port scanning, these algorithms can trigger timely alerts for further investigation. This aids in ensuring the integrity and security of data transmissions [40].
- *Network segmentation*: Unsupervised learning can be applied to segment the communication network based on the similarity of devices or user behavior. By creating distinct segments, the network can apply tailored security policies and access controls, reducing the impact of potential security breaches. In addition, unsupervised learning can be employed to cluster sensitive data into different groups, ensuring that privacy-sensitive information is stored securely and accessed only by authorized users. This technique helps minimize the risk of data exposure and unauthorized access [41].
- *Intrusion detection*: Unsupervised learning can be employed to continuously monitor network activity and identify potential intrusion attempts in real time. Unsupervised anomaly detection algorithms, such as isolation forest and One-Class SVM, are designed to identify rare or abnormal instances in the data. By analyzing network behavior, these models can detect suspicious activities that may not match known attack patterns [42].
- *Dynamic security policy adaptation*: Unsupervised learning can dynamically adapt security policies based on real-time network activity. This flexibility allows the network to respond swiftly to emerging threats [43].
- *Fingerprinting and device identification*: Device fingerprinting combined with unsupervised learning can identify and distinguish devices connected to a network based on their communication traits without the need for labeled training data [44]. The process of device fingerprinting involves extracting relevant features from the network traffic, such as packet size, timing, protocols used, data payloads, and communication ports. By analyzing network traffic features and using clustering algorithms, devices with similar characteristics can be grouped, allowing for more accurate device identification and enhanced security [45].

- *Data clustering for privacy preservation*: Clustering algorithms are commonly based on unsupervised learning. These include K-means, hierarchical clustering, and density-based spatial clustering of applications with noise (DBSCAN), and they can be employed to cluster sensitive data into groups, ensuring that privacy-sensitive information is stored securely and accessed only by authorized users [46]. Clustering works on the principle of maximizing intra-cluster similarity and minimizing inter-cluster similarity.

The ability of unsupervised learning to autonomously identify patterns and anomalies make them valuable tools for network security and privacy. These techniques enable efficient data analysis, real-time threat detection, and automated decision-making, contributing to a more secure and privacy-conscious communication environment. These solutions encompass a wide range of clustering, density estimation, feature extraction, and dimensionality reduction techniques, each tailored to address specific security and privacy challenges in communication networks, among others.

10.3.3 Reinforcement Learning

Reinforcement learning (RL) offers a promising approach for enhancing the security and privacy of communication systems through dynamic decision-making and adaptive strategies [47]. RL algorithms are designed to enable agents to learn optimal actions in an environment by interacting with it and receiving feedback in the form of rewards or penalties [48]. In the context of communication systems, RL can be leveraged to improve network security and privacy by continuously adapting to evolving threats and dynamically adjusting security measures.

- *Adaptive access control*: RL can be used to optimize access control policies based on real-time network conditions and user behavior. By observing network activities, the RL agent can learn to grant or deny access to specific resources, applications, or devices based on their historical interactions and the ongoing security posture of the system [49]. RL agents can adapt to changes in user behavior and network conditions, effectively mitigating potential security risks.
- *Intrusion detection and prevention*: RL can be applied to design intelligent intrusion detection and prevention systems. By modeling the network environment as a Markov decision process (MDP), the RL agent learns to detect anomalous behavior or potential attacks [50]. Upon identifying suspicious activities, the RL agent can initiate appropriate countermeasures or inform security analysts.
- *Reactive security policies*: RL agents can adaptively update security policies based on changes in the network environment. As the threat landscape evolves, potential gaps in security policies can be found and adjusted by leveraging RL [51]. RL facilitates multiagent collaboration to enhance security in complex

communication networks. Agents can learn to cooperate and communicate to detect and respond to sophisticated or coordinated attacks that involve multiple vectors. This adaptability ensures that security measures remain effective against emerging threats.

- *Autonomous threat hunting*: RL can empower autonomous threat hunting agents that continuously search for potential vulnerabilities or signs of intrusion within the network. Such agents dynamically explore the network, identify weak points and potential risks that require immediate attention [52]. RL can also be employed to detect zero-day threats that exploit unknown vulnerabilities. The RL agent learns from the behavior of known threats and applies this knowledge to identify new attack patterns that may indicate the presence of a zero-day exploit.
- *Secure network slicing*: In 5G and beyond networks, RL can optimize network slicing in terms of security, ensuring that the critical network slices are isolated and protected from other slices that may implement weaker security feature. RL agents can learn to dynamically allocate resources to ensure that critical resources, such as bandwidth and processing power, are allocated efficiently while considering different security requirements across different services and applications [53]. This includes prioritizing security-critical tasks and ensuring that sufficient resources are available for security monitoring and analysis [54].

In all these applications, RL agents learn from interactions with the environment, receiving rewards for actions that improve security and privacy and penalties for actions that compromise them. The RL training process involves extensive simulations and real-world experiments to optimize the agent's decision-making capabilities [55]. RL's ability to adapt to changing conditions and learn from experience makes it a valuable tool for enhancing the security and privacy of communication systems in the face of evolving cyber threats. As RL techniques continue to advance, they hold immense potential for revolutionizing the field of network security and privacy.

10.3.4 Generative Adversarial Networks

GANs have emerged as a groundbreaking technique to improve the security and privacy of communication systems. GANs have shown remarkable success in various fields, and their potential in enhancing communication system security is no exception [56]. These advanced AI models comprise two neural networks, the generator and the discriminator networks, working collaboratively in a competitive process. The generator network generates fake data samples and the discriminator network's role is to distinguish between real and fake data samples [56]. Through this adversarial process, GANs can learn to generate highly realistic data that closely resembles real data, making them well-suited for various security and privacy applications:

- *Data augmentation for anomaly detection*: GANs can be used to augment the data set used for training anomaly detection systems. By generating synthetic data samples that mimic normal network behavior, GANs help create a more comprehensive and diverse data set [57]. This enables anomaly detection models to better distinguish between normal and abnormal network activities, improving the accuracy of intrusion detection.
- *Adversarial testing*: GANs can play a critical role in evaluating the resilience of security measures by conducting adversarial testing. By generating adversarial examples, which are data samples intentionally designed to deceive security systems, GANs enable security systems to assess the robustness of their defenses [58]. This approach helps identify vulnerabilities and weaknesses in security algorithms, allowing for targeted improvements.
- *Stealth Communication*: GANs enable the design of covert communication. By generating data samples resembling benign network traffic, GANs create stealthy channels that are difficult for unauthorized users to identify. Such covert communication channels are valuable in scenarios where secure and private data exchange is essential. In addition, GANs can be applied to stenography, a technique used to hide secret information within innocuous-looking data [59]. By training the generator to embed secure data within images or other media, GANs can protect sensitive information from interception during transmission.

GANs' ability to generate realistic and diverse data samples provides unique opportunities for enhancing security and privacy in communication systems. However, the use of GANs in security applications also raises challenges, such as ensuring that the generated data is sufficiently close to real data and avoiding potential bias or adversarial attacks on the GAN itself. As GAN technology continues to advance, addressing these challenges will be critical to realizing the full potential of GANs for secure and private communication systems.

10.3.5 Federated Learning

FL has emerged as a cutting-edge AI solution to address the challenges of security and privacy in communication systems. It enables collaborative model training across distributed devices while preserving data privacy and confidentiality [60]. The training process takes place locally on individual devices, and only model updates are shared with the central server [61]. More details and applications of FL for enhancing security and privacy in communication systems are elaborated in continuation.

- *Decentralized model training*: Decentralized model training is a pivotal concept in FL, empowering secure and privacy-preserving AI model development across distributed devices [62]. In traditional machine learning approaches, all data are centralized and sent to a central server for model training. This centralized approach raises concerns about data privacy, as raw data must be transmitted and stored on the central server, potentially exposing sensitive information. FL addresses these concerns by distributing the model training process across decentralized devices, such as smartphones or edge devices. The model updates from all the devices are securely aggregated without revealing individual data points [63]. Techniques such as secure multiparty computation are used to ensure that the aggregated updates are privacy-preserving [64]. Instead of sending raw data to a central server, each device performs local model training using its locally available data. This ensures that sensitive data never leaves the device, eliminating the risk of data exposure during transmission.
- *Differential privacy*: Differential privacy is an integral component of FL to enhance user privacy and ensure data protection during the model aggregation process. The challenge however lies in preventing malicious actors from deducing specific user information from these different locally trained models and aggregated model updates. Differential privacy addresses this concern by introducing randomness into the model updates before aggregation. This process adds carefully calibrated noise to the updates, making it mathematically challenging for any malicious entity, including the central server, to discern individual user contributions to the model [65]. However, it is important to carefully balance the amount of noise added to the model for privacy protection and the model's accuracy. The aggregated model maintains its utility and effectiveness for making accurate predictions while ensuring the privacy of user data.

10.3.6 Ensemble Learning

Ensemble learning is an advanced AI technique that has emerged as a promising solution for enhancing the security and privacy of communication systems. It leverages the power of multiple models working together to achieve better performance and robustness than any individual model by itself [66]. This collaborative approach is particularly advantageous for addressing complex security challenges in communication networks. Ensemble learning techniques include bagging, boosting, and random forest [66]. Bagging involves training multiple instances of the same base model on different subsets of the training data. These models are then combined by averaging their predictions, reducing variance, and enhancing their overall performance. Boosting focuses on iteratively improving the performance of weak models by assigning higher weight to misclassified samples.

AdaBoost and gradient boosting are popular boosting algorithms [67]. Random forest is an extension of bagging that employs decision trees as base models. It reduces overfitting and increases diversity by introducing randomness during tree construction. In the context of security and privacy for communication systems, ensemble learning offers several benefits.

- *Zero-day vulnerability identification*: Zero-day vulnerabilities are security flaws in software that are unknown to the vendor and, consequently, lack patches or defenses. Detecting these vulnerabilities before they are exploited is critical for maintaining network security [68]. For identifying zero-day vulnerabilities, ensemble learning amalgamates the outputs of multiple vulnerability assessment tools [69]. These tools may encompass static code analysis, dynamic analysis, and behavioral analysis techniques. The outputs of the different vulnerability assessment tools are merged within the ensemble learning framework. This involves combining the individual tool's outputs, which may include information on potential zero-day vulnerabilities, false positives, and normal behavior. Ensemble learning employs aggregation strategies to make a final decision regarding the presence of a zero-day vulnerability [70]. It enhances the chances of detecting previously unknown vulnerabilities by capitalizing on the diversity of detection strategies. Common aggregation methods include majority voting, weighted averaging, and stacking.
- *Data anonymization*: Data anonymization is a critical practice for preserving privacy when sharing sensitive information. Ensemble learning techniques provide a robust approach to enhance data anonymization while maintaining data utility [71]. This process involves transforming original data into a form that hides individual identities while still enabling useful analysis. Ensemble learning adds an extra layer of sophistication to this process. When applied to data anonymization, ensemble learning employs a variety of methods and combines their outcomes to achieve a more reliable and comprehensive anonymization result. Ensemble learning ensures the diversity of anonymization methods employed. These methods include generalization, suppression, perturbation, or adding noise to data [72]. The outputs of individual anonymization models are aggregated to generate a final data set. This aggregation helps in minimizing the limitations and biases of any single method. Ensemble models also integrate privacy-preserving metrics to evaluate the level of data anonymization achieved. These metrics assess the risk of reidentification while ensuring that the transformed data maintains its analytical usefulness. Ensemble learning not only focuses on privacy but also aims at enhancing the utility of anonymized data [73]. By considering a diverse range of strategies, the resulting data maintains its original characteristics, making it valuable for analytical purposes.

10.4 AI-Driven Physical Layer Security Solutions

Physical layer security (PLS) techniques have evolved independently from traditional cryptographic security mechanisms, aiming to enhance the confidentiality, availability, and privacy of wireless communication services. PLS involves a deliberate exploration of protocol and environmental attributes, or physical layer (PHY) communications characteristics to bolster the security of transmitted messages across communication channels [74]. This extra layer of security complements the conventional security measures deployed at higher protocol layers. The potential advantages associated with adopting PLS strategies, as opposed to traditional cryptographic approaches, are discussed below.

1) PLS functions autonomously, in contrast to cryptography, which operates at higher layers and involves greater computational complexity [75]. This self-sufficiency ensures the ability to maintain secure and dependable communication between various transceivers, even when eavesdroppers with substantial computing resources are present.
2) PLS strategies can be seamlessly integrated into existing systems such as 5G networks or future 6G networks, often requiring minimal additional hardware resources [76]. This simplifies the incorporation of PLS with other security methodologies, effectively enhancing the confidentiality and privacy of wireless communication networks [77].

Various PHY layer attributes hold the potential to be harnessed for PLS. These attributes encompass factors such as channel state information (CSI), power allocation, and signal characteristics. Substantial research endeavors have been directed towards the creation of a variety of PLS methodologies that use these inherent PHY attributes with the ultimate goal of enhancing the security of wireless communication networks [78, 79].

The wireless communication channel can be leveraged for PLS. The inherent unpredictability within such channel, stemming from noise, multipath propagation, mobility, and other communication dynamics can be strategically managed to authenticate users and curtail the interception of data by malicious nodes. Leveraging AI within the realm of PLS introduces a transformative dimension that can amplify the efficacy and efficiency of these techniques [80]. By employing AI in PLS the system can not only detect anomalies but also adapt to evolving threats, thereby establishing a new foundation for securing cellular communication networks [81]. At the core of this augmentation lies the ability of AI algorithms to enhance the accuracy of key PLS attributes, such as CSI, power allocation, and signal classification. AI can identify correlations between these attributes and security vulnerabilities, facilitating the development of predictive models that anticipate potential breaches. This proactive stance allows network administrators to take preemptive measures to thwart security threats before they materialize.

Diverse methods have been explored to improve the attributes of secure channels while enhancing the received signal strength at the designated receivers. Conversely, alternative strategies have been put forth to deliberately weaken the signal received by potential attackers. This weakening is achieved by deliberately compromising the quality of the attacker's channel [82]. Some of these techniques are discussed in continuation.

10.4.1 Intelligent Beamforming

Intelligent beamforming through AI enables enhancing the PLS of wireless communication networks. This advanced technique optimizes signal transmission, reception, and directionality, thereby fortifying the security and privacy of wireless communication [83].

Beamforming refers to the strategic focusing of transmitted or received signals in a specific direction. AI algorithms can optimize beamforming by dynamically adjusting antenna weights to counteract eavesdropping attempts or interference [84]. Adaptive beamforming that adjusts the precoding weights in real time to accommodate changing network conditions and potential threats. The transmission or reception direction is steered toward the intended user while simultaneously minimizing signal leakage toward unintended recipients. To this end, RL can be effectively employed to continuously refine the directionality of beams [85]. By analyzing real-time CSI feedback, the AI system can intelligently adjust beamforming patterns to ensure that the transmitted signals are confined to the intended receiver. Moreover, AI algorithms can optimize beamforming patterns to create nulls or regions of minimal signal strength in the directions of expected eavesdroppers. This null-steering technique ensures that even if an eavesdropper manages to intercept some signal energy, the received signal strength will be weak, rendering the intercepted data useless. This adaptability ensures that even in dynamic and unpredictable environments, the system can proactively adjust the beamforming patterns to maintain secure and private communication.

Numerous research efforts and work items have been dedicated to exploring the integration of AI for optimizing beamforming and antenna array management in the context of secure 5G and 6G communications. The authors of Deep Learning Driven 3D Robust Beamforming for Secure Communication of UAV Systems [86] propose a DL-driven 3D robust beamforming method for secure communication in unmanned aerial vehicle (UAV) systems. The main contribution of the paper is the development of a neural network that optimizes the beamformer for confidential signal and artificial noise (AN) based on partial CSI of both the legitimate

UAV and the eavesdropping UAV. The goal is to maximize the average secrecy rate of the system. The proposed method outperforms benchmarks in terms of secrecy rate and beam steering flexibility, as demonstrated through simulations. The work presented in [87] proposes a new relay system that uses a deep neural network to design a secure beamforming vector that adapts to the current channel of the legitimate user and is resistant to eavesdropping. The model takes into account the statistical characteristics of the channel and learns the CSI to design secure beamforming, effectively preventing eavesdropping. The proposed model achieves a bit error rate (BER) close to the Shannon limit for the intended receiver demodulating the received signals, while ensuring that eavesdroppers cannot demodulate the transmitted signals. Simulation results demonstrate that the proposed model meets the predefined BER requirements and effectively enhances the security of the system. The authors of [88] focus on the secure beamforming problem in massive multiple-input multiple-output (MIMO) systems. The work presents an RL approach to optimize the system performance without requiring channel estimation, thus avoiding the issue of imperfect CSI in massive MIMO systems. The RL algorithm dynamically optimizes the system performance by observing the state of the cache and acquiring feedback from the delay of packets to reduce the delay of the system without using CSI. The paper presents a theoretical analysis and numerical results using the asynchronous advantage actor–critic (A3C) algorithm to solve the RL problem and compare the results with a randomized policy in a time-variant wireless environment.

10.4.2 AI-Based Radio Frequency Fingerprinting Techniques

AI-based radio frequency (RF) fingerprinting techniques have been receiving increasing attention in the realm of wireless communication networks. They harness the power of AI-driven algorithms to significantly enhance the security and privacy of wireless communications by employing intelligent methods for signal transmission, reception, and identification. In essence, RF fingerprinting exploits the unique characterization of wireless devices based on their transmitted signals. The uniqueness of RF fingerprints arises from the analog element imperfections created in the manufacturing process and different aging processes. Such elements in the transmitter include digital-to-analog converters, band-pass filters, frequency mixers, and amplifiers [89]. By leveraging RF fingerprinting techniques empowered by AI, there is a great potential to enhance the PLS by identifying potential spoofing or impersonation attacks that emit wireless signals with different RF fingerprints than enrolled (legitimate) wireless devices.

The transmitter identification system based on RF fingerprinting typically involves three stages: signal collection, feature (fingerprint) extraction, and classification. The signal collection stage requires capturing raw in-phase and quadrature I/Q samples of the received signal. Then, fingerprints can be extracted in different parts of the signal, particularly in the steady-state part [90] and in the transient part [91, 92]. The classification stage is then performed based on the extracted features to identify different transmitters. The key to the success of RF fingerprinting is the extraction of useful features that can distinguish different transmitters. With the advances in deep neural networks, which have shown tremendous success in many technological domains. The feature extraction stage can be integrated with the classification stage for learning high-level feature representations to maximize the overall classification performance.

In recent research endeavors, RF fingerprinting techniques based on deep neural networks have shown remarkable potential. For instance, the authors of [93] propose a deep recurrent neural network-based approach for device identification and authentication using RF fingerprints. The method achieves high accuracy in distinguishing between authorized and unauthorized devices, making it a promising tool for enhancing wireless security. Reference [94] describes a differential constellation trace figure (DCTF) for PHY layer feature extraction and a convolutional neural network (CNN)-based classification scheme to identify mobile phones operating in a 2G GSM network. Their theoretical analysis shows that the differential process of GSM signals can effectively reflect the characteristics of RF fingerprints of different phones and that the CNN can be trained to achieve state-of-the-art performance. A CNN has also been integrated into a PHY layer authentication method for identifying satellite transmitters [95].

10.4.3 AI-Assisted Power Control

At the PHY layer, wireless communication relies on the transmission of electromagnetic signals, which naturally radiate beyond the intended recipients, making such signals susceptible to eavesdropping. AI-assisted power control addresses this challenge by dynamically optimizing the transmission powers of network nodes, such as base stations and user equipment, in response to changing network conditions and security threats [96]. AI models can trained to estimate the secrecy rate, representing the data rate that eavesdroppers cannot decode transmitted information [97, 98]. By continuously monitoring network parameters, AI systems can make real-time decisions to maximize the secrecy rate while maintaining the desired quality of service for legitimate users.

The work presented in [99] focuses on enhancing the secrecy rate of femto-edge users (FEUs) in dense femtocell deployments within heterogeneous networks. FEUs are susceptible to security attacks due to their distributed and dynamic nature. The objective is to improve data transmission security and reliability. The study formulates a power control problem for terahertz-enabled femto-base stations (FBSs) with FEUs and an eavesdropper. Addressing the complex nonconvex optimization problem, the paper employs a MDP to transform it into a multiagent deep RL challenge. Multiagent Q-learning is used to enhance the learning and reduce the output dimensionality. A deep deterministic policy gradient (DDPG) is then applied for power control. Simulation results demonstrate that the proposed technique enhances the average secrecy rate of the network compared to existing state-of-the-art approaches. The framework provided in [100] presents an enhanced secure scheme tailored for wireless communication systems facing intelligent threats in the form of eavesdropping, jamming, and spoofing. The conventional secure approach relies on a Q-learning-based algorithm to achieve the Nash equilibrium (NE) in a zero-sum game between the transmitter and the attacker. Beamforming is employed at the transmitter to suppress eavesdropping, while the receiver uses filtering to counter jamming and spoofing. This combined approach curtails the advantages the attacker can gain in the game. In addition, a Q-learning-based power control strategy is employed to attain a new NE. Simulation outcomes show the efficacy of the proposed scheme for mitigating intelligent attacks, surpassing conventional methods in terms of secrecy performance. Reference [101] proposes a secrecy-aware energy-efficient strategy for a two-tier heterogeneous network (HetNet). This HetNet comprises a macrocell operating in the sub-6 GHz spectrum and multiple millimeter wave (mmWave) picocells. Within each picocell, several users coexist alongside an eavesdropper that intercepts picocell user signals. The scheme tackles a complex joint optimization problem involving power control, channel allocation, and beamforming. More precisely, it employs the dueling and double Q architecture from a dueling double-deep Q-network (D3QN) to optimize power control, channel allocation, and beamforming vectors. The optimization is shaped by adhering to minimum secrecy rate and signal-to-interference-plus-noise ratio (SINR) thresholds while driving the secrecy energy-efficiency (SEE) maximization for picocells. The SEE performance of the proposed solution outperforms several benchmarks including multiagent RL (MARL), multiagent deep Q-network (MA–DQN), the joint beamforming-based secrecy energy efficiency maximization (JBF–SEEM) scheme, and the one time pad-based encrypted data transmission (O–EDT).

10.5 Case Study: UAV-Assisted PLS for Terrestrial Wireless Communications Networks

While security considerations should ideally be integrated into wireless communication system design, the evolving nature of cellular standards and use cases, especially for upcoming 6G networks, means that not all threats can be anticipated. Therefore, security must be an ongoing consideration both during the design phase and throughout operation. UAVs, commonly known as drones, are increasingly used in various applications, including surveillance, disaster management, and wireless communication support [102]. This case study explores the application of AI-assisted PLS in a UAV-enhanced wireless communication network scenario. The extensive deployment of UAVs necessitates effective coordination through radio signaling. For instance, UAVs must convey their presence and establish communication with other UAVs, terrestrial networks, and ground users [103]. Multiple working groups within the Third Generation Partnership Project (3GPP) are actively developing specifications for the seamless integration of UAVs into cellular networks [104]. A network-connected UAV can serve various roles, including that of a UE, a communication relay, or an aerial base station. Table 10.2 captures prevention, detection, and recovery techniques that are enabled by UAVs against common wireless security attacks.

AI plays a pivotal role in enhancing the performance of UAV-assisted PLS strategies in terms of attack prevention, detection, and recovery as presented in Table 10.2. This includes deploying AI algorithms, such as DRL, to dynamically optimize UAV relay trajectories by taking into consideration real-time threat assessment data and adapting UAV trajectories to limit data rates within

Table 10.2 UAV-enabled prevention, detection, and recovery strategies against common wireless security attacks.

Category	Prevention strategies	Detection strategies	Recovery strategies
Eavesdropping	• **Emit artificial noise via UAVs to establish secure zones** • Employ UAV relays for safe information transfer.	• Utilize UAVs equipped with RF and non-RF sensors for eavesdropper localization. • Detect local oscillator leakage from eavesdropper RF front-ends.	• Restore communication using UAV relays. • Optimize UAV trajectories to limit data rates within attack ranges.
Jamming	• Utilize UAV relaying and UAV base stations multipoint transmissions. • Proactively reposition UAVs for beamforming and dynamic relaying.	• Deploy UAVs with RF environment sensing capabilities • Utilize UAV swarms for efficient jammer localization.	• Establish hot zones with redundant UAV transmission and reception points. • Utilize UAV swarm mobility to expedite communication recovery.
Spoofing	• Facilitate secondary authorization through UAVs.	• Enhance RF environment awareness via distributed UAVs. • Employ cooperative UAVs for localization and anomaly identification.	• Implement multi-hop transmission with UAVs providing focused, higher-power transmissions near affected nodes.

potential attack ranges [74]. Moreover, AI-enhanced sensor fusion algorithms, such as CNNs, can be used for more accurate sensor data analysis to detect local oscillator leakage from eavesdroppers' RF front-ends [105], enhancing the reliability of eavesdropper localization. UAVs equipped with AI algorithms can continuously monitor the RF environment and analyze complex RF data to detect abnormal patterns indicative of jamming attacks [106]. Random Forest models can be trained to differentiate between normal RF environment behavior and anomalies caused by spoofing attacks [107]. Such models can process various features extracted from the data to make real-time anomaly predictions. Swarm intelligence algorithms, such as ant colony optimization, can assist in the deployment of UAV swarms to establish hot zones [108]. These algorithms optimize the distribution of UAVs to maintain communication despite jamming. Q-learning can help UAVs adapt their transmission strategies to counteract spoofing attacks [109]. It can optimize multihop routes and power levels to maintain secure communication paths. These examples showcase how various AI algorithms can be applied to address specific challenges in UAV-assisted PLS. A summary of these examples and applications of AI-assisted UAV-enabled PLS techniques is presented in Table 10.3. Depending on the specific requirements and constraints of the scenario, different AI techniques may be chosen to achieve optimal results.

Modeling a dynamic wireless communication environment, we have conducted simulations to assess the security of the downlink transmission between a ground base station and mobile users in the presence of an eavesdropper attack while the UAV is deployed to act as a mobile aerial relay. The users are in constant motion, traveling unidirectionally, and encounter a single stationary eavesdropper. We are simulating two cases where a single UAV is deployed to relay the transmission between the base station and two and four users, respectively. The simulation scenarios for these cases are illustrated in Figure 10.2. In the context of a terrestrial receiver, the signal received from the UAV encompasses both line-of-sight (LoS) and non-LoS (NLoS) components, along with multiple reflections that induce multipath fading [110]. Based on empirical field measurements [110], the likelihood of achieving LoS connections, denoted as $\rho_{\text{LoS}}(\theta)$, between the terrestrial receiver and the UAV can be statistically represented as

$$\rho_{\text{LoS}}(\theta) = \frac{1}{1 + C\exp\left[-B(\theta - C)\right]}. \tag{10.1}$$

Parameters C and B are constants that depend on the specific communication environment, such as rural, urban, or dense urban. We have $\rho_{\text{NLoS}}(\theta) = 1 - \rho_{\text{LoS}}(\theta)$.

The path loss, on the other hand, depends on several factors, including the heights of the UAV transmitter (h_t) and ground receiver (h_r) as well as the

Table 10.3 AI applications for UAV-assisted PLS.

PLS strategy	Description	AI algorithms
UAV path planning	Optimize UAV flight paths for secure data transmission.	RL, DRL, and Q-learning for path optimization.
Dynamic relay positioning	Adjust UAV positions to improve signal quality and security.	Genetic algorithms (GA) for optimizing relay positions.
Resource allocation	Optimize power allocation, bandwidth, and antenna patterns.	RL, DRL, and Q-learning for path optimization.
Spectrum sensing	Detect changes in the RF spectrum indicative of attacks.	ML classifiers such as support vector machines (SVM) for spectrum analysis.
Anomaly detection	Identify abnormal patterns in communication data.	Random forest for detecting unusual RF patterns.
Relay swarming	Employ UAV swarms for resilient communication.	Swarm intelligence: Particle swarm optimization (PSO) and ACO for swarm behavior.
Resource sharing and load balancing	Distribute communication tasks among UAVs efficiently.	Distributed AI algorithms for load balancing.
Multi-path routing	Establish multiple data transmission paths for redundancy.	PSO for route optimization.
Jammer identification	Detect and locate signal jammers to minimize their impact on communication.	AI-based RF fingerprinting for jammer identification.

horizontal distance (d) that separates them. The spatial expectation of path loss can be written as

$$\text{PL}(\theta, d, h_t, h_r) = \sum_{\varsigma} \rho_{\varsigma}(\theta)\text{PL}_{\varsigma}(d, h_t, h_r). \tag{10.2}$$

The path loss is typically higher for NLoS connections compared to LoS links because of signal attenuation stemming shadowing and indirect signal paths. To quantify this difference, we use parameter $\eta = \frac{\eta_{\text{LOS}}}{\eta_{\text{NLOS}}}$, ($\eta < 1$) to represent the excessive attenuation factor of NLoS. The small-scale fading of air-to-ground (A2G) channels is described using the Nakagami-m distribution [111].

The degree to which the system's confidentiality is at risk can be quantified using the secrecy rate. This rate signifies the rate of transmission at which the eavesdropper cannot decode any information. It is computed as the difference between the channel capacities of the legitimate and the wiretap links, as outlined in [112].

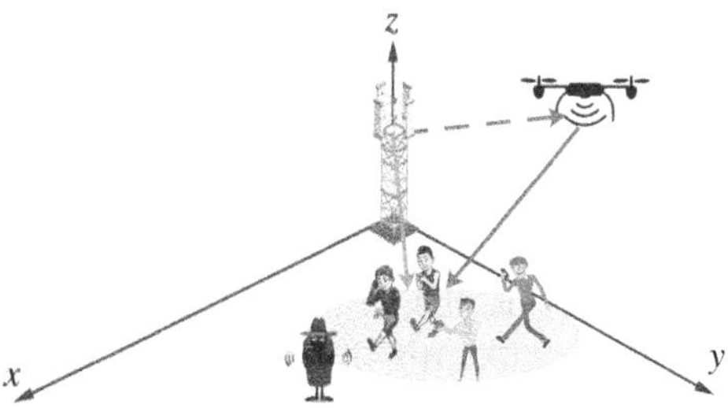

Figure 10.2 System model for UAV-assisted PLS simulation scenario.

The secrecy rate is obtained as

$$\begin{aligned} R^{\mathrm{Sec}} &= \max\left(\left(R^{\mathrm{LEG}} - R^{\mathrm{EVS}}\right), 0\right), \\ \text{where } R^{\mathrm{LEG}} &= \max\left(R^{BU}, R^{UR}\right), \\ \text{and } R^{\mathrm{EVS}} &= \max\left(R^{BE}, R^{UE}\right). \end{aligned} \tag{10.3}$$

The legitimate rate R^{LEG} is defined as the maximum rate achieved between base station and the UAV R^{BU} and the UAV and the ground receiver R^{UR}. The R^{BU} transmission rate is calculated as

$$\begin{aligned} R^{BU} &= B\log_2\left(1 + \mathrm{SNR}^{BU}\right), \\ \mathrm{SNR}^{BU} &= \frac{P_{T}g(d^{BU})^{-\alpha_{AG}}\left(\varpi_{\mathrm{BU}}^{\mathrm{LOS}} + \varpi_{\mathrm{BU}}^{\mathrm{NLOS}}\right)}{\sigma^2}, \end{aligned} \tag{10.4}$$

B represents the system bandwidth, P_T stands for the ground base station transmit power, g is the Nakagami-m small-scale fading parameter, d_{BU} captures the distance between the transmitter and UAV, and $\alpha_{\mathrm{AG}} \approx 2$ approximates the A2G pathloss exponent. The formula uses the expressions $\varpi_{\mathrm{BU}}^{\mathrm{LOS}} = \rho_{\mathrm{LOS}}(\theta_{\mathrm{BU}})10^{-\frac{\eta_{\mathrm{LOS}}}{10}}$ and $\varpi_{\mathrm{BU}}^{\mathrm{NLOS}} = \rho_{\mathrm{NLOS}}(\theta_{\mathrm{BU}})10^{-\frac{\eta_{\mathrm{NLOS}}}{10}}$, where θ_{BU} is the elevation angle between the base station and the UAV relay and σ^2 is the noise power. Equivalently, the R_{UR} rate between the UAV and the ground UE is calculated as

$$\begin{aligned} R^{UR} &= B\log_2\left(1 + \mathrm{SNR}^{UR}\right), \\ \mathrm{SNR}^{UR} &= \frac{P_{U}g(d^{UR})^{-\alpha_{AG}}\left(\varpi_{\mathrm{UR}}^{\mathrm{LOS}} + \varpi_{\mathrm{UR}}^{\mathrm{NLOS}}\right)}{\sigma^2}, \end{aligned} \tag{10.5}$$

where P_U represents the aerial relay transmit power, d^{UR} is the distance between the UAV and the ground users, $\varpi_{\mathrm{UR}}^{\mathrm{LOS}} = \rho_{\mathrm{LOS}}(\theta_{\mathrm{UR}})10^{-\frac{\eta_{\mathrm{LOS}}}{10}}$, $\varpi_{\mathrm{UR}}^{\mathrm{NLOS}} = \rho_{\mathrm{LOS}}(\theta_{\mathrm{UR}})10^{-\frac{\eta_{\mathrm{NLOS}}}{10}}$, and θ_{UR} is the elevation angle between the UAV and the UE.

The rate between the transmitter and the eavesdropper,

$$R^{BE} = B\log_2\left(1 + \frac{P_T l(d^{BE})^{-\alpha_G}}{\sigma^2}\right), \tag{10.6}$$

determines the eavesdropping rate R^{EVS}. In this equation, variable l represents the channel power gain of the Rayleigh small-scale fading for ground transmissions, d^{BE} captures the distance between the base station and the eavesdropper and

$\alpha_G > 2$ represents the ground path loss exponent. Finally, R^{UE} is the rate of transmission between the aerial relay and the eavesdropper:

$$\begin{aligned} R^{UE} &= B\log_2\left(1 + \text{SNR}^{UE}\right), \\ \text{SNR}^{UE} &= \frac{P_U g (d^{UE})^{-\alpha_{AG}} \left(\varpi_{\text{UE}}^{\text{LOS}} + \varpi_{\text{UE}}^{\text{NLOS}}\right)}{\sigma^2}, \end{aligned} \tag{10.7}$$

where d^{UE} is the distance between the UAV relay and the eavesdropper, $\varpi_{\text{UE}}^{\text{LOS}} = \rho_{\text{LOS}}(\theta_{\text{UE}})10^{-\frac{\eta_{\text{LOS}}}{10}}$, $\varpi_{\text{UE}}^{\text{NLOS}} = \rho_{\text{LOS}}(\theta_{\text{UE}})10^{-\frac{\eta_{\text{NLOS}}}{10}}$, and θ_{UE} is the elevation angle between UAV relay and the eavesdropper.

The objective of this simulation is to maximize the secrecy rate of the downlink transmissions, denoted as R^{Sec}, by strategically choosing the position for the UAV and managing the allocation of transmission power for the UAV is subject to position and power constraints. The secrecy rate optimization problem over T time slots is formulated as

$$\begin{aligned} (P1):\ & \max_{x_u, y_u, P_u} \frac{1}{T}\sum_{t=1}^{T} R[t]^{\text{Sec}}, \\ s.t.\ & (x_u[t], y_u[t]) \leq (L_x, L_y), \\ & 0 \leq P_u[t] \leq P_{\max}, \end{aligned} \tag{10.8}$$

where x_u and y_u are the aerial relay positioning parameters at a constant height, (L_x, L_y) captures the maximum 2D coordinates of the available flight area, and $P_{\max}$ is the maximum transmission power.

The optimization problem described in (10.8) involves the simultaneous optimization of UAV positioning and transmit power optimization. The challenge arises from the nonconvex nature of the objective function with respect to parameters x_u, y_u, and P_u in addition to the complexity introduced by multiple constraints. Consequently, this problem is classified as NP-hard, as discussed in prior works [76, 113, 114]. Therefore, we opt for an RL approach to solve this problem, specifically a DQL algorithm. This choice enables us to efficiently and accurately tackle the problem, leading to enhanced secrecy rates within the network. The MDP for the DQL agent is composed of the state space $\mathcal{S}$, the action space $\mathcal{A}$, the reward space $\mathcal{R}$, and the transition probability space $\mathcal{T}$, i.e. $(\mathcal{S}, \mathcal{A}, \mathcal{R}, \mathcal{T})$. Each state s_t at a given time slot t comprises two key elements: $s_t = \Delta x, \Delta y$. These correspond to the distances along the x and y axes between the UAV and the center of the mobile users cluster. These measurements are vital for characterizing the spatial channel gain relationships and positions and, hence, the SNR within the system. At each time slot t, an action a_t comprises three components: δ_x, δ_y, and δ_p. Specifically, δ_x and δ_y denote adjustments in the horizontal spatial dimensions along the x and y directions, respectively, whereas δ_p represents the variation in transmit power. These actions dictate the UAV's movement and the transmit power adaptation during each time step. The reward

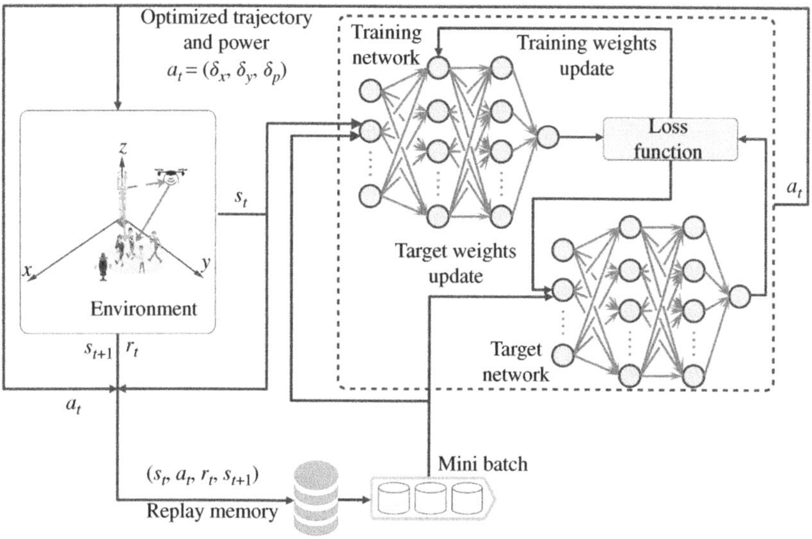

Figure 10.3 The proposed DQN block diagram.

$r_t(s_t, a_t)$ is obtained by the DQL agent after taking action a_t in state s_t at time slot t. The reward is designed to encourage actions that lead to higher secrecy rates. It is defined based on the instantaneous SNR difference between the legitimate SNR and the malicious SNR are as follows:

$$r_t(s_t, a_t) = \left(\max\left(\mathrm{SNR}^{BU}, \mathrm{SNR}^{UR}\right) - \max\left(\mathrm{SNR}^{BE}, \mathrm{SNR}^{UE}\right)\right). \tag{10.9}$$

The proposed DQN architecture for solving this optimization problem is illustrated in Figure 10.3. We conduct a numerical analysis to evaluate the effectiveness of our proposed DQN-based approach for optimizing UAV positioning and power control for maximizing the secrecy rate in the presence of an eavesdropping attack. In our scenario, we deploy a UAV relay that operates consistently at an altitude of 80 m. This UAV is equipped with an omnidirectional antenna to communicate with ground mobile users. The UAV's location is optimized using the proposed DQL algorithm, initially starting from the cluster center. Figure 10.4 shows the achieved average secrecy rate over different x positions of the mobile users cluster for the two cases of 2 and 4 UEs and the case of not employing the UAV relay. It can be observed that the UAV relaying solution has a positive effect on the secrecy rate. The utilization of the aerial relay ensures that the secrecy rate of mobile users remains consistently high across wide positions. However, there is a rapid decline in the rate to zero when the mobile user clusters and the eavesdropper come into close proximity. Nevertheless, the rate also exhibits a rapid recovery thereafter, which is not the case when the aerial relay is not

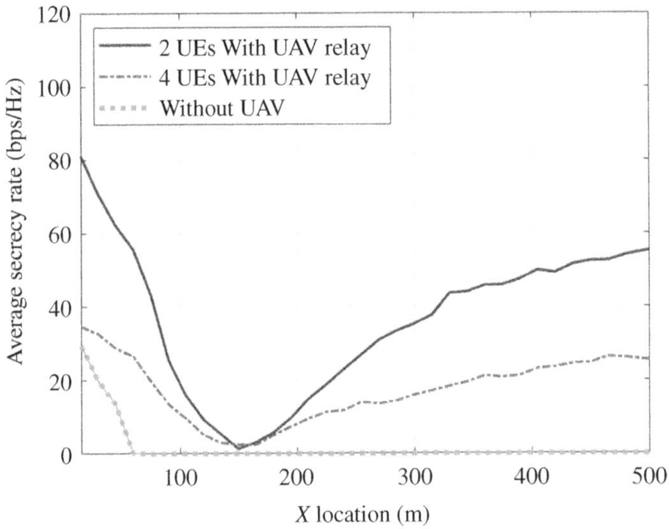

Figure 10.4 DQL algorithm for optimizing DL power for UAV-assisted PLS.

present. We also notice that the achieved secrecy rate for the two user cases is higher than the one achieved for the four UEs. This is so because of the limited coverage area of a single UAV, which prevents the UAV from optimally optimizing the legitimate rate of all users in the cluster.

10.6 Practical Considerations and Challenges of Implementing AI-Based Security Solutions

Implementing AI-based security solutions for wireless communication networks comes with several challenges, including scalability, performance optimization, and privacy.

10.6.1 Scalability and Performance Optimization of AI Models

Scalability challenges: Wireless communication networks generate enormous volumes of data. Traditional ML models may struggle to efficiently handle such data loads. Complex AI models and, particularly, DL models require significant computational resources [115]. As the network scales and the volume of data grows, so does the complexity of the models, making scalability a concern. Training large models on extensive data sets can be time consuming and resource intensive [116]. The proliferation of edge devices and the IoT add additional complications. These devices often have limited computational resources, making it challenging to deploy resource-intensive AI models directly on them [117].

In wireless networks, low-latency response is essential, particularly for security-related decisions. Scalability efforts should not compromise the ability to provide real-time responses to security threats [118]. To that end, it is recommended to distribute AI processing to edge devices whenever possible [119] and implement lightweight models that can run on resource-constrained devices, thus enabling real-time decision-making at the edge without overburdening central systems [120]. Another solution is to leverage hardware accelerators like GPUs, TPUs, or specialized AI chips to boost model inference speed and handle larger volumes of data [121] or implement data reduction techniques such as data summarization or feature selection to reduce the amount of data to be processed without sacrificing model accuracy.

Performance optimization challenges: Security solutions for wireless communication networks must operate in real time to detect and respond to threats promptly. This imposes strict performance requirements on AI models, including low inference latency. In mobile and battery-powered devices, such as smartphones or IoT sensors, energy-efficient operation is critical. AI models must be optimized to minimize power consumption while meeting security requirements [122]. There is often a tradeoff between model complexity and accuracy. While complex models may achieve high accuracy, they may not be suitable for resource-constrained devices due to their computational demands. These challenges can be addressed by employing model compression techniques, such as knowledge distillation [123], to create smaller and more efficient models that approximate the behavior of larger ones, designing AI architectures with low-latency considerations, optimizing model inference for real-time applications, using hardware components designed for energy efficiency in edge devices, pruning unimportant connections or neurons in neural networks to reduce model size and computational overhead [124], or quantizing DL models to lower precision (e.g., from 32-bit floating-point to 8-bit fixed-point) [125].

10.6.2 Privacy Considerations of AI-Enhanced Wireless Network Security

In the rapidly evolving landscape of wireless communication networks, the integration of AI solutions, which is expected within the 6G networks, has ushered in transformative benefits, enhancing efficiency, performance, and user experience. However, as with any technological advancement, AI-based wireless network optimizers are not immune to security vulnerabilities [126]. In fact, they introduce a new dimension of challenges, including the potential for various attacks that exploit the characteristics and mechanisms of AI algorithms [127]. These attacks target the vulnerabilities inherent in AI models and their deployment within wireless communication networks. They capitalize on the unique aspects of AI, such as pattern recognition, data-driven decision-making, and adaptive learning, to undermine the integrity, privacy, and availability of network operations [128]. Understanding these attack vectors is important when introducing AI for managing wireless communication networks.

One prominent category of attacks against AI-enhanced network management is adversarial attacks. In this context, adversaries deliberately manipulate input data to mislead AI models, causing them to make incorrect predictions or classifications [129]. This can lead to severe consequences, especially in safety-critical applications including autonomous vehicles or medical diagnosis [130]. Addressing adversarial attacks requires a multifaceted approach. Research is dedicated to developing robust AI models that can resist such attacks by enhancing their generalization capabilities and augmenting their decision-making processes to account for potential manipulation [131]. In addition, adversarial training techniques involve exposing AI models to adversarial examples during training, making them more resilient in real-world scenarios. Preprocessing input data by employing techniques such as dimensionality reduction and noise addition can make models less sensitive to adversarial perturbations. Employing ensemble methods, where multiple models work together to make predictions, can enhance the overall robustness of the system [132]. Adversarial attacks are less likely to fool a diverse set of models simultaneously. Also, using transfer learning with models that are pretrained on large datasets can improve generalization and enhance resistance against adversarial attacks.

Another related attack is data poisoning, wherein attackers inject malicious or misleading data into the training dataset used to train AI models. This tampering can compromise the accuracy and reliability of models [133]. The consequences of data poisoning attacks include biased predictions, poor performance, security breaches, and privacy risks. Counteracting data poisoning attacks demands a combination of preventive and corrective measures [130], including regularly inspecting and cleansing training data sets to identify and remove any potentially poisoned instances. One can also employ anomaly detection techniques to flag suspicious data points or robust training techniques that are less susceptible to malicious data [134]. Similarly, ensemble methods and outlier rejection mechanisms or input validation mechanisms that assess the integrity and consistency of incoming data can help mitigate the effect of poisoned data. Data diversification from various sources and continuous evaluation of the AI model's performance on new data can be effective for minimizing the chances of data poisoning attacks or these being unnoticed.

Membership inference attacks and model inversion attacks represent sophisticated threats to the privacy and security of machine learning models in wireless communication networks. These attacks leverage vulnerabilities of AI models to either identify if specific data points were part of the training data set (membership inference) or reverse-engineer sensitive training data from model outputs (model inversion) [135]. Membership inference attacks leverage the target model's outputs to infer membership information [136]. Such attacks often train a shadow model, which mimics the target model's behavior without access to its internal parameters or data. This shadow model is trained on a data set that includes some data points from the target model's training data and some synthetic or publicly available data. By comparing the responses of the shadow model to the target model's outputs, attackers can deduce whether a given data point was used in the training of the target model. Model inversion attacks manipulate the model's outputs, often by providing crafted queries, to obtain partial or complete information about the training data [137]. By observing how the model responds to these queries, attackers can infer attributes or details about the training data and compromise it confidentiality [130]. Successful membership inference and model inversion attacks can breach user privacy by revealing their involvement in sensitive tasks, such as medical diagnoses, financial transactions, or communication patterns. Mitigating these attacks can be done through different techniques including privacy-preserving training, regular model auditing, data anonymization, input validation, and model output noise.

10.7 Conclusions and Outlook

This chapter has underscored the potential impact that AI can have in fortifying the security defenses of communication networks. Many traditional security measures are rule-based and static. AI introduces a dynamic and adaptable approach that responds in real time to emerging threats, thus reducing the window of opportunity for cyberattacks. AI algorithms can effectively detect anomalies and patterns in the data. They excel in recognizing complex and evolving threat vectors. AI techniques can handle massive data volumes and analyze vast network parameters simultaneously. Their adaptability ensures that security solutions are capable of adapting to new threats and that they can evolve alongside threats. This means that as malicious actors develop new tactics, AI security systems can learn and counteract them.

The adoption of AI in wireless communication networks is an ongoing journey. Future research will focus on refining AI algorithms, making them more efficient, accurate, and adaptable for this new context. This includes the development of specialized models that are tailored to the unique demands of wireless network security. Researchers will strive to reduce false positives, enhance real-time threat detection, and optimize resource allocation. As the use of AI in wireless networks grows, interoperability between AI-driven components and solutions will become more relevant. Standardization efforts will thus be essential to create a cohesive and effective security framework. In the future, AI algorithms will likely move closer to the edge of the network, enabling faster response times. This can lead to more distributed AI models that can make real-time security decisions directly within the network infrastructure. Finally, research to make AI-based security systems, including future 6G networks, which are more resilient and adaptive to adversarial attacks will be vital. Ensuring that AI systems can defend themselves against attempts to manipulate or compromise the data they rely on is essential for long-term network security.

References

1 Zhou, L., Yeh, K.-H., Hancke, G. et al. (2018). Security and privacy for the industrial Internet of Things: an overview of approaches to safeguarding endpoints. *IEEE Signal Processing Magazine* 35 (5): 76–87.

2 Keshk, M., Turnbull, B., Sitnikova, E. et al. (2021). Privacy-preserving schemes for safeguarding heterogeneous data sources in cyber-physical systems. *IEEE Access* 9: 55077–55097.

3 Dapel, M.E., Asante, M., Uba, C.D., and Agyeman, M.O. (2023). Artificial intelligence techniques in cybersecurity management. *Cybersecurity in the Age of Smart Societies: Proceedings of the 14th International Conference on Global Security, Safety and Sustainability*, London, September 2022, 241–255. Springer.

4 Suryavanshi, A., Apoorva, G., TN, M.B. et al. (2023). The integration of blockchain and AI for web 3.0: a security perspective. *2023 4th International Conference on Innovative Trends in Information Technology (ICITIIT)*, 1–8. IEEE.

5 Abdalla, A.S., Upadhyaya, P.S., Shah, V.K., and Marojevic, V. (2022). Toward next generation open radio access networks: what O-RAN can and cannot do! *IEEE Network* 36 (6): 206–213.

6 Aiden, M.K., Sabharwal, S.M., Chhabra, S., and Al-Asadi, M. (2023). AI and blockchain for cyber security in cyber-physical system. *AI Models for Blockchain-Based Intelligent Networks in IoT Systems: Concepts, Methodologies, Tools, and Applications*, 203–230. Springer.

7 Liu, Z., Thapa, N., Shaver, A. et al. (2020). Anomaly detection on IoT network intrusion using machine learning. *2020 International Conference on Artificial Intelligence, Big Data, Computing and Data Communication Systems (icABCD)*, 1–5. IEEE.

8 Das, P., Illa, M., Pokhariyal, R. et al. (2023). Role of neural network, fuzzy, and IoT in integrating artificial intelligence as a cyber security system. *2023 2nd International Conference on Electronics and Renewable Systems (ICEARS)*, 652–658. IEEE.

9 Hussain, F., Hussain, R., Hassan, S.A., and Hossain, E. (2020). Machine learning in IoT security: current solutions and future challenges. *IEEE Communications Surveys & Tutorials* 22 (3): 1686–1721.

10 Abdalla, A.S. (2023). Physical layer security with unmanned aerial vehicles for advanced wireless networks. PhD dissertation. Bagley College of Engineering, Mississippi State University.

11 Ye, N., Kang, B., Qi, B. et al. (2023). Towards high-security ubiquitous IoT networks using AI. *2023 IEEE International Conference on Acoustics, Speech, and Signal Processing Workshops (ICASSPW)*, 1–5.

12 Leka, E., Hoxha, E., and Rexha, G. (2023). Security and privacy concerns associated with the Internet of Things (IoT) and the role of adapting blockchain and machine learning-a systematic literature review. *2023 46th MIPRO ICT and Electronics Convention (MIPRO)*, 1690–1696.

13 Shiu, Y.-S., Chang, S.Y., Wu, H.-C. et al. (2011). Physical layer security in wireless networks: a tutorial. *IEEE Wireless Communications* 18 (2): 66–74.

14 Zou, Y. and Wang, G. (2015). Intercept behavior analysis of industrial wireless sensor networks in the presence of eavesdropping attack. *IEEE Transactions on Industrial Informatics* 12 (2): 780–787.

15 Kapetanovic, D., Zheng, G., and Rusek, F. (2015). Physical layer security for massive MIMO: an overview on passive eavesdropping and active attacks. *IEEE Communications Magazine* 53 (6): 21–27.

16 Pirayesh, H. and Zeng, H. (2022). Jamming attacks and anti-jamming strategies in wireless networks: a comprehensive survey. *IEEE Communications Surveys & Tutorials* 24 (2): 767–809.

17 Zhen, J. and Srinivas, S. (2003). Preventing replay attacks for secure routing in ad hoc networks. *International Conference on Ad-Hoc Networks and Wireless*, 140–150. Springer.

18 Chen, Y., Trappe, W., and Martin, R.P. (2007). Detecting and localizing wireless spoofing attacks. *2007 4th Annual IEEE Communications Society Conference on Sensor, Mesh and Ad Hoc Communications and Networks*, 193–202. IEEE.

19 Mallik, A. (2019). Man-in-the-middle-attack: understanding in simple words. *Cyberspace: Jurnal Pendidikan Teknologi Informasi* 2 (2): 109–134.

20 Twardokus, G. and Rahbari, H. (2023). Towards protecting 5G sidelink scheduling in C-V2X against intelligent DoS attacks. *IEEE Transactions on Wireless Communications* 22 (11): 7273–7286.

21 Khouzani, M., Sarkar, S., and Altman, E. (2012). Maximum damage malware attack in mobile wireless networks. *IEEE/ACM Transactions on Networking* 20 (5): 1347–1360.

22 Martins, G., Bhatia, S., Koutsoukos, X. et al. (2015). Towards a systematic threat modeling approach for cyber-physical systems. *2015 Resilience Week (RWS)*, 1–6. IEEE.

23 Khera, Y., Kumar, D., and Garg, N. (2019). Analysis and impact of vulnerability assessment and penetration testing. *2019 International Conference on Machine Learning, Big Data, Cloud and Parallel Computing (COMITCon)*, 525–530. IEEE.

24 Kasper, T., Oswald, D., and Paar, C. (2011). Wireless security threats: eavesdropping and detecting of active RFIDs and remote controls in the wild. *SoftCOM 2011, 19th International Conference on Software, Telecommunications and Computer Networks*, 1–6. IEEE.

25 Lichtman, M., Jover, R.P., Labib, M. et al. (2016). LTE/LTE-A jamming, spoofing, and sniffing: threat assessment and mitigation. *IEEE Communications Magazine* 54 (4): 54–61.

26 Al-Shareeda, M.A. and Manickam, S. (2022). Man-in-the-middle attacks in mobile ad hoc networks (MANETs): analysis and evaluation. *Symmetry* 14 (8): 1543.

27 Feng, Z., Ning, J., Broustis, I. et al. (2011). Coping with packet replay attacks in wireless networks. *2011 8th Annual IEEE Communications Society Conference on Sensor, Mesh and Ad Hoc Communications and Networks*, 368–376. IEEE.

28 Pizzi, S., Suraci, C., Iera, A. et al. (2020). A sidelink-aided approach for secure multicast service delivery: from human-oriented multimedia traffic to machine type communications. *IEEE Transactions on Broadcasting* 67 (1): 313–323.

29 Abdalla, A.S. and Marojevic, V. (2023). End-to-end O-RAN security architecture, threat surface, coverage, and the case of the open fronthaul. *arXiv preprint arXiv:2304.05513*.

30 Cunningham, P., Cord, M., and Delany, S.J. (2008). Supervised learning. In: *Machine Learning Techniques for Multimedia: Case Studies on Organization and Retrieval* (ed. M. Cord and P. Cunningham), 21, 978-3-540-75171-7–49. Berlin, Heidelberg: Springer-Verlag.

31 Kruczkowski, M. and Niewiadomska-Szynkiewicz, E. (2014). Comparative study of supervised learning methods for malware analysis. *Journal of Telecommunications and Information Technology* 4: 24–33.

32 Tsai, C.-F., Hsu, Y.-F., Lin, C.-Y., and Lin, W.-Y. (2009). Intrusion detection by machine learning: a review. *Expert Systems with Applications* 36 (10): 11994–12000.

33 Raikar, M.M., Meena, S., Mulla, M.M. et al. (2020). Data traffic classification in software defined networks (SDN) using supervised-learning. *Procedia Computer Science* 171: 2750–2759.

34 Jia, W., Shukla, R.M., and Sengupta, S. (2019). Anomaly detection using supervised learning and multiple statistical methods. *2019 18th IEEE International Conference on Machine Learning and Applications (ICMLA)*, 1291–1297. IEEE.

35 Gross, R., Sweeney, L., De La Torre, F., and Baker, S. (2008). Semi-supervised learning of multi-factor models for face de-identification. *2008 IEEE Conference on Computer Vision and Pattern Recognition*, 1–8. IEEE.

36 Weinand, A., Sattiraju, R., Karrenbauer, M., and Schotten, H.D. (2019). Supervised learning for physical layer based message authentication in URLLC scenarios. *2019 IEEE 90th Vehicular Technology Conference (VTC2019-fall)*, 1–7. IEEE.

37 Towhid, M.S. and Shahriar, N. (2022). Encrypted network traffic classification using self-supervised learning. *2022 IEEE 8th International Conference on Network Softwarization (NetSoft)*, 366–374. IEEE.

38 Barlow, H.B. (1989). Unsupervised learning. *Neural Computation* 1 (3): 295–311.

39 Soni, J., Prabakar, N., and Upadhyay, H. (2020). Visualizing high-dimensional data using t-distributed stochastic neighbor embedding algorithm. In: *Principles of Data Science* (ed. H.R. Arabnia, K. Daimi, R. Stahlbock, et al.), 189–206. Cham: Springer International Publishing. ISBN: 978-3-030-43981-1.

40 Yu, C., Lan, J., Xie, J., and Hu, Y. (2018). QoS-aware traffic classification architecture using machine learning and deep packet inspection in SDNs. *Procedia Computer Science* 131: 1209–1216.

41 Debbabi, F., Jmal, R., Fourati, L.C., and Ksentini, A. (2020). Algorithmics and modeling aspects of network slicing in 5G and beyonds network: survey. *IEEE Access* 8: 162748–162762.

42 De Araujo-Filho, P.F., Pinheiro, A.J., Kaddoum, G. et al. (2021). An efficient intrusion prevention system for can: hindering cyber-attacks with a low-cost platform. *IEEE Access* 9: 166855–166869.

43 Henriques, J., Caldeira, F., Cruz, T., and Simes, P. (2022). An automated closed-loop framework to enforce security policies from anomaly detection. *Computers & Security* 123: 102949.

44 Meidan, Y., Bohadana, M., Shabtai, A. et al. (2017). Detection of unauthorized IoT devices using machine learning techniques. *arXiv preprint arXiv:1709.04647*.

45 Nguyen, T.T. and Armitage, G. (2008). A survey of techniques for internet traffic classification using machine learning. *IEEE Communications Surveys & Tutorials* 10 (4): 56–76.

46 da Silva, J.C., Giannella, C., Bhargava, R. et al. (2005). Distributed data mining and agents. *Engineering Applications of Artificial Intelligence* 18 (7): 791–807.

47 Abdalla, A.S. and Marojevic, V. (2022). ARIS for safeguarding MISO wireless communications: a deep reinforcement learning approach. *2022 5th International Conference on Advanced Communication Technologies and Networking (CommNet)*, 1–6.

48 Li, Y. (2017). Deep reinforcement learning: an overview. *arXiv preprint arXiv:1701.07274*.

49 Liu, Z. and Elhanany, I. (2006). RL-MAC: A reinforcement learning based MAC protocol for wireless sensor networks. *International Journal of Sensor Networks* 1 (3-4): 117–124.

50 Hamadi, R., Ghazzai, H., and Massoud, Y. (2023). Reinforcement learning based intrusion detection systems for drones: a brief survey. *2023 IEEE International Conference on Smart Mobility (SM)*, 104–109. IEEE.

51 Jin, Q. and Wang, L. (2019). Intranet user-level security traffic management with deep reinforcement learning. *2019 International Joint Conference on Neural Networks (IJCNN)*, 1–8. IEEE.

52 Radoglou-Grammatikis, P., Liatifis, A., Grigoriou, E. et al. (2021). TRUSTY: A solution for threat hunting using data analysis in critical infrastructures. *2021 IEEE International Conference on Cyber Security and Resilience (CSR)*, 485–490. IEEE.

53 Boateng, G.O., Sun, G., Mensah, D.A. et al. (2022). Consortium blockchain-based spectrum trading for network slicing in 5G RAN: a multi-agent deep reinforcement learning approach. *IEEE Transactions on Mobile Computing* 22 (10): 5801–5815.

54 Abdalla, A.S., Moore, J., Adhikari, N., and Marojevic, V. (2024). ZTRAN: Prototyping zero trust security xApps for open radio access network deployments. *IEEE Wireless Communications* 31 (2): 66–73.

55 Abdalla, A.S. and Marojevic, V. (2022). DDPG learning for aerial RIS-assisted MU-MISO communications. *2022 IEEE 33rd Annual International Symposium on Personal, Indoor and Mobile Radio Communications (PIMRC)*, 701–706.

56 Creswell, A., White, T., Dumoulin, V. et al. (2018). Generative adversarial networks: an overview. *IEEE Signal Processing Magazine* 35 (1): 53–65.

57 Lim, S.K., Loo, Y., Tran, N.-T. et al. (2018). DOPING: Generative data augmentation for unsupervised anomaly detection with GAN. *2018 IEEE international Conference on Data Mining (ICDM)*, 1122–1127. IEEE.

58 Duy, P.T., Khoa, N.H., Nguyen, A.G.-T., and Pham, V.-H. (2021). DIGFuPAS: Deceive IDs with GAN and function-preserving on adversarial samples in SDN-enabled networks. *Computers & Security* 109: 102367.

59 Zhang, K.A., Cuesta-Infante, A., Xu, L., and Veeramachaneni, K. (2019). SteganoGAN: High capacity image steganography with GANs. *arXiv preprint arXiv:1901.03892.*

60 Li, L., Fan, Y., Tse, M., and Lin, K.-Y. (2020). A review of applications in federated learning. *Computers & Industrial Engineering* 149: 106854.

61 Li, T., Sahu, A.K., Talwalkar, A., and Smith, V. (2020). Federated learning: challenges, methods, and future directions. *IEEE Signal Processing Magazine* 37 (3): 50–60.

62 Abdel-Basset, M., Hawash, H., and Moustafa, N. (2021). Toward privacy preserving federated learning in internet of vehicular things: challenges and future directions. *IEEE Consumer Electronics Magazine* 11 (6): 56–66.

63 Yang, Q. (2021). Toward responsible AI: an overview of federated learning for user-centered privacy-preserving computing. *ACM Transactions on Interactive Intelligent Systems (TiiS)* 11 (3-4): 1–22.

64 Ziller, A., Trask, A., Lopardo, A. et al. (2021). PySyft: A library for easy federated learning. In: *Federated Learning Systems: Towards Next-Generation AI* (ed. M.H. ur Rehman and M.M. Gaber), 111–139. Cham: Springer International Publishing. ISBN: 978-3-030-70604-3.

65 Wei, K., Li, J., Ding, M. et al. (2020). Federated learning with differential privacy: algorithms and performance analysis. *IEEE Transactions on Information Forensics and Security* 15: 3454–3469.

66 Dong, X., Yu, Z., Cao, W. et al. (2020). A survey on ensemble learning. *Frontiers of Computer Science* 14: 241–258.

67 Sagi, O. and Rokach, L. (2018). Ensemble learning: a survey. *Wiley Interdisciplinary Reviews: Data Mining and Knowledge Discovery* 8 (4): e1249.

68 Abri, F., Siami-Namini, S., Khanghah, M.A. et al. (2019). Can machine/deep learning classifiers detect zero-day malware with high accuracy? *2019 IEEE International Conference on Big Data (Big Data)*, 3252–3259. IEEE.

69 Nkongolo, M., Van Deventer, J.P., Kasongo, S.M. et al. (2022). A cloud based optimization method for zero-day threats detection using genetic algorithm and ensemble learning. *Electronics* 11 (11): 1749.

70 Al-rimy, B.A.S., Maarof, M.A., Prasetyo, Y.A. et al. (2018). Zero-day aware decision fusion-based model for crypto-ransomware early detection. *International Journal of Integrated Engineering* 10 (6): 82–88.

71 Peethambaran, G., Naikodi, C., and Suresh, L. (2020). An ensemble learning approach for privacy—quality—efficiency trade-off in data analytics. *2020 International Conference on Smart Electronics and Communication (ICOSEC)*, 228–235. IEEE.

72 Chaudhary, N., Gupta, V., Sandhir, K. et al. (2022). Privacy preserving ensemble learning classification model for mental healthcare. *2022 7th*

International Conference on Parallel, Distributed and Grid Computing (PDGC), 513–518. IEEE.

73 Dasika, A.K. and Paruchuri, P. (2020). An ensemble learning approach to improve tracking accuracy of multi sensor fusion. *Neural Information Processing: 27th International Conference, ICONIP 2020*, Bangkok, Thailand (18–22 November 2020), Proceedings, Part V 27, 704–712. Springer.

74 Abdalla, A.S. and Marojevic, V. (2023). Multi-agent learning for secure wireless access from UAVs with limited energy resources. *IEEE Internet of Things Journal* 10 (24): 22356–22370.

75 Mukherjee, A., Fakoorian, S.A.A., Huang, J., and Swindlehurst, A.L. (2014). Principles of physical layer security in multiuser wireless networks: a survey. *IEEE Communications Surveys & Tutorials* 16 (3): 1550–1573.

76 Abdalla, A.S. and Marojevic, V. (2021). Securing mobile multiuser transmissions with UAVs in the presence of multiple eavesdroppers. *IEEE Transactions on Vehicular Technology* 70 (10): 11011–11016.

77 Mucchi, L., Jayousi, S., Caputo, S. et al. (2021). Physical-layer security in 6G networks. *IEEE Open Journal of the Communications Society* 2: 1901–1914.

78 Abdalla, A.S., Shang, B., Marojevic, V., and Liu, L. (2020). Performance evaluation of aerial relaying systems for improving secrecy in cellular networks. *2020 IEEE 92nd Vehicular Technology Conference (VTC2020-Fall)*, 1–5.

79 Abdalla, A.S., Shang, B., Marojevic, V., and Liu, L. (2020). Securing mobile IoT with unmanned aerial systems. *2020 IEEE 6th World Forum on Internet of Things (WF-IoT)*, 1–6.

80 Zhao, L., Zhang, X., Chen, J., and Zhou, L. (2020). Physical layer security in the age of artificial intelligence and edge computing. *IEEE Wireless Communications* 27 (5): 174–180.

81 Sharma, H. and Kumar, N. (2023). Deep learning based physical layer security for terrestrial communications in 5G and beyond networks: a survey. *Physical Communication* 57: 102002.

82 Gu, Y., Wu, Z., Yin, Z., and Zhang, X. (2019). The secrecy capacity optimization artificial noise: a new type of artificial noise for secure communication in MIMO system. *IEEE Access* 7: 58353–58360.

83 Yang, H., Xiong, Z., Zhao, J. et al. (2020). Deep reinforcement learning-based intelligent reflecting surface for secure wireless communications. *IEEE Transactions on Wireless Communications* 20 (1): 375–388.

84 Abdalla, A.S., Behfarnia, A., and Marojevic, V. (2023). UAV trajectory and multi-user beamforming optimization for clustered users against passive eavesdropping attacks with unknown CSI. *IEEE Transactions on Vehicular Technology* 72 (11): 14426–14442.

85 Yang, H., Lam, K.-Y., Nie, J. et al. (2021). 3D beamforming based on deep learning for secure communication in 5G and beyond wireless networks. *2021 IEEE Globecom Workshops (GC Wkshps)*, 1–6. IEEE.

86 Dong, R., Wang, B., and Cao, K. (2021). Deep learning driven 3D robust beamforming for secure communication of UAV systems. *IEEE Wireless Communications Letters* 10 (8): 1643–1647.

87 Xing, J., Lv, T., and Zhang, X. (2019). Cooperative relay based on machine learning for enhancing physical layer security. *2019 IEEE 30th Annual International Symposium on Personal, Indoor and Mobile Radio Communications (PIMRC)*, 1–6.

88 Zhang, X. and Sun, S. (2019). Dynamic optimization for secure MIMO beamforming using large-scale reinforcement learning. *2019 IEEE Wireless Communications and Networking Conference (WCNC)*, 1–6.

89 Soltanieh, N., Norouzi, Y., Yang, Y., and Karmakar, N.C. (2020). A review of radio frequency fingerprinting techniques. *IEEE Journal of Radio Frequency Identification* 4 (3): 222–233.

90 Candore, A., Kocabas, O., and Koushanfar, F. (2009). Robust stable radiometric fingerprinting for wireless devices. *2009 IEEE International Workshop on Hardware-Oriented Security and Trust*, 43–49. IEEE.

91 Smith, N., Smith, L., Kodipaka, S. et al. (2021). Real-time location fingerprinting for mobile devices in an indoor prison setting. *Signal Processing, Sensor/Information Fusion, and Target Recognition XXX*, volume 11756, 267–275. SPIE.

92 Islam, F., Farmer, J., Dahal, A. et al. (2022). Wifi fingerprinting based room level classification: combining short term Fourier transform and imbalanced learning method. *Signal Processing, Sensor/Information Fusion, and Target Recognition XXXI*, volume 12122, 269–277. SPIE.

93 Merchant, K. and Nousain, B. (2019). Enhanced RF fingerprinting for IoT devices with recurrent neural networks. *MILCOM 2019 - 2019 IEEE Military Communications Conference (MILCOM)*, 590–597. IEEE.

94 Wang, S., Peng, L., Fu, H. et al. (2020). A convolutional neural network-based RF fingerprinting identification scheme for mobile phones. *IEEE INFOCOM 2020-IEEE Conference on Computer Communications Workshops (INFOCOM WKSHPS)*, 115–120. IEEE.

95 Oligeri, G., Sciancalepore, S., Raponi, S., and Di Pietro, R. (2022). PAST-AI: Physical-layer authentication of satellite transmitters via deep learning. *IEEE Transactions on Information Forensics and Security* 18: 274–289.

96 Abdalla, A.S., Behfarnia, A., and Marojevic, V. (2022). Aerial base station positioning and power control for securing communications: a deep Q-network approach. *2022 IEEE Wireless Communications and Networking Conference (WCNC)*, 2470–2475.

97 Wyner, A.D. (1975). The wire-tap channel. *The Bell System Technical Journal* 54 (8): 1355–1387.

98 Maurer, U. (1993). Secret key agreement by public discussion from common information. *IEEE Transactions on Information Theory* 39 (3): 733–742.

99 Luo, Y., Yang, Y., An, S., and Feng, Z. (2021, 2021). Secure and reliable D2D communications with active attackers: a game-theoretic perspective and machine learning approaches. In: *Machine Learning and Intelligent Communications* (ed. M. Guan and Z. Na), 519–533. Cham: Springer International Publishing. ISBN: 978-3-030-66785-6.

100 Li, C., Zhou, W., Yu, K. et al. (2019). Enhanced secure transmission against intelligent attacks. *IEEE Access* 7: 53596–53602.

101 Sharma, H., Kumar, N., and Tekchandani, R.K. (2023). SecBoost: Secrecy-aware deep reinforcement learning based energy-efficient scheme for 5G HetNets. *IEEE Transactions on Mobile Computing* 23 (2): 1401–1415.

102 Abdalla, A.S., Powell, K., Marojevic, V., and Geraci, G. (2020). UAV-assisted attack prevention, detection, and recovery of 5G networks. *IEEE Wireless Communications* 27 (4): 40–47.

103 Abdalla, A.S. and Marojevic, V. (2022). Security threats and cellular network procedures for unmanned aircraft systems: challenges and opportunities. *IEEE Communications Standards Magazine* 6 (4): 104–111.

104 Abdalla, A.S. and Marojevic, V. (2021). Communications standards for unmanned aircraft systems: the 3GPP perspective and research drivers. *IEEE Communications Standards Magazine* 5 (1): 70–77.

105 Liu, Y., Wang, J., Li, J. et al. (2021). Machine learning for the detection and identification of internet of things devices: a survey. *IEEE Internet of Things Journal* 9 (1): 298–320.

106 Gamage, C., Dinalankara, R., Samarabandu, J., and Subasinghe, A. (2023). A comprehensive survey on the applications of machine learning techniques on maritime surveillance to detect abnormal maritime vessel behaviors. *WMU Journal of Maritime Affairs* 22 (4): 447–477.

107 Cheng, J., Li, M., Tang, X. et al. (2018). Flow correlation degree optimization driven random forest for detecting DDoS attacks in cloud computing. *Security and Communication Networks* 2018 (1): 6459326.

108 Manikandan, K. and Sriramulu, R. (2022). Optimized path planning strategy to enhance security under swarm of unmanned aerial vehicles. *Drones* 6 (11): 336.

109 Alabadi, M. and Albayrak, Z. (2020). Q-learning for securing cyber-physical systems: a survey. *2020 International Congress on Human-Computer Interaction, Optimization and Robotic Applications (HORA)*, 1–13. IEEE.

110 Al-Hourani, A., Kandeepan, S., and Lardner, S. (2014). Optimal lap altitude for maximum coverage. *IEEE Wireless Communications Letters* 3 (6): 569–572.

111 Abdalla, A.S. and Marojevic, V. (2022). Aerial RIS for MU-MISO: joint base station beamforming and RIS phase shifter optimization. *2022 IEEE International Conference on Sensing, Communication, and Networking (SECON Workshops)*, 19–24.

112 Yang, N., Wang, L., Geraci, G. et al. (2015). Safeguarding 5G wireless communication networks using physical layer security. *IEEE Communications Magazine* 53 (4): 20–27.

113 Zhang, Y., Mou, Z., Gao, F. et al. (2020). UAV-enabled secure communications by multi-agent deep reinforcement learning. *IEEE Transactions on Vehicular Technology* 69 (10): 11599–11611.

114 Li, J., Gao, H., Lv, T. and Lu, Y. (2018). Deep reinforcement learning based computation offloading and resource allocation for MEC. *2018 IEEE Wireless Communications and Networking Conference (WCNC)*, 1–6.

115 Chen, J. and Ran, X. (2019). Deep learning with edge computing: a review. *Proceedings of the IEEE* 107 (8): 1655–1674.

116 Hu, W., Fey, M., Ren, H. et al. (2021). OGB-LSC: A large-scale challenge for machine learning on graphs. *arXiv preprint arXiv:2103.09430*.

117 Al-Emran, M., Malik, S.I., and Al-Kabi, M.N. (2020). A survey of Internet of Things (IoT) in education: opportunities and challenges. In: *Toward Social Internet of Things (SIoT): Enabling Technologies, Architectures and Applications: Emerging Technologies for Connected and Smart Social Objects* (ed. A.E. Hassanien, R. Bhatnagar, N.E.M. Khalifa, and M.H.N. Taha), 197–209. Cham: Springer International Publishing. ISBN: 978-3-030-24513-9.

118 Abdalla, A.S. and Marojevic, V. (2020). Machine learning-assisted UAV operations with the UTM: requirements, challenges, and solutions. *2020 IEEE 92nd Vehicular Technology Conference (VTC2020-Fall)*, 1–5.

119 Long, Y., Chakraborty, I., Srinivasan, G., and Roy, K. (2021). Complexity-aware adaptive training and inference for edge-cloud distributed AI systems. *2021 IEEE 41st International Conference on Distributed Computing Systems (ICDCS)*, 573–583. IEEE.

120 Sudharsan, B., Patel, P., Breslin, J. et al. (2021). Toward distributed, global, deep learning using IoT devices. *IEEE Internet Computing* 25 (3): 6–12.

121 Wang, H., Qu, Z., Zhou, Q. et al. (2021). A comprehensive survey on training acceleration for large machine learning models in IoT. *IEEE Internet of Things Journal* 9 (2): 939–963.

122 Mungoli, N. (2023). Scalable, distributed AI frameworks: leveraging cloud computing for enhanced deep learning performance and efficiency. *arXiv preprint arXiv:2304.13738*.

123 Gou, J., Yu, B., Maybank, S.J., and Tao, D. (2021). Knowledge distillation: a survey. *International Journal of Computer Vision* 129: 1789–1819.

124 Han, S., Pool, J., Tran, J., and Dally, W. (2015). Learning both weights and connections for efficient neural network. *Advances in Neural Information Processing Systems 28.*

125 Jin, Q., Ren, J., Zhuang, R. et al. (2022). F8Net: Fixed-point 8-bit only multiplication for network quantization. *arXiv preprint arXiv:2202.05239.*

126 Rahman, T.F., Abdalla, A.S., Powell, K. et al. (2021). Network and physical layer attacks and countermeasures to AI-enabled 6G O-RAN. *arXiv preprint arXiv:2106.02494.*

127 Ma, C., Li, J., Ding, M. et al. (2020). On safeguarding privacy and security in the framework of federated learning. *IEEE Network* 34 (4): 242–248.

128 Sachdev, R. (2020). Towards security and privacy for edge AI in IoT/IoE based digital marketing environments. *2020 5th International Conference on Fog and Mobile Edge Computing (FMEC)*, 341–346. IEEE.

129 Bae, H., Jang, J., Jung, D. et al. (2018). Security and privacy issues in deep learning. *arXiv preprint arXiv:1807.11655.*

130 Rahman, M.M., Arshi, A.S., Hasan, M.M. et al. (2023). Security risk and attacks in AI: a survey of security and privacy. *2023 IEEE 47th Annual Computers, Software, and Applications Conference (COMPSAC)*, 1834–1839. IEEE.

131 Yamin, M.M., Ullah, M., Ullah, H., and Katt, B. (2021). Weaponized AI for cyber attacks. *Journal of Information Security and Applications* 57: 102722.

132 Benzaïd, C. and Taleb, T. (2020). AI for beyond 5G networks: a cyber-security defense or offense enabler? *IEEE Network* 34 (6): 140–147.

133 Hu, Y., Kuang, W., Qin, Z. et al. (2021). Artificial intelligence security: threats and countermeasures. *ACM Computing Surveys (CSUR)* 55 (1): 1–36.

134 Li, B., Qi, P., Liu, B. et al. (2023). Trustworthy AI: from principles to practices. *ACM Computing Surveys* 55 (9): 1–46.

135 Kuzlu, M., Catak, F.O., Zhao, Y. et al. (2023, 2023). Security and privacy concerns in next-generation networks using artificial intelligence-based solutions: a potential use case. In: *Wireless Networks: Cyber Security Threats and Countermeasures* (ed. H. Jahankhani and A. El Hajjar), 205–226. Cham: Springer International Publishing. ISBN: 978-3-031-33631-7.

136 Hamidouche, M., Bellafqira, R., Quellec, G., and Coatrieux, G. (2022). White-box membership attack against machine learning based retinopathy classification. *arXiv preprint arXiv:2206.03584.*

137 Khosravy, M., Nakamura, K., Pasquali, A. et al. (2022). When AI facilitates trust violation: an ethical report on deep model inversion privacy attack. *2022 International Conference on Computational Science and Computational Intelligence (CSCI)*, 929–935. IEEE.

Index

Artificial Intelligence for Future Networks, First Edition.
Edited by Mohammad A. Matin, Sotirios K. Goudos, and George K. Karagiannidis.

j

k

l

m

t

u

Printed and bound by CPI Group (UK) Ltd, Croydon, CR0 4YY
17/09/2026
14964311-0006